THE 8051/8052 MICROCONTROLLER

Architecture, Assembly Language, and Hardware Interfacing

Craig Steiner

Universal Publishers
Boca Raton, Florida
USA • 2005

The 8051/8052 Microcontroller

Architecture, Assembly Language, and Hardware Interfacing

ISBN: 1-58112-459-7
Author: Craig Steiner
Cover Photo and Design: Erika Oliden González

Universal Publishers
Boca Raton, FL USA
www.universal-publishers.com

For Erika

Preface

Despite its relatively old age, the MCS-51 (8052) line of microcontrollers remains one of the most popular in use today. Many derivative microcontrollers have since been produced that are based on—and are compatible with—the 8052. Thus the ability to program an 8052 is an important skill for anyone that plans to develop microcontroller-based products.

This book was written to help the student, hobbyist, and professional understand the 8052 architecture and learn to write assembly language software as well as the general concepts involved in constructing an 8052-based device at the hardware level. Even if you'll be developing your program in 'C', a working knowledge of the underlying architecture as well as the microcontroller's assembly language is extremely important in writing efficient code that will separate a typical 'C' programmer from a far more versatile embedded 8052 software developer.

This book is intended to be both instructional as well as an easy-to-use reference. The various chapters of the book will explain the 8052 architecture step by step so that someone just beginning to use the 8052 may get a firm grasp of the architecture. The appendices are a useful reference that will assist both the novice programmer as well as the experienced developer long after the architecture has been mastered.

No specific knowledge of the 8052 is required but the book does assume the following:

1. A general and conceptual knowledge of programming.
2. An understanding of decimal, hexadecimal, and binary number systems.
3. A general knowledge of electronics and integrated circuits, though no prior knowledge of the 8052 IC is assumed.

This book will first briefly explain what a microcontroller is, then focus on the details of the 8052 microcontroller, describe its architecture and explain its assembly language. We'll then provide a design for a working single-board computer (SBC) and discuss each section of the design at the hardware level and cover the software and hardware development tools that make it possible to create and test the firmware developed by the user. The book will close by implementing software that can access each part of the SBC's hardware and describe a "monitor program" which can be used to control it.

There is some overlap, however, in that the architecture cannot be fully explained without some use of assembly language while the assembly language cannot be fully explained without first understanding the architecture. For this reason there will be sporadic references to assembly language instructions in chapters prior to assembly language being formally explained. In these cases, the references to assembly language instructions will be brief, well-commented, and, hopefully, understandable in the context of the architecture that is being explained. If an assembly language instruction used in the first few chapters isn't clear—or if you're anxious to read more about the instruction—feel free to jump ahead to the chapter on assembly language and then come back to where you were. Don't feel that you must read the entire book in a strictly sequential fashion.

I hope you find this book useful and educational. I would cordially invite you to visit my website,

http://www.8052.com, where you will find additional resources, example source code, and a message forum with a user community of over 12,000 people that exchange ideas, comments, questions, and answers on the 8052 microcontroller.

Additionally, I would invite you to submit any comments, suggestions, or corrections regarding the material in this book to me personally so that future editions may be further improved. A list of errata and corrections for the book will be found at **http://www.8052.com/book**.

I would like to extend my thanks to Atmel and Dallas Semiconductor which both provided me with sample microcontrollers that were extremely useful in the design of the SBC. And, of course, I would like to thank everyone who assisted or encouraged me during the writing of this book. While not en exhaustive list, I would like to specifically thank the following people for their assistance: Gerard Steiner, Steve Taylor, Agustin Dominguez, Erik Malund, Michael Karas, and my wife Erika Oliden. Whether it was proof-reading the manuscript, providing feedback on circuit diagrams, or simply tolerating the long hours I put in writing this book, their help and support was invaluable.

Craig Steiner
craigsteiner@8052.com
http://www.8052.com

Table of Contents

viii

8052 ARCHITECTURE

CHAPTER 1: INTRODUCTION TO 8052 MICROCONTROLLERS

Many developers new to microcontrollers, including the 8052, come from a PC/Windows or Macintosh environment. While most programming concepts will transfer over to the 8052 environment with no problem, there are some issues that may need clarification as you enter the microcontroller world. Before delving into the details of microcontrollers and, specifically, the 8052, we'll address some common stumbling blocks.

1.1 What is a Microcontroller?

A **microcontroller** (often abbreviated **MCU**) is a single integrated circuit that executes a user program, normally for the purpose of *controlling* some device—hence the name micro*controller*.

Microcontrollers are normally found in devices such as microwave ovens, automobiles, keyboards, CD players, cell phones, VCRs, security systems, time & attendance clocks, electronic toys, etc. These are devices that require some amount of computing power but don't require so much as to justify the use of a more complex and expensive 486 or Pentium system which generally requires a large amount of supporting circuitry and memory. A microwave oven just doesn't need that much processing power.

Microcontroller-based systems are generally physically smaller, more reliable, and cheaper than full-blown PC-based systems. They are ideal for the types of applications described above where unit cost and size are very important considerations. In such applications it is almost always desirable to produce designs that utilize the smallest number of integrated circuits, occupy the smallest amount of physical space, require the least amount of energy, and cost as little as possible.

1.1.1 Microcontroller Program Storage

The program for a microcontroller is normally stored on a memory IC—called an EPROM—or in the microcontroller IC itself.

An **EPROM** (Electrically Programmable Read Only Memory) is a special type of IC that does nothing more than store program code or other data that is not lost when power is removed. Traditionally, software for a microcontroller is assembled or compiled on a PC and is subsequently programmed (or "burned") into an EPROM which is then physically inserted into the circuitry of the hardware. The microcontroller accesses the program stored in the EPROM and executes it. This allows the program to be made available to the microcontroller without the need for a hard drive, floppy drive, nor any of the other circuitry and software necessary to manage such devices.

In recent years a growing number of microcontrollers have offered the capability of having programs loaded internally into the microcontroller IC itself. Rather than having a circuit that includes both a microcontroller and an external EPROM, it is now possible to have a just a microcontroller which stores the program code internally.

1.1.2 "Loading" a Microcontroller Program

The manner in which software is transferred from the PC to the hardware depends on whether the design uses an external EPROM or is using a more modern microcontroller that allows the program to be loaded into the microcontroller IC itself.

Programming an EPROM requires special hardware called a **device programmer** (see section 16.1). An EPROM programmer is a device that connects to the PC via the serial, parallel, or USB port. The EPROM is placed into a connector ("socket") and special software transfers the program from the PC to the device programmer which in turn "burns" the program onto the chip. Once the program is burned into the chip, the EPROM is removed from the device programmer and inserted into the physical hardware in which it is to run. EPROMs can subsequently be erased and reprogrammed by exposing them to ultraviolet light for a brief period of time. Devices called **EPROM erasers** exist which illuminate EPROMs with ultraviolet light for the purpose of erasing them.

Programming a microcontroller that stores the program within the microcontroller itself generally requires a serial or parallel port be available for downloading updates to the program. Most of these devices can be programmed by traditional device programmers, as described in the previous paragraph, but if a circuit is being designed from scratch it is usually a good idea to plan for the possibility of programming the microcontroller without removing the IC from the circuit itself—this is especially true of surface-mount parts that may be difficult to remove from the circuit. The datasheet for the microcontroller being used should provide the information necessary to design the circuit for this "in-circuit programming" capability. Datasheets for microcontrollers are available from the websites of each of the companies that produce them.

1.2 What is an 8051 or 8052?

The **8052** is an 8-bit microcontroller originally developed by Intel in the late 1970s. It includes an instruction set of 255 operation codes (**opcodes**), 32 input/output lines, three user-controllable timers, an integrated and automatic serial port, and 256 bytes of on-chip RAM. The **8051** is similar but has only two timers and 128 bytes of on-chip RAM.

The 8052 was designed such that control of the MCU and input/output between the MCU and external devices is accomplished primarily using **Special Function Registers** (**SFR**s, see chapter 3). Each SFR has an address between 128 and 255. Additional functions can be added to new *derivative* MCUs by adding additional SFRs while remaining compatible with the original 8052. This allows the developer to use the same software development tools with any device that is "8052-compatible."

Over time, other semiconductor firms adopted the "8052 core" for their microcontrollers, using the same instruction set and underlying SFRs. This allowed the 8052 architecture to become an industry-wide standard. Now, more than 20 years later, dozens of semiconductor companies produce hundreds of microcontrollers that are based on the original 8052 core. The additional features that each semiconductor-firm offers in their MCUs are accessed by utilizing new SFRs in addition to the standard 8052 SFRs that are found in all 8052-compatible MCUs.

While most microcontrollers based on the 8052 core will include the standard SFRs from the 8052 core, some derivatives may only implement a subset of them. They may also change the function of some bits. This is generally not the case, but it is something to look out for when using derivative chips.

In this book, the term "**8052**" will refer to any MCU that is compatible with the original 8052. As a minimum it will support the 8052 instruction set, support the 8052's 26 SFRs, provide three user timers, and have at least 256 bytes of internal RAM.

The term "**8051**" will refer to any 8052-compatible MCU that doesn't meet the specifications in the previous paragraph, rather mirroring Intel's 8051 microcontroller which was a more limited version of the 8052. As a minimum, an 8051 must support the 21 SFRs supported by the original 8051 and support the standard 8052 instruction set. Generally an 8051 will have two user timers and 128 bytes of internal RAM, although some devices have as little as 64 bytes.

1.2.1 Derivative Chips

The term "**derivative chip**" refers to any 8051 or 8052-compatible microcontroller. There are currently hundreds of derivatives produced by dozens of semiconductor firms.

A derivative will generally—but not always—be able to execute a standard 8052 program without modification. A derivative chip must be based on the 8052 instruction set and support the appropriate SFRs (at least 21 SFRs for an 8051 or 26 for an 8052). Some derivatives operate at different speeds than a standard 8052 so some adjustment to program timing may be necessary when moving to a given derivative from a standard device.

Software development tools designed for the 8052 can always be used to develop software for any derivative chip as long as the programmer explicitly defines any new SFRs that are supported by the derivative chip that they are using.

1.3 Using Windows or Linux

Some people ask whether or not a microcontroller is Windows-compatible or if they can load Linux on their microcontroller. No, a microcontroller cannot run Windows nor can it run Linux. Nor is a microcontroller Windows-compatible, per se. Your microwave oven doesn't run Linux and your automobile is not Windows-compatible.

It is important to remember what microcontrollers are used for. A microcontroller is not a personal computer. It is an integrated circuit that will run short, specialized programs to control hardware devices such as those mentioned earlier. A microwave oven and an automobile simply don't need complex operating systems to perform their designated functions.

That said, Windows or Linux can be used to develop programs *for* microcontrollers. Many Windows products exist that allow 8052 programs to be written in the Windows environment even though the software will ultimately be run by a microcontroller. Once the software is executed by a microcontroller it doesn't matter whether it was originally developed under Windows, Linux, or any other operating system.

CHAPTER 2: TYPES OF MEMORY

The 8052 has three very general types of memory. To program the 8052 effectively it is necessary to have a basic understanding of these memory types. They are: On-Chip Memory, External Code Memory, and External RAM.

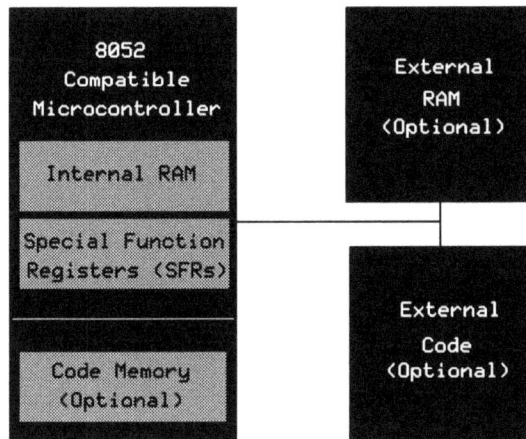

Code Memory is code (or program) memory that is used to store the actual program. This often resides off-chip in the form of an EPROM. Many modern derivative chips allow program storage on the MCU itself (on-chip code memory) and some modern derivative chips do not even support the concept of having code memory located off-chip.

External RAM is RAM memory that resides off-chip. This is typically in the form of standard static RAM.

On-Chip Memory refers to any memory (code, RAM, or other) that physically exists in the MCU itself. On-chip memory can be of several types which will be discussed in section 2.3.

2.1 Code Memory

Code memory is the memory that holds the actual 8052 program that is to be run. This memory is conventionally limited to a maximum of 64K and comes in many shapes and sizes: Code memory may be found on-chip, either burned into the microcontroller as ROM or EPROM, or loaded into flash program memory in newer derivatives. Code may also be stored completely off-chip in an external ROM or, more commonly, an external EPROM. Various combinations of these memory types may also be used—that is to say, it is often possible to have 4K of code memory on-chip and 64k of code memory off-chip in an EPROM, depending on the derivative chip being used.

When the program is stored on-chip, the 64K maximum is often reduced to 1k, 2k, 4k, 8k, 16k, or 32k. This varies depending on the derivative chip in question. Each derivative offers specific capabilities and one of the distinguishing factors is the amount of on-chip code memory the part offers.

Code memory of a "classic" 8052 system is usually off-chip EPROM. This is especially true in low-cost development systems and in systems developed by students in which learning how to interface with off-chip memory is part of the exercise.

Since code memory is restricted to 64K, 8052 programs are limited to 64K. Some compilers offer ways to get around this limit when used with specially wired hardware and a technique known as "memory banking." However, without such special compilers and hardware configurations, programs are limited to 64K.

Some manufacturers such as Dallas Semiconductor, Philips, and Analog Devices have special 8052 derivatives that can address several megabytes of memory.

2.1.1 Memory Architecture

In any given system, the memory architecture can normally be described as either "Harvard" or "Von Neumann". In the simplest terms, this describes whether code and data are contained in two separate memory areas or share a common memory space.

Von Neumann architecture is a system in which code (compiled program, instructions, and constants) and data (variables, registers, etc.) are stored in the same memory space. This is the architecture used in PCs where programs are loaded on-demand from hard drives and subsequently executed from memory. This architecture is very flexible but has the potential disadvantage that a poorly written program can actually destroy itself—or destroy other programs—by erroneously writing data to an area of memory that holds program code. In Windows, this type of problem commonly results in the infamous "General Protection Fault" error message.

Harvard architecture is a system in which code and data are stored in separate memory space. This architecture has the advantage that a program cannot accidentally overwrite itself. This architecture is commonly used in embedded systems where the program is seldom modified nor is it loaded and unloaded.

8052 systems generally are designed with a Harvard architecture. Thus there may be 64k of data memory available (often in the form of an external RAM IC) and 64k of code memory (often in the form of an external EPROM). An 8052 program running under this architecture cannot modify itself intentionally or accidentally.

The 8052 can be used in a Von Neumann architecture by connecting the MCU to external RAM in a certain way—this will be discussed in section 14.2.5. In practice, this is seldom done.

2.2 External RAM

Although 8052s contain a small amount of on-chip internal RAM (see section 2.3.1), external RAM is also supported.

As the name suggests, external RAM is any random access memory that is found off-chip. Since the memory is off-chip the assembly language instructions to access it are slower and less flexible. For example, to increment an internal RAM location by 1 requires only 1 instruction and 1 instruction cycle. To increment a value stored in external RAM requires 4 instructions and 7 instruction cycles. In this case, external memory is seven times slower and requires four times as much program memory to manipulate.

What external RAM loses in speed and flexibility it gains in quantity. While internal RAM is normally limited to 256 bytes (128 with 8051s), the 8052 supports external RAM up to 64K.

Generally, the 8052 may only address 64k of RAM. To expand RAM beyond this limit requires programming and hardware tricks.

2.3 On-Chip Memory

As mentioned at the beginning of this chapter, the 8052 includes a certain amount of on-chip memory. Aside from on-chip code memory, on-chip memory is really one of two types: **Internal RAM** and **Special Function Register** (SFR) memory. The layout of the 8052's internal memory is presented in the following memory map.

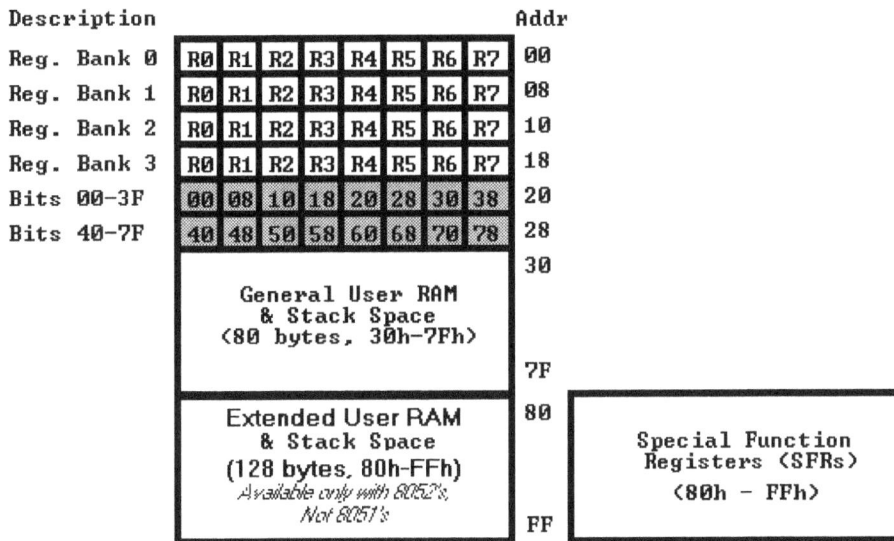

Description		Addr
Reg. Bank 0	R0 R1 R2 R3 R4 R5 R6 R7	00
Reg. Bank 1	R0 R1 R2 R3 R4 R5 R6 R7	08
Reg. Bank 2	R0 R1 R2 R3 R4 R5 R6 R7	10
Reg. Bank 3	R0 R1 R2 R3 R4 R5 R6 R7	18
Bits 00-3F	00 08 10 18 20 28 30 38	20
Bits 40-7F	40 48 50 58 60 68 70 78	28
		30
	General User RAM & Stack Space (80 bytes, 30h-7Fh)	
		7F
	Extended User RAM & Stack Space (128 bytes, 80h-FFh) *Available only with 8052's, Not 8051's*	80
	Special Function Registers (SFRs) (80h - FFh)	
		FF

2.3.1 Internal RAM

As the figure above shows, the 8052 has a bank of 256 bytes of internal RAM. This internal RAM is found on-chip within the 8052 so it is the fastest RAM available. It is also the most flexible in terms of reading, writing, and modifying its contents. Internal RAM is volatile so when the 8052 is powered-up, this memory is undefined.

The 256 bytes of internal RAM are subdivided as shown in the memory map above. The first eight bytes (00h - 07h) are "register bank 0". By manipulating the Program Status Word (PSW) SFR, a program may choose to use register banks 0, 1, 2, or 3. These alternative register banks are located in internal RAM in addresses 08h through 1Fh. "Register banks" are discussed in section 2.3.1.2. For now it is sufficient to know that they are located in—and are part of—internal RAM.

Bit memory also is located in and is a part of internal RAM. Bit memory is covered in section 2.3.1.3. For now just keep in mind that bit memory actually resides in internal RAM from addresses 20h through 2Fh.

Also note that all of internal RAM is general-use, byte-wide memory, regardless of whether it is used by register banks or bit memory. This means that internal RAM allocated to register banks 1, 2, and 3 and bit memory may be used for other purposes if the program will not be using those register banks and/or bit variables.

The remaining 208 bytes of internal RAM, from addresses 30h through FFh, may be used by user variables that need to be accessed frequently or at high-speed. This area is also utilized by the microcontroller as a storage area for the operating stack. This significantly limits the 8052's stack size since the area reserved for the stack is only 208 bytes—and usually much less since the 208 bytes of internal RAM must be shared between the stack and user variables.

> While the 8052 has 256 bytes of internal RAM, the 8051 only has 128 bytes. This further restricts the amount of RAM available on-chip for user variables and the operating stack.

2.3.1.1 Stack

The stack is a "last in, first out" (LIFO) storage area that exists in internal RAM. It is used by the 8052 to store values that the user program manually pushes onto the stack as well as to store the return addresses for calls and interrupt service routines.

The stack is defined and controlled by an SFR called the stack pointer, or SP. SP, as a standard 8-bit SFR, holds a value between 0 and 255 that represents the internal RAM address of the *end* of the current stack. If a value is removed from the stack, it will be taken from the internal RAM address pointed to by SP and SP will subsequently be decremented by 1. If a value is pushed onto the stack, SP will first be incremented and the value will be inserted in internal RAM at the address now pointed to by SP.

SP is initialized to 07h when an 8052 is first powered-up. This means the first value to be pushed onto the stack will be placed at internal RAM address 08h (07h + 1), the second will be placed at 09h, etc.

> By default, the 8052 initializes the stack pointer (SP) to 07h. This means that the stack will start at address 08h and expand upwards. If a program will be using the alternate register banks (banks 1, 2 or 3) it must initialize the stack pointer to an address above the highest register bank that it will be using, otherwise the stack will overwrite the alternate register banks. Similarly, if the program will be using bit variables it is usually a good idea to initialize the stack pointer to some value greater than 2Fh to guarantee that the bit variables are protected from the stack.

2.3.1.2 Register Banks

The 8052 includes eight "R" registers which are used in many of its instructions. These "R" registers are numbered from 0 through 7 (R0, R1, R2, R3, R4, R5, R6, and R7) and are generally used to assist in manipulating values and moving data from one memory location to another. For example, to add the value of R4 to the accumulator, the following assembly language instruction would be used:

```
ADD A,R4
```

Thus if the accumulator (A) contained the value 6 and R4 contained the value 3, the accumulator would contain the value 9 after this instruction was executed. However, as the memory map shows, the "R" register R4 is really part of internal RAM. Specifically, R4 is address 04h of internal RAM. Thus the above instruction is functionally equivalent to the following operation:

```
ADD A,04h
```

This instruction adds the value found in internal RAM address 04h to the value of the accumulator, leaving the result in the accumulator. Since R4 is really internal RAM address 04h, the above instruction produces the same result as the previous ADD instruction.

But as the memory map also shows, the 8052 has four distinct register banks. When the 8052 is first initialized, register bank 0 (addresses 00h through 07h) is used by default. However, the user program may instruct the 8052 to use one of the alternate register banks; i.e., register banks 1, 2, or 3. In that case, R4 will no longer be the same as internal RAM address 04h. For example, if the program instructs the 8052 to use register bank 1, register R4 will now be synonymous with internal RAM address 0Ch. If the program selects register bank 2, R4 will be synonymous with 14h, and if it selects register bank 3 it will be synonymous with address 1Ch.

The register bank is selected by setting or clearing the bits RS0 and RS1 in the Program Status Word (PSW) Special Function Register.

```
MOV PSW,#00h        ;Sets register bank 0
MOV PSW,#08h        ;Sets register bank 1
MOV PSW,#10h        ;Sets register bank 2
MOV PSW,#18h        ;Sets register bank 3
```

The above instructions will make more sense after the topic of Special Function Registers is covered in section 2.3.2 and in chapter 3.

The concept of register banks adds a great level of flexibility to the 8052, especially when dealing with interrupts (more about interrupts in chapter 10). But always remember that the register banks reside in the first 32 bytes of internal RAM.

If only the first register bank (i.e. bank 0) is being used, internal RAM locations 08h through 1Fh may be used for anything the programmer desires. If the program uses register banks 1, 2, or 3, be very careful about using addresses below 20h as the program may end up overwriting the value of "R" registers in other register banks.

2.3.1.3 Bit Memory

The 8052, being a communications and control-oriented microcontroller that often has to deal with "on" and "off" situations, gives the developer the ability to access a number of bit variables directly with simple instructions to set, clear, and compare these bits. These variables may be either 1 or 0.

There are 128 bit variables available to the developer, numbered 00h through 7Fh. The developer may make use of these variables with instructions such as SETB and CLR. For example, to set bit number 24h (hex), the following instruction may be used:

```
SETB 24h
```

It is important to note that bit memory, like the register banks above, is really a part of internal RAM. The 128 bit variables occupy the 16 bytes of internal RAM from 20h through 2Fh. Thus, if the value FFh is written to internal RAM address 20h, bits 00h through 07h have effectively been set. That is to say that the instruction:

```
MOV 20h,#0FFh
```

is equivalent to the instructions:

```
SETB 00h
SETB 01h
SETB 02h
SETB 03h
SETB 04h
SETB 05h
SETB 06h
SETB 07h
```

As illustrated above, bit memory isn't really a new type of memory but rather a subset of internal RAM. Since the 8052 provides special instructions to access these 16 bytes of memory on a bit-by-bit basis it is useful to think of it as a separate type of memory, yet always keep in mind that it is just a subset of internal RAM—and that operations performed on internal RAM can change the values of the bit variables.

If a program does not use bit variables, internal RAM locations 20h through 2Fh may be used for anything the developer wishes. If bit variables are to be used, however, be very careful about using addresses from 20h through 2Fh as the value of the corresponding bits will be overwritten in the process.

By default, the 8052 initializes the stack pointer (SP) to 07h when the microcontroller is booted. This means that the stack will start at address 08h and expand upwards. If bit variables will be used it is usually a good idea to initialize the stack pointer to some value greater than 2Fh to guarantee that the bit variables are protected from the stack.

While bit memory 00h through 7Fh are for developer-defined functions, bit memory 80h and above are used to access certain SFRs (see section 2.3.2) on a bit-by-bit basis. For example, if output lines P0.0 through P0.7 are all clear (0) and the program needs to raise P0.1, the following instruction may be executed:

```
MOV P0,#02h
```

12

The following instruction also accomplishes the task of raising P0.1 by treating the line as an SFR bit variable:

```
SETB 81h
```

Both of these instructions accomplish the same thing but the SETB command will turn on the P0.1 line without affecting the status of any of the other P0 output lines. The MOV command effectively turns off all the other output lines which, in some cases, may not be acceptable.

When working with bit addresses of 80h and above, remember that they refer to the bits of corresponding SFRs that are divisible by eight. This is a complicated way of saying that bits 80h through 87h refer to bits 0 through 7 of SFR 80h, bits 88h through 8Fh refer to bits 0 through 7 of SFR 88h, bits 90h through 97h refer to bits 0 through 7 of 90h, etc.

2.3.2 Special Function Register (SFR) Memory

Special Function Registers (**SFR**s) are areas of memory that control specific functionality of the 8052. For example, four SFRs permit access to the 8052's 32 input/output lines (8 lines per SFR). Another SFR allows the program to read or write to the 8052's serial port. Other SFRs allow for the configuration of the serial port baud rate, control of and access to timers, and configure the 8052's interrupt system.

When programming, SFRs have the *illusion* of being internal RAM because SFR addresses are 0FFh and below, just like internal RAM. For example, the following instruction would write the value "1' to internal RAM location 50h:

```
MOV 50h,#01h
```

Similarly, to send the value "1" to the 8052's serial port, the program would write that value to the "SBUF" SFR which has an SFR address of 99 Hex. Thus to write the value "1" to the serial port the program would execute the instruction:

```
MOV 99h,#01h
```

While it may appear as though the SFR is part of internal RAM, this is not the case. When using this method of memory access (called "direct address", more on that in chapter 5), any address between 00h and 7Fh refers to an internal RAM memory address; any address between 80h and FFh refers to an SFR control register.

SFRs are used to control the way in which the 8052 functions. Each SFR has a specific purpose and format that will be discussed later. Not all addresses above 80h are assigned to SFRs. However, this area may NOT be used as additional RAM memory even if a given address has not been assigned to an SFR.

Since all direct access to addresses 80h through FFh refers to SFRs, direct addressing cannot be used to access internal RAM addresses 80h through FFh. The upper 128 bytes of internal RAM must be accessed using "indirect addressing" which will be explained in chapter 5.

CHAPTER 3: SPECIAL FUNCTION REGISTERS (SFRs)

The 8052 is a flexible microcontroller with a relatively large number of modes of operation. A program may inspect or change the operating mode of the 8052 by manipulating the values of the **Special Function Registers** (SFRs).

SFRs are accessed as if they were normal internal RAM. The only difference is that internal RAM is addressed in "direct mode" with addresses 00h through 7Fh while SFR registers are accessed in the range of 80h through FFh.

Each SFR has an address (80h through FFh) and a name. The following chart provides a graphical representation of the 8052's SFRs, their names, and their address.

80	P0	SP	DPL	DPH				PCON	87
88	TCON	TMOD	TL0	TL1	TH0	TH1			8F
90	P1								97
98	SCON	SBUF							9F
A0	P2								A7
A8	IE								AF
B0	P3								B7
B8	IP								B9
C0									C7
C8	T2CON		RCAP2L	RCAP2H	TL2	TH2			CF
D0	PSW								D7
D8									DF
E0	ACC								E7
E8									EF
F0	B								F7
F8									FF

Although the address range of 80h through FFh offers 128 possible addresses, only 26 SFRs are defined for a standard 8052 (21 for an 8051). All other addresses in the SFR range (80h through FFh) are considered invalid. Writing to or reading from these registers may produce undefined values or behavior.

SFR addresses that have not been assigned to an SFR should not be accessed. Doing so may provoke undefined behavior and may cause the program to be incompatible with other derivatives that use the given SFR for some other purpose.

3.1 Referencing SFRs

When writing code in assembly language, SFRs may be referenced either by their name or their address.

For example, the chart above indicates that the SBUF SFR is at address 99h. To write the value 24h to the

SBUF SFR, the following instruction could be used:

```
MOV 99h,#24h
```

This instruction moves the value 24h into address 99h. Since the value 99h is in the range of 80h-FFh, it refers to an SFR. Since 99h refers to the SBUF SFR, this instruction will accomplish the goal of writing the value 24h to the SBUF SFR.

While the above instruction certainly works, it is not necessarily easy to remember the address of each SFR when writing software. Luckily, 8052 assemblers allow the developer to use the name of the SFR rather than its numeric address. The above instruction would more commonly be written as:

```
MOV SBUF,#24h
```

This is much easier to read since it is obvious the SBUF SFR is being accessed. The assembler will automatically convert this to its numeric address at assemble-time.

3.2 Referencing Bits of SFRs

Individual bits of SFRs are referenced in one of two ways. The general convention is to name the SFR followed by a period and the bit number. For example, SCON.0 refers to bit 0 (the least significant bit) of the SCON SFR. SCON.7 refers to bit 7 (the most significant bit) of SCON.

These bits also have names: SCON.0 is RI and SCON.7 is SM0. When talking about SFR bits they are often referred to by their name. It is also acceptable for a program to refer to SFR bits by the name of the bit as long as the SFR is bit-addressable.

3.3 Bit-Addressable SFRs

All SFRs that have addresses divisible by eight (such as 80h, 88h, 90h, 98h, etc.) are bit-addressable. This means that individual bits of these SFRs can be set and cleared by using the SETB and CLR instruction.

The SFRs whose names appear in the left column in the table at the beginning of this chapter are SFRs that may be accessed using bit operations such as SETB and CLR. The other SFRs cannot be accessed using bit operations.

3.4 SFR Types

The four SFRs that appear in bold type—P0, P1, P2, and P3—are related to the I/O ports. The 8052 has four I/O ports of 8 bits, for a total of 32 I/O lines. Whether a given I/O line is high or low, and the value read from each line, is controlled by these SFRs.

The SFRs that appear in italics are registers that in some way control the operation or the configuration of some aspect of the 8052. For example, TCON controls the timers, SCON controls the serial port.

The remaining SFRs are "other SFRs." These SFRs can be thought of as auxiliary SFRs in the sense that they don't directly configure the microcontroller, but the 8052 cannot operate without them. For example, once the serial port has been configured using SCON, the program may read or write to the serial port using the SBUF register.

3.5 SFR Descriptions

This section will endeavor to briefly explain each of the standard SFRs found in the above SFR chart. It is not the intention of this section to fully explain the functionality of each SFR—that information will be covered in other chapters. The purpose of this section is to provide a general idea of the purpose of each SFR.

P0 (Port 0, Address 80h, Bit-Addressable): This is input/output port 0. Each bit of this SFR corresponds to one of the pins on the microcontroller. For example, bit 0 of port 0 is pin P0.0, bit 7 is pin P0.7. Writing a value of 1 to a bit of this SFR will set a high level on the corresponding I/O pin whereas a value of 0 will bring it to a low level. A port pin must be brought high in order for it to be read.

> While the 8052 has four I/O ports (P0, P1, P2, and P3), if the hardware uses external RAM or external code memory, the program may not use P0 or P2. This is because the 8052 uses ports P0 and P2 to address the external memory.

SP (Stack Pointer, Address 81h): This is the stack pointer of the microcontroller. SP indicates where the next value to be taken from the stack will be read from in internal RAM. If a value is pushed onto the stack, the value will be written to the address of SP + 1. That is to say that if SP holds the value 07h, a PUSH instruction will push the value onto the stack at address 08h. This SFR is modified by all instructions that modify the stack, such as PUSH, POP, LCALL, RET, RETI, and whenever interrupts are triggered by the microcontroller.

> SP, on startup, is initialized to 07h. This means the stack will start at 08h and subsequently expand upwards in internal RAM. Since alternate register banks 1, 2, and 3, as well as the user bit variables, occupy internal RAM from addresses 08h through 2Fh, it is necessary to initialize SP to some higher value if the alternate register banks and/or bit memory will be used. It's not a bad idea to initialize SP to 2Fh as the first instruction of every programs unless there is absolutely no possibility that either the register banks or the bit variables will be used.

DPL/DPH (Data Pointer Low/High, Addresses 82h/83h): DPL and DPH work together to represent a 16-bit value called the **data pointer**. The data pointer is used in operations related to external RAM and some instructions involving code memory. Since it is an unsigned two-byte integer value, it can represent values from 0000h to FFFFh (0 through 65,535 decimal).

> The DPTR register is DPH and DPL taken together as a 16-bit value. In reality, DPTR must almost always be dealt with one byte at a time. For example, to push DPTR onto the stack, first DPL is pushed and then DPH. It is not possible to simply "push DPTR" onto the stack all at once. Additionally, there is an instruction to "increment DPTR." When this instruction in executed, the two bytes are operated upon as a 16-bit value. DPTR is a useful storage location for the occasional 16-bit value that may be manipulated in a program—especially if the value must be incremented frequently.

PCON (Power Control, Address 87h): The Power Control SFR is used to control the 8052's power control modes. Certain operation modes allow the 8052 to go into a type of sleep mode that requires much less power. These modes of operation are controlled through PCON. Additionally, one of the bits in PCON is used to double the baud rate of the 8052's serial port.

TCON (Timer Control, Address 88h, Bit-Addressable): The Timer Control SFR is used to configure and modify the way in which the 8052's two basic timers operate. This SFR controls whether each of the two timers is running or stopped and contains a flag to indicate when a timer has overflowed. Additionally, some non-timer related bits are located in the TCON SFR. These bits are used to configure the way in which the external interrupts are activated and also contain the external interrupt flags that are set when an external interrupt has occurred.

TMOD (Timer Mode, Address 89h): The Timer Mode SFR is used to configure the mode of operation of each of the two timers. Using this SFR the program may configure each timer to act as a 16-bit timer, an 8-bit auto-reload timer, a 13-bit timer, or two separate timers. Additionally, the timers may be configured to only count when an external pin is activated or to count "events" that are indicated on an external pin.

TL0/TH0 (Timer 0 Low/High, Addresses 8Ah/8Bh): These two SFRs, taken together, represent timer 0. Their exact behavior depends on how the timer is configured by the TMOD SFR.

TL1/TH1 (Timer 1 Low/High, Addresses 8Ch/8Dh): These two SFRs, taken together, represent timer 1. Their exact behavior depends on how the timer is configured by the TMOD SFR.

P1 (Port 1, Address 90h, Bit-Addressable): This is input/output port 1. Each bit of this SFR corresponds to one of the pins on the microcontroller. For example, bit 0 of port 1 is pin P1.0, bit 7 is pin P1.7. Writing a value of 1 to a bit of this SFR will set a high level on the corresponding I/O pin whereas a value of 0 will bring it to a low level. A port pin must be brought high in order for it to be read.

SCON (Serial Control, Address 98h, Bit-Addressable): The Serial Control SFR is used to configure the behavior of the 8052's on-chip serial port. This SFR controls how the baud rate of the serial port is determined, whether the serial port is activated to receive data, and contains flags that are set when a byte is successfully sent or received.

To use the 8052's on-chip serial port it is generally necessary to initialize SCON, TCON (or T2CON), and TMOD. This is because SCON controls the serial port but in most cases the program will wish to use one of the timers to establish the serial port's baud rate. In these cases it is necessary to configure TCON and TMOD, or T2CON.

SBUF (Serial Control, Address 99h): The Serial Buffer SFR is used to send and receive data via the serial port. Any value written to SBUF will be sent out the serial port's TXD pin. Likewise, any value the 8052 receives via the serial port's RXD pin will be delivered to the user program via SBUF. In other words, SBUF acts as if it were really two registers: one which serves as the output port when written to and the other which serves as the input port when read from.

P2 (Port 2, Address A0h, Bit-Addressable): This is input/output port 2. Each bit of this SFR corresponds to one of the pins on the microcontroller. For example, bit 0 of port 2 is pin P2.0, bit 7 is pin P2.7. Writing a value of 1 to a bit of this SFR will set a high level on the corresponding I/O pin whereas a value of 0 will bring it to a low level. A port pin must be brought high in order for it to be read.

While the 8052 has four I/O ports (P0, P1, P2, and P3), if the hardware uses external RAM or external code memory, the program may not use P0 or P2. This is because the 8052 uses ports P0 and P2 to address the external memory.

IE (Interrupt Enable, Address A8h): The Interrupt Enable SFR is used to enable and disable interrupts. The low 7 bits of the SFR are used to enable/disable specific interrupts whereas the highest bit is used to enable or disable *all* interrupts. Thus, if the high bit of IE is clear, all interrupts are disabled regardless of whether an individual interrupt is enabled by setting a lower bit.

P3 (Port 3, Address B0h, Bit-Addressable): This is input/output port 3. Each bit of this SFR corresponds to one of the pins on the microcontroller. For example, bit 0 of port 3 is pin P3.0, bit 7 is pin P3.7. Writing a value of 1 to a bit of this SFR will set a high level on the corresponding I/O pin whereas a value of 0 will bring it to a low level. A port pin must be brought high in order for it to be read.

IP (Interrupt Priority, Address B8h, Bit-Addressable): The Interrupt Priority SFR is used to specify the relative priority of each interrupt. An interrupt may either be of low (0) priority or high (1) priority. An interrupt may only preempt interrupts of lower priority. For example, if the 8052 is configured so that all interrupts are of low priority except the serial interrupt, the serial interrupt will always be able to interrupt the system even if another interrupt is currently executing. However, if a serial interrupt is executing no other interrupt will be able to interrupt it since the serial interrupt routine has the highest priority.

T2CON (Timer Control 2, Address C8h, Bit-Addressable): The Timer Control 2 SFR is used to configure and control the way in which timer 2 operates. This SFR is only available on 8052s.

TL2/TH2 (Timer 2 Low/High, Addresses CCh/CDh): These two SFRs, taken together, represent timer 2. Their exact behavior depends on how the timer is configured in the T2CON SFR. This SFR is only available on 8052s, not on 8051s.

RCAP2L/RCAP2H (Timer 2 Capture Low/High, Addresses CAh/CBh): These two SFRs, taken together, represent the timer 2 capture register. It may be used as a reload value for timer 2, or to capture the value of timer 2 under certain circumstances. The exact purpose and function of these two SFRs depends on the configuration of T2CON. This SFR is only available on 8052s, not on 8051s.

PSW (Program Status Word, Address D0h, Bit-Addressable): The Program Status Word is used to store a number of important bits that are set and cleared by 8052 instructions. The PSW SFR contains the carry flag, the auxiliary carry flag, the overflow flag, and the parity flag. Additionally, the PSW register contains the register bank flags that are used to set the active register bank.

When writing an interrupt service routine, it is a very good idea to always save the PSW SFR on the stack and restore it when the interrupt is complete. Many 8052 instructions modify the bits of PSW. If the interrupt routine does not guarantee that PSW is the same upon exit as it was upon entry, the program is bound to behave unpredictably—and it will be tricky to debug since the behavior will tend to be apparently random.

ACC (Accumulator, Address E0h, Bit-Addressable): The accumulator is one of the most-used SFRs on the 8052 since it is involved in so many instructions. The accumulator exists as an SFR at address E0h which means the instruction MOV A,#20h is the same as MOV 0E0h,#20h. However, it is a good idea to use the first method since special instructions exist that allow the first option to be assembled in only two bytes whereas the second option requires three bytes.

B (B Register, Address F0h, Bit-Addressable): The "B" register is explicitly used in the multiply and divide instructions. The B register is also commonly used by programmers as an auxiliary register to store temporary values.

3.6 Other SFRs for Derivative Chips

The table at the beginning of this chapter is a summary of all the SFRs that exist in a standard 8052. Almost all derivative microcontrollers of the 8052 support these basic SFRs in order to maintain compatibility with the underlying MCS-51 standard.

A common practice when semiconductor firms wish to develop a new 8052 derivative is to add additional SFRs to support new functions that exist in the new chip. For example, the Dallas Semiconductor DS80C320 is upward compatible with the 8052. This means that any program that runs on a standard 8052 should run without modification on the DS80C320. This also means that all the SFRs defined above also apply to the Dallas component.

However, since the DS80C320 provides many new features that the standard 8052 does not, there must be some way to control and access these new features. This is accomplished by adding additional SFRs to those covered in this chapter. For example, since the DS80C320 supports two serial ports (as opposed to just one on the 8052), the SFRs SBUF2 and SCON2 have been added. In addition to all the SFRs listed above, the DS80C320 also recognizes these two new SFRs as valid and uses their values to determine the mode of operation of the secondary serial port. Obviously, these new SFRs have been assigned to SFR addresses that were unused in the original 8052. In this manner new 8052 derivative chips may be developed which will run existing 8052 programs.

If a program utilizes new SFRs that are specific to a given derivative chip and not included in the above SFR list, the program will not run properly on a standard 8052 where that SFR does not exist. Thus, only use non-standard SFRs if the program is only expected to run on a specific derivative microcontroller. Likewise, if code is written that uses non-standard SFRs and is subsequently shared with a third-party, be sure to let that party know that the code is using non-standard SFRs to save them the headache of discovering that due to strange program behavior at run-time.

CHAPTER 4: BASIC REGISTERS

A number of 8052 registers can be considered "basic." Very little can be done without them and a detailed explanation of each one is warranted to make sure the reader understands these registers before getting into more complicated areas of development.

4.1 Accumulator

Those that have worked with other assembly languages will be familiar with the concept of an accumulator register.

The accumulator, as its name suggests, is used as a general register to accumulate the results of a large number of instructions. It can hold an 8-bit (1-byte) value and is the most versatile register the 8052 has due to the sheer number of instructions that make use of it. More than half of the 8052's 255 instructions manipulate or use the accumulator in some way.

For example, when adding the number 10 and 20, the resulting 30 will be stored in the accumulator. Once a value is stored in the accumulator the program may continue processing the value or it may be stored in another register or in memory.

4.2 "R" registers

The "R" registers are sets of eight registers that are named R0 through R7.

These registers are used as auxiliary registers in many operations. To continue with the above example, perhaps the values 10 and 20 are being added. The original number 10 may be stored in the accumulator while the value 20 may be stored in, say, register R4. To perform the addition the following instruction would be executed:

```
ADD A,R4
```

After executing this instruction the accumulator will contain the value 30.

The "R" registers may be thought of as very important auxiliary, or "helper", registers. The accumulator alone would not be very useful if it were not for these "R" registers.

The "R" registers are also used to store values temporarily. For example, to add the values in R1 and R2 together and then subtract the values of R3 and R4, the following code could be used:

```
MOV A,R3      ;Move the value of R3 into the accumulator
ADD A,R4      ;Add the value of R4
MOV R5,A      ;Store the resulting value temporarily in R5
MOV A,R1      ;Move the value of R1 into the accumulator
ADD A,R2      ;Add the value of R2
CLR C         ;Make sure the carry bit is clear before we subtract
SUBB A,R5     ;Subtract the value of R5 (which now contains R3 + R4)
```

R5 is used to temporarily hold the sum of R3 and R4. Of course, this isn't the most efficient way to calculate (R1+R2) - (R3 +R4) but it does illustrate the use of the "R" registers as a way to store values temporarily.

The concept of register banks adds a great level of flexibility to the 8052, especially when dealing with interrupts.

4.3 "B" Register

The "B" register is very similar to the accumulator in the sense that it may hold an 8-bit (1-byte) value. The "B" register is only used explicitly by two 8052 instructions: MUL AB and DIV AB. To quickly and easily multiply or divide the accumulator by another number, the other number is stored in in "B" and the corresponding instruction is executed.

Aside from the MUL and DIV instructions, the "B" register is often used as another temporary storage register, much like a ninth "R" register.

4.4 Program Counter (PC)

The Program Counter (PC) is a 16-bit register that tells the 8052 the address of the next instruction to execute in code memory. When the 8052 is initialized, the program counter always starts at 0000h and is incremented each time an instruction is executed. It is important to note that the program counter isn't always incremented by one. Since some instructions are 2 or 3 bytes in length, it will be incremented by 2 or 3 in these cases.

The program counter is special in that there is no way to directly modify its value. That is to say, there is no instruction along the lines of PC=2430h. On the other hand, if the instruction LJMP 2430h is executed the value of PC will be set to 2430h automatically—effectively accomplishing the same thing.

It is also interesting to note that while the value of the program counter may be changed (by executing a jump instruction, etc.), it is not possible to read its value directly. That is, there is no direct way to query the 8052 to find the address in code memory that is about to be executed.

> Although the microcontroller itself does not provide a method for reading the program counter, there is a trick that can be used to read the value of PC. This is covered in section 22.1.

4.5 Data Pointer (DPTR)

The data pointer (DPTR) is the 8052's only user-accessible 16-bit (2-byte) register. The accumulator, "R" registers, and "B" register are all 8-bit. The program counter that was just described is a 16-bit value but isn't directly user-accessible as a working register.

DPTR, as the name suggests, is used to point to data. It is used by a number of instructions that allow the 8052 to access external memory. When the 8052 accesses external memory it accesses the memory at the address pointed to by DPTR.

While DPTR is most often used to point to data in external memory or code memory, many developers take advantage of the fact that it's the only true 16-bit register available. It is often used to store 2-byte values that have nothing to do with memory locations, especially if the value needs to frequently be incremented by one.

4.6 Stack Pointer (SP)

The stack pointer, like all registers except DPTR and PC, may hold an 8-bit (1-byte) value. The stack pointer is used to indicate where the next value to be removed from the stack should be taken from in internal RAM.

When a value is pushed onto the stack, the 8052 first increments the value of SP and then stores the value at the resulting memory location. When a value is popped off the stack, the 8052 returns the value from the memory location indicated by SP, and then decrements the value of SP.

This order of operation is important. When the 8052 is powered-up or reset, SP will be initialized to 07h. If a value is immediately pushed onto the stack the value will be stored in internal RAM address 08h. This is consistent with the explanation in the previous paragraph: first the 8052 will increment the value of SP (from 07h to 08h) and then will store the pushed value at that memory address (08h).

SP is modified directly by six instructions: PUSH, POP, ACALL, LCALL, RET, and RETI. It is also used intrinsically whenever an interrupt is triggered.

CHAPTER 5: ADDRESSING MODES

As is the case with all processors, the 8052 utilizes several memory addressing modes. An "addressing mode" refers to the form in which a given memory location or data value is being accessed (or "addressed"). The addressing modes supported by the 8052 are listed below with an example of each:

Immediate Addressing	`MOV A,#20h`
Direct Addressing	`MOV A,30h`
Indirect Addressing	`MOV A,@R0`
External Direct	`MOVX A,@DPTR`
External Indirect	`MOVX A,@R0`
Code Indirect	`MOVC A,@A+DPTR`

Each of these addressing modes provides important flexibility to the developer.

5.1 Immediate Addressing

Immediate addressing is so-named because the value to be stored in memory immediately follows the opcode in code memory. That is to say, the instruction itself dictates what value will be stored in memory.

For example:

```
MOV A,#20h
```

This instruction uses immediate addressing because the accumulator (A) will be loaded with the value that immediately follows; in this case 20h (hexadecimal). The pound sign (#) indicates that this is direct addressing and the number that follows is to be loaded into the first register.

Immediate addressing is very fast since the value to be loaded is included in the instruction. However, since the value to be loaded is fixed at compile-time it is not very flexible. It is used to load the same known value every time the instruction executes.

5.2 Direct Addressing

Direct addressing is so-named because the value to be stored in memory is obtained by directly retrieving it from another memory location which is specified in the instruction.

For example:

```
MOV A,30h
```

This instruction will read the data out of internal RAM address 30h (hexadecimal) and store it in the accumulator (A). This syntax is very similar to immediate addressing discussed in the previous section, but direct addressing omits the pound sign. The lack of a pound sign indicates that this is direct addressing as opposed to immediate addressing.

Direct addressing is generally fast since, although the value to be loaded isn't included in the instruction, it is quickly accessible since it is stored in the 8052's internal RAM. It is also much more flexible than immediate addressing since the value to be loaded is whatever is found at the given address—the value of which may change.

It is important to note that when using direct addressing any instruction that refers to an address between 00h and 7Fh is referring to internal RAM while any address between 80h and FFh is referring to the SFR control registers that control the 8052 itself. For this reason, direct addressing cannot be used to access the upper 128 bytes of internal RAM which share the same addresses as the SFRs. The upper half of internal RAM can only be accessed with the next addressing mode, "indirect addressing."

5.3 Indirect Addressing

Indirect addressing is a very powerful addressing mode that provides an exceptional level of flexibility. Indirect addressing is also the only way to access the upper 128 bytes of internal RAM found on an 8052. This addressing mode appears as follows:

```
MOV A,@R0
```

This instruction causes the 8052 to read the value of the R0 register. The 8052 will then load the accumulator (A) with the value from internal RAM that is found at the address pointed to by R0.

For example:

```
MOV R0,#40h
MOV A,@R0
```

In this example, the instruction will find that the register R0 holds the value 40h so it will load the accumulator without the value currently stored in internal RAM address 40h. If internal RAM address 40h contains 67h, the accumulator will be loaded with the number 67h.

Indirect addressing always refers to internal RAM—it never refers to an SFR. In a prior example it was mentioned that SFR 99h can be used to write a value to the serial port. Thus one may think that the following would be a valid solution to write the value '1' to the serial port:

```
MOV R0,#99h        ;Load the SFR address of the serial port
MOV @R0,#01h       ;Send 01 to the serial port -- WRONG!!
```

This is not valid. Since indirect addressing always refers to internal RAM, these two instructions would write the value 01h to internal RAM address 99h on an 8052. On an 8051 these two instructions would produce an undefined result since the 8051 only has 128 bytes of internal RAM.

5.4 External Direct

External memory is accessed using a suite of instructions that operate with "external direct" addressing. It is so called because the address to be accessed is contained directly in the DPTR register, but it accesses external memory.

There are only two commands that use External Direct addressing mode:

```
MOVX A,@DPTR          ;Read from external RAM
MOVX @DPTR,A          ;Write to external RAM
```

Both commands utilize DPTR. With these instructions, DPTR must first be loaded with the address of external memory that is to be read or written. Once DPTR holds the correct external memory address, the first command will move the contents of that external memory address to or from the accumulator.

For example, to read the contents of external RAM address 1516h, the following instructions would be executed:

```
MOV DPTR,#1516h       ;Select the external address to read
MOVX A,@DPTR          ;Move the contents of external RAM into accumulator
```

The second command will do the opposite: it allows the program to write the value of the accumulator to the external memory address pointed to by DPTR. For example, to write the contents of the accumulator to external RAM address 1516, the following instructions would be executed:

```
MOV DPTR,#1516h       ;Select the external address to write
MOVX @DPTR,A          ;Move the contents of accumulator into external RAM
```

Technically, accessing external memory with the MOVX @DPTR instructions is *indirect* addressing since the address to be accessed is referred to indirectly by the DPTR register. However, to access a specific external RAM memory location the most direct way is to load DPTR with the address in question and access it with the MOVX instruction. Thus while this approach is technically "indirect," we've called it "external direct" since it is the most direct method of accessing a specific external RAM memory location and to differentiate it from the following addressing mode (external indirect) which is very similar to the "indirect addressing" mentioned in section 5.3 in its use of R0 and R1.

5.5 External Indirect

External memory can also be accessed using a form of indirect addressing called "external indirect." This form of addressing is usually only used in relatively small projects that have a very small amount of external RAM. An example of this addressing mode is:

```
MOVX @R0,A
```

The value of R0 is first read and the value of the accumulator is written to that address in external RAM. Since the value of R0 can only be 00h through FFh the project would effectively be limited to 256 bytes of external RAM. There are relatively simple hardware/software tricks that can be implemented to access more than 256 bytes of memory using external indirect addressing; however, it is usually easier to use external direct addressing if the project has more than 256 bytes of external RAM.

5.6 Code Indirect

Two additional 8052 instructions allow the developer to access the program code itself. This is useful for accessing data tables, strings, etc. The two instructions are:

```
MOVC A,@A+DPTR
MOVC A,@A+PC
```

For example, to access the data stored in code memory at address 2021h, the following instructions would be implemented:

```
MOV DPTR,#2021h      ;Set DPTR to 2021h
CLR A                ;Clear the accumulator (set to 00h)
MOVC A,@A+DPTR       ;Read code memory address 2021h into the accumulator
```

The MOVC A,@A+DPTR moves the value contained in the code memory address that is pointed to by adding DPTR to the accumulator; since the accumulator was zeroed by the CLR A instruction, the MOVX instruction will load the accumulator with the contents pointed to by DPTR. The MOVC A,@A+DPTR instruction is also useful in constructing what are known as "jump tables", but this will be further explained in section 11.31.

CHAPTER 6: PROGRAM FLOW

When an 8052 is first initialized or reset, the program counter is reset to 0000h. The 8052 then begins to execute instructions sequentially in memory unless a program instruction causes the PC to be otherwise altered. There are various types of instructions that can modify the value of the PC: conditional branching instructions, direct jumps and calls, and "returns" from subroutines. Interrupts, when enabled, can also cause the program flow to deviate from its otherwise sequential path.

6.1 Conditional Branching

The 8052 contains a suite of instructions which, as a group, are referred to as "conditional branching" instructions. These instructions cause program execution to follow a non-sequential path if a certain condition is true. Take, for example, the JB instruction. This instruction means "jump if bit set." An example of the JB instruction might be:

```
        JB 45h,HELLO
        NOP
HELLO:....
```

In this case, the 8052 will analyze the contents of bit 45h. If the bit is set, program execution will jump immediately to the label HELLO, skipping the NOP instruction. If the bit is not set, the conditional branch fails and program execution continues, as usual, with the NOP instruction that follows.

Conditional branching is the fundamental building block of program logic since all decisions are accomplished with these instructions. Conditional branching can be thought of as the "IF... THEN" structure of 8052 assembly language.

An important aspect of conditional branching is that the program may only branch to instructions located within 128 bytes prior to or 127 bytes after the address that *follows* the conditional branch instruction.

6.2 Direct Jumps

While conditional branching is extremely important, it is often necessary to make a direct branch to a given memory location without basing it on a specific logical decision. This is equivalent to "GOTO" in the BASIC programming language. In this case the program flow continues at the specified memory address without considering any conditions. This is accomplished using "direct jump" instructions. Consider the example:

```
        LJMP NEW_ADDRESS
                .
                .
                .
NEW_ADDRESS:  ....
```

The LJMP instruction in this example means "long dump." When the 8052 executes this instruction, the program counter is loaded with the address of NEW_ADDRESS and program execution continues sequentially from there.

The obvious difference between the direct jump and call instructions and conditional branching is that with direct jumps and calls, program flow *always* changes. With conditional branching, program flow only changes if a certain condition is true.

In addition to LJMP, there are two other instructions that cause a direct jump to occur: SJMP and AJMP. These two instructions perform the exact same function as LJMP in that they always cause program flow to continue at the address indicated by the instruction. However, these instructions differ from LJMP in that they are not capable of jumping to any address. They both have limitations as to the range of the jumps.

1. The SJMP command, like the conditional branching instructions, can only jump to an address within +127/-128 bytes of the address following the SJMP command.
2. The AJMP command can only jump to an address that is in the same 2k block of memory as the address following the AJMP command. That is to say, if the AJMP command is at code memory location 650h, it can only jump to addresses 0000h through 07FFh (0 through 2047, decimal).

These instructions, while less flexible than LJMP due to the restricted range of the addresses which they can jump to, are useful in that LJMP requires three bytes of code memory whereas both SJMP and AJMP require only two. Although this may sound like a trivial difference, a surprising amount of code memory may be conserved by using AJMP and SJMP whenever possible.

Some assemblers will select the most appropriate jump instruction automatically. That is, they'll automatically change LJMPs to SJMPs when it is possible to do so. This is a very powerful capability that may be a feature worth looking for in an assembler if projects with tight memory restrictions are to be developed.

6.3 Direct Calls

Another operation that will be familiar to seasoned programmers is the LCALL instruction. This is similar to a "GOSUB" in BASIC.

When the 8052 executes an LCALL instruction, it immediately pushes the current program counter on the stack and then continues executing code at the address indicated by the LCALL instruction.

Similar in format to the AJMP instruction that was described in the previous section, the ACALL instruction provides a way to perform the equivalent of an "LCALL" with a 2-byte instruction (instead of 3) as long as the target routine is within the same 2k block of memory.

6.4 Returns from Routines

Another structure that can cause program flow to change is the "return from subroutine" instruction, known as RET in 8052 assembly language. The RET instruction, when executed, returns to the address following the instruction that called the given subroutine. More accurately, it returns to the last address that is stored on the stack.

The RET command is direct in the sense that it always changes program flow without basing it on a condition, but is variable in the sense that where program flow continues can be different each time the RET instruction is executed depending on from where the subroutine was called originally.

6.5 Interrupts

An interrupt is a special feature that allows the 8052 to break from its normal program flow to execute an immediate task, providing the illusion of "multi-tasking." The word "interrupt" can often be substituted with the word "event."

An interrupt is triggered whenever a corresponding event occurs. When the event occurs the 8052 temporarily puts "on hold" the normal execution of the main program and executes a special section of code referred to as the "interrupt service routine" (ISR). The ISR performs whatever special functions are required to handle the event and then returns control to the 8052 at which point program execution continues as if it had never been interrupted.

The topic of interrupts is somewhat tricky and very important. For that reason chapter 10 will cover this topic in detail.

CHAPTER 7: INTERNAL TIMING

In order to understand and make better use of the 8052 it is necessary to understand some underlying concepts regarding timing.

The 8052 operates with timing derived from an external crystal or a clock signal generated by some other system. A crystal is a component that allows an electronic oscillator to run at a very precise and known frequency. One can find crystals of virtually any frequency depending on the application requirements. One of the most commonly used crystal frequencies used with the 8052 is 11.0592 megahertz.

The reason that such an unusual crystal frequency is popular has to do with the generation of baud rates which will be discussed in full in the serial communication chapter (chapter 9). For the remainder of this discussion it will be assumed that an 11.0592 MHz crystal is being used.

Microcontrollers use their oscillators to synchronize operations. The 8052 operates using what are called "instruction cycles." A single instruction cycle is the minimum amount of time in which a single instruction can be executed, although many instructions take multiple cycles.

A cycle is, in reality, 12 clock cycles from the crystal. If an instruction takes one instruction cycle to execute, it will take 12 clocks of the crystal to execute. Since the crystal oscillates 11,059,200 times per second and one instruction cycle is 12 clock cycles, the number of instruction cycles the 8052 can execute per second may be calculated as follows:

```
11,059,200 / 12 = 921,600
```

This means that the 8052 can execute 921,600 single-cycle instructions per second. Since a large number of 8052 instructions are single-cycle instructions it is often said that the 8052 can execute roughly one million instructions per second (1 MIPS), although in reality it is less—and, depending on the instructions being used, an estimate of about 600,000 instructions per second is more realistic. For example, if a program is using exclusively 2-cycle instructions, one would find that the 8052 executes 460,800 instructions per second.

The traditional 8052 also has two comparatively slow instructions (MUL AB and DIV AB) that require a full four cycles to execute—if the program were to execute nothing but those instructions the performance would be approximately 230,400 instructions per second.

Many derivative chips change instruction timing. For example, some optimized versions of the 8052 execute instructions in 4 oscillator cycles instead of 12; such a chip would be effectively three times faster than the 8052 when used with the same 11.0592 MHz crystal.

Since all the instructions require different amounts of time to execute, a very obvious question is how to keep track of time in a time-critical application if there is no reference to time in the outside world. It is for precisely this reason that the 8052 includes timers which allow the program to time events with high precision—which is the topic of the next chapter.

CHAPTER 8: TIMERS

The 8052 is equipped with three timers, each of which may be controlled, set, read, and configured individually. The timers have three general functions: 1) keeping time and/or calculating the amount of time between events, 2) counting the events themselves, or 3) generating baud rates for the serial port.

The three timer uses are distinct so they will be discussed separately. The first two uses will be discussed in this chapter while the use of timers for baud rate generation will be discussed in the chapter relating to serial communication (chapter 9).

> While the 8052 has three timers, the 8051 only has two. Timers 0 and 1 work and are configured in exactly the same way and will be discussed first. Timer 2, available only on 8052-compatible derivatives, is configured differently and will be discussed later in this chapter.

8.1 How does a timer count?

The answer to this question is very simple: a timer counts up. It doesn't matter whether the timer is being used as a timer, a counter, or a baud rate generator—timers are always incremented by the microcontroller.

> Some derivatives allow the program to configure whether the timers count up or down. However, since this option only exists on some derivatives it is beyond the scope of this discussion which is aimed at the standard 8052. It is only mentioned here in the event that a program absolutely needs a timer to count backwards.

8.2 Using Timers to Measure Time

One of the primary uses of timers is to measure time. When a timer is used for this purpose it is also called an **interval timer** since it is measuring the time of the interval between two events.

8.2.1 How long does a timer take to count?

When a timer is in interval timer mode and correctly configured, it will increment by 1 every instruction cycle. As explained in the previous chapter, a single instruction cycle consists of 12 crystal pulses. Thus the timer will be incremented:

```
11,059,200 / 12 = 921,600 times per second
```

Unlike instructions—some of which require one instruction cycle, others two, and others four—the timers are consistent. They will always be incremented once per instruction cycle. Thus if a timer has counted from 0 to 50,000 it may be calculated that:

```
50,000 / 921,600 = .0543 seconds
```

.0543 seconds have passed. In plain English, about one-twentieth of a second.

Clearly it's not very useful to know .0543 seconds have passed. If an event is to be executed once per second the program would have to wait for the timer to count from 0 to 50,000 18.45 times. How can the program wait for "half of a time?" It can't. So another important calculation is used.

It is possible to determine how many times the timer will be incremented in .05 seconds by performing some simple multiplication:

```
.05 * 921,600 = 46,080 cycles
```

This means that it will take .05 seconds (1/20th of a second) to count from 0 to 46,080. If it is known that it takes 1/20th of a second to finish this count and an event is to be executed every second, the program simply waits for the timer to count from 0 to 46,080 twenty times; the event is then executed, the timers are reset, and the program waits for the timer to count up another 20 times. The event will effectively be executed once per second, accurate to within thousandths of a second.

In this manner we have a system with which to measure time. All that is needed is to explain how to control the timers and initialize them to provide the program with the necessary information.

8.2.2 Timer SFRs

The 8052 has two timers which function essentially the same way: timer 0 and timer 1. These two timers share two SFRs (TMOD and TCON) which control the timers, and each timer also has two SFRs dedicated solely to maintaining the value of the timer itself, TH0/TL0 for timer 0 and TH1/TL1 for timer 1. Timer 2 functions slightly different and will be discussed in section 8.4.

The SFRs used to control and manipulate the timers are presented in the following table.

SFR Name	Description	SFR Address	Bit Addressable
TH0	Timer 0 High Byte	8Ch	No
TL0	Timer 0 Low Byte	8Ah	No
TH1	Timer 1 High Byte	8Dh	No
TL1	Timer 1 Low Byte	8Bh	No
TCON	Timer Control	88h	Yes
TMOD	Timer Mode	89h	No

Timer 0 has two SFRs dedicated exclusively to itself: TH0 and TL0. TL0 is the low-byte of the value of the timer while TH0 is the high-byte of the value of the timer. That is to say, when Timer 0 has a value of 0, both TH0 and TL0 will contain 0. When Timer 0 has the value 1000 (decimal), TH0 will hold the high byte of the value (3 decimal) and TL0 will contain the low byte of the value (232 decimal). Reviewing low/high byte notation, recall that the high byte must be multiplied by 256 and added to the low byte to calculate the final value. In this case:

```
(TH0 * 256) + TL0 = 1000
(3 * 256) + 232 = 1000
```

Timer 1 works the exact same way but its SFRs are TH1 and TL1.

Since there are only two bytes devoted to the value of each timer, the maximum value is 65,535. If a timer contains the value 65,535 and is subsequently incremented it will reset—or overflow—back to 0.

8.2.3 Timer Mode (TMOD) SFR

The first control SFR, TMOD (Timer Mode, address 89h), is used to control the mode of operation of both timers. Each bit of the SFR gives the microcontroller specific information concerning how to run a timer. The high four bits (bits 4 through 7) relate to timer 1 whereas the low four bits (bits 0 through 3) perform the exact same functions for timer 0.

The individual bits of TMOD have the following functions:

Bit	Name	Explanation	Timer
7	GATE1	When this bit is set the timer will only run when INT1 (P3.3) is high. When this bit is clear the timer will run regardless of the state of INT1.	1
6	C/T1	When this bit is set the timer will count events on T1 (P3.5). When this bit is clear the timer will be incremented every instruction cycle.	1
5	T1M1	Timer mode bit 1 (see below)	1
4	T1M0	Timer mode bit 0 (see below)	1
3	GATE0	When this bit is set the timer will only run when INT0 (P3.2) is high. When this bit is clear the timer will run regardless of the state of INT0.	0
2	C/T0	When this bit is set the timer will count events on T0 (P3.4). When this bit is clear the timer will be incremented every instruction cycle.	0
1	T0M1	Timer mode bit 1 (see below)	0
0	T0M0	Timer mode bit 0 (see below)	0

As shown in the table above, four bits (two for each timer) are used to specify a mode of operation. The modes of operation are:

TxM1	TxM0	Timer Mode	Description
0	0	0	13-bit timer
0	1	1	16-bit timer
1	0	2	8-bit auto-reload
1	1	3	Split timer

8.2.3.1 13-bit Timer Mode (mode 0)

Timer mode "0" is a 13-bit timer. This is a relic that was kept around in the 8052 to maintain compatibility with its predecessor, the 8048. The 13-bit timer mode is not normally used in new development.

When the timer is in 13-bit mode, TLx will count from 0 to 31. When TLx is incremented from 31, it will "reset" to 0 and increment THx. Thus only 13 bits are being used: bits 0-4 of TLx and bits 0-7 of THx. This also means, in essence, the timer can only contain 8192 values. If the 13-bit timer is set to 0, it will overflow back to zero 8192 instruction cycles later.

There really is very little reason to use this mode and it is only mentioned so that the reader won't be surprised if he or she end ups analyzing archaic code that has been passed down through the generations and finds this mode being used.

8.2.3.2 16-bit Timer Mode (mode 1)

Timer mode "1" is a 16-bit timer and is very commonly used. It functions just like 13-bit mode except that all 16 bits are used.

TLx is incremented from 0 to 255. When TLx is incremented from 255, it resets to 0 and causes THx to be incremented by 1. Since this is a full 16-bit timer, the timer may contain up to 65,536 distinct values. If a 16-bit timer is set to 0, it will overflow back to 0 after 65,536 instruction cycles.

8.2.3.3 8-bit Auto-Reload Mode (mode 2)

Timer mode "2" is 8-bit auto-reload mode. In this mode, THx holds the **reload value** and TLx is the timer itself. TLx counts up and when it reaches 255 and is subsequently incremented, instead of resetting to 0 (as in the case in modes 0 and 1) it will be reset to the value stored in THx.

For example, if TH0 holds the value FDh and TL0 holds the value FEh, their values would appear as follows if they were monitored for seven instruction cycles.

Instruction Cycle	TH0 Value	TL0 Value
1	FDh	FEh
2	FDh	FFh
3	FDh	FDh
4	FDh	FEh
5	FDh	FFh
6	FDh	FDh
7	FDh	FEh

As the table shows, the value of TH0 never changes. In fact, when mode 2 is used, THx is almost always set to a known value during program initialization and TLx is the SFR that is constantly incremented. THx is left unchanged throughout program execution.

The benefit of auto-reload mode is that, perhaps, the program needs the timer to always have a value from 200 to 255. If mode 0 or 1 is used the program would have to check in code to see if the timer had overflowed and, if so, reset the timer to 200. This uses up valuable execution time. When mode 2 is used the microcontroller takes care of this. Once the program has configured a timer in mode 2, it doesn't have to worry about checking to see if the timer has overflowed nor does it need to worry about resetting the value—the microcontroller hardware will do it all automatically. Auto-reload mode is also extremely useful in causing overflows at a fixed frequency.

The auto-reload mode is often used for establishing a baud rate which will be explained in the section on serial communications (chapter 9).

8.2.3.4 Split Timer Mode (mode 3)

Timer mode "3" is a split-timer mode. When timer 0 is placed in mode 3, it essentially becomes two separate 8-bit timers. That is to say, timer 0 is TL0 and timer 1 is TH0. Both timers count from 0 to 255 and overflow back to 0. All the control bits related to timer 1 will be tied to TH0 and those related to timer 0 will be tied to TL0.

While timer 0 is in split mode, the real timer 1 (i.e. TH1 and TL1) can be put into modes 0, 1 or 2 normally—however, the real timer 1 can't be stopped or started since the bits that do that are now linked to TH0. The real timer 1 may be stopped by configuring timer 1 to be in mode 3—in this case, the real timer 1 will not increment.

The only reason for using split timer mode is if two separate timers are needed and, additionally, a baud rate generator. In such a case the real timer 1 can be used as a baud rate generator and TH0/TL0 used as two separate timers.

8.2.4 Timer Control (TCON) SFR

One additional SFR controls the two timers and provides valuable information about them. The TCON SFR has the following structure:

Timer Control (TCON): SFR Address 88h

Bit	Name	Bit Address	Explanation	Timer
7	TF1	8Fh	Timer 1 Overflow. This bit is set by the microcontroller when timer 1 overflows.	1
6	TR1	8Eh	Timer 1 Run. When this bit is set timer 1 is turned on. When this bit is clear timer 1 is off.	1
5	TF0	8Dh	Timer 0 Overflow. This bit is set by the microcontroller when timer 0 overflows.	0
4	TR0	8Ch	Timer 0 Run. When this bit is set timer 0 is turned on. When this bit is clear timer 0 is off.	0

Only the most significant four bits of TCON are described in the previous table. The remaining four bits (TCON.0 through TCON.3) are related to interrupts and will be discussed in chapter 10.

A new piece of information in this chart is the column "bit address." That is because this SFR is bit-addressable, which means that if the program needs to set the bit TF1—which is the most significant bit of TCON—the following instruction could be executed:

```
MOV TCON, #80h
```

But since the SFR is bit-addressable the following instruction could be executed instead:

```
SETB TF1
```

This has the benefit of setting the high bit of TCON without changing the value of any of the other bits of the SFR. Usually when a timer is started or stopped it isn't necessary or desirable to modify the other bits in TCON so the program normally takes advantage of the fact that the SFR is bit-addressable and modifies the bit directly.

8.2.5 Initializing a Timer

Now that the timer-related SFRs have been explained, it is possible to write code that will initialize the timer and start it running. The first step is to determine which mode the timer should operate in. In this case the program needs a 16-bit timer that runs continuously; that is to say, it is not dependent on any external pins.

First, the TMOD SFR must be initialized. Since the program will be using timer 0 it will be using the low 4 bits of TMOD. The first two bits, GATE0 and C/T0 are both 0 since the timer is to run independently of the external pins. 16-bit mode is timer mode 1 so T0M1 must be cleared and T0M0 must be set. Effectively, the only bit that should be turned on is bit 0 of TMOD. Thus to initialize the timer the following instruction is executed:

```
MOV TMOD,#01h
```

Timer 0 is now in 16-bit timer mode. However, the timer is not running. To start the timer running the TR0 bit must be set. This is accomplished with the instruction:

```
SETB TR0
```

Upon executing these two instructions timer 0 will immediately begin counting, being incremented once every instruction cycle.

8.2.6 Reading the Timer

There are two common methods of reading the value of a 16-bit timer; which method a program uses depends on the specific application. The program may either read the actual value of the timer as a 16-bit number or it may simply detect when the timer has overflowed.

8.2.6.1 Reading the value of a Timer

If a timer is in an 8-bit mode—that is, either 8-bit auto-reload mode or in split timer mode—then reading the value of the timer is simple. The program simply reads the 1-byte value of the timer and it's done.

However, when the program is working with a 13-bit or 16-bit timer the chore is a little more complicated. Consider what would happen if the program were to read the low byte of the timer as 255 then read the high byte of the timer as 15. In this case, what actually happened was that the timer value was 14/255 (high byte 14, low byte 255) but the program read 15/255.

This is because the program reads the low byte as 255 but as this instruction was executed, another instruction cycle passed—enough for the timer to be incremented again, at which point the value rolled over from 14/255 to 15/0. As a result, the program has erroneously read the timer as 15/255 instead of 14/255.

The solution is fairly simple. The program reads the high byte of the timer, then reads the low byte, then reads the high byte again. If the high byte read the second time is not the same as the high byte read the first time, the cycle is repeated. In code, this would be written as:

```
REPEAT:
  MOV A,TH0           ;Store value of TH0 in accumulator
  MOV R0,TL0          ;Store value of TL0 in R0
  CJNE A,TH0,REPEAT   ;If value of TH0 has changed, repeat
```

In this case, the accumulator is loaded with the high byte of timer 0. R0 is then loaded with the low byte of timer 0. Finally, the program checks to see if the high byte read out of timer 0—which is now stored in the accumulator—is the same as the current timer 0 high byte. If it isn't, the timer has just "rolled over" and its value must be re-read—which is accomplished by going back to REPEAT. When the loop exits, the low byte of the timer will be in R0 and the high byte in the accumulator.

Another alternative is to simply suspend the timer (i.e. CLR TR0), read the timer value, and then reactivate the timer (i.e. SETB TR0). In this case, the timer isn't running so no special tricks are necessary. Of course, this implies that the timer will be stopped for a few instruction cycles which has the potential of causing the timer to "lose time." Whether or not this is acceptable depends on the specific application.

8.2.6.2 Detecting Timer Overflow

Often it is only necessary to know that the timer has overflowed. That is to say, the program is not particularly interested in the *value* of the timer but rather *when* the timer has overflowed back to 0 (or to the THx reset value in the case of auto-reload mode).

Whenever a timer overflows from its highest value the microcontroller automatically sets the TFx bit in the TCON register. This is useful since rather than checking the exact value of the timer the program may just check if the TFx bit is set. If the TF0 bit is set, it means that timer 0 has overflowed; if TF1 is set, it means that timer 1 has overflowed.

This approach can be used to cause the program to execute a fixed delay. In section 8.2.1 it was calculated that it takes the 8052 $1/20^{th}$ of a second to count from 0 to 46,080. Thus, to use the TFx flag to indicate when $1/20^{th}$ of a second has passed, the timer must be set initially to 65,536 minus 46,080 (19,456). If the timer is set to 19,456, $1/20^{th}$ of a second later the timer will overflow. The following code executes a pause of $1/20^{th}$ of a second:

```
1   MOV THO,#76      ;High byte of 19,456 (76 * 256 = 19,456)
2   MOV TL0,#00      ;Low byte of 19,456 (19,456 + 0  = 19,456)
3   MOV TMOD,#01     ;Put timer 0 in 16-bit mode
4   CLR TF0          ;Make sure TF0 bit is clear initially
5   SETB TR0         ;Make Timer 0 start counting
6   JNB TF0,$        ;If TF0 not set, jump back to this instruction
```

Lines 1 and 2 initialize the value of timer 0 to 19,456. Lines 3 through 5 configure timer 0 to 16-bit mode and turn it on. The last instruction in line 6 reads "Jump back to the same instruction if TF0 is not set." The "$" operand means, in most assemblers, the address of the current instruction.

As long as the timer has not overflowed, the TF0 bit will not be set and the program will keep executing the same instruction. After $1/20^{th}$ of a second, timer 0 will overflow, the TF0 bit will be set, and program execution will then break out of the loop, thus concluding the delay.

8.2.7 Timing the length of events

The 8052 provides another useful feature that can be used to time the length of events.

Take, for example, a program that is measuring electricity consumption in an office by measuring how long the light is turned on each day. When the light is turned on, the program needs to measure time. When the light is turned off, it needs to *not* measure time. One option would be to connect the light switch to one of the pins, constantly read the pin, and turn the timer on or off based on the state of that pin. While this would work, the 8052 offers an easier method of accomplishing this.

Looking again at the TMOD SFR, there is a bit called GATE0. Until now, that bit has always been cleared since the program needed the timer to run regardless of the state of the external pins. However, now it would be useful if an external pin could control whether the timer was running or not. It can.

All that needs to be done is to connect the light switch to pin INT0 (P3.2) of the 8052 and set the bit GATE0. When GATE0 is set, timer 0 will only run if P3.2 is high. When P3.2 is low (i.e., the light switch is off) the timer will automatically be stopped.

Thus, with no control code whatsoever the external pin P3.2 can control whether or not the timer is running while the microcontroller program is taking care of other tasks.

8.3 Using Timers as Event Counters

Up to this point, timers have been used exclusively for the purpose of keeping track of time. The 8052 also allows timers to be used to count events.

This can be useful in many applications. For example, consider a sensor placed across a road that sends a

42

pulse every time a car passes over it which could be used to determine the volume of traffic on the road. The sensor could be attached to one of the 8052's I/O lines with the program constantly monitoring it to detect when it pulsed high and then increment the counter when it went back to a low state. If the sensor were connected to P1.0 the code to count passing cars would look something like this:

```
JNB P1.0,$    ;If a car hasn't raised the signal, keep waiting
JB P1.0,$     ;The line is high, car is on the sensor right now
INC COUNTER   ;The car has passed completely, so we count it
```

It's only three lines of code, but what if the program needed to be doing other processing at the same time? It can't be stuck in the `JNB P1.0,$` loop waiting for a car to pass if it needs to be doing other things. And what if the program were doing other things when a car passed over? It's possible that the car will raise the signal and that the signal will fall low again before the program checks the line status; this would result in the car not being counted. Of course, there are ways to get around this limitation but the code quickly becomes big and complex.

Since the 8052 offers a way to use the timers to count events it is not necessary to bother with such messy code. Counting events is actually surprisingly easy—only a single additional bit must be configured.

Assuming that timer 0 will be used to count the number of cars that pass, the bit table for the TCON SFR shows that there is a bit called "C/T0" (TMOD.2). As the explanation indicates, clearing this bit will cause timer 0 to be incremented every instruction cycle—this is what was already explained earlier in this chapter to measure time. If this bit is set, however, timer 0 will monitor the P3.4 line. Instead of being incremented every instruction cycle, timer 0 will count events on the P3.4 line. So in the case of the "car-counting" example, the program would set the C/T0 bit, the sensor would be connected to P3.4, and the program would let the 8052 do the work. Whenever the program wants to know how many cars have passed, it just reads the value of timer 0 which will contain the current car count.

What exactly is an event? What does timer 0 actually "count" when it is in counter mode? Speaking at the electrical level, the 8052 counts one-to-zero transitions on the P3.4 line. This means that when a car first runs over the sensor, it will raise the input to a high ("1") condition. At that point the 8052 will not count anything since it is a 0-1 transition. When the car has passed, the sensor will fall back to a low ("0") state. This is a 1-0 transition and at that instant the counter will be incremented by 1.

It is important to note that the 8052 checks the P3.4 line only once each instruction cycle (12 clock cycles). This means that if P3.4 is low, goes high, and goes back low in 6 clock cycles it will not be detected by the 8052. More specifically, it means the 8052 event counter is only capable of counting events that occur at a maximum of 1/24th the rate of the crystal frequency. If the crystal frequency is 12.000 MHz it can count a maximum of 500,000 events per second (12.000 MHz * 1/24 = 500,000). If the event being counted occurs more than 500,000 times per second, it will not be accurately counted by the 8052 without using additional external circuitry, a faster crystal, or a derivative chip that implements an instruction cycle of less than 12 crystal cycles.

8.4 Using Timer 2

The 8052 has a third timer, timer 2. Keep in mind that the 8051 and some lower-end derivative chips don't include timer 2. Timer 2 functions slightly different than timers 0 and 1 and, for that reason, it is being addressed separately from the first two.

8.4.1 Timer 2 Control (T2CON) SFR

The operation of timer 2 is controlled almost entirely by the T2CON SFR, at address C8h, described in the table on the next page. Note that since this SFR is evenly divisible by 8 that it is bit-addressable.

Timer 2 Control (T2CON): SFR Address C8h

Bit	Name	Bit Address	Explanation
7	TF2	CFh	Timer 2 Overflow. This bit is set when T2 overflows. When T2 interrupt is enabled, this bit will cause the interrupt to be triggered. This bit will not be set if either TCLK or RCLK bits are set.
6	EXF2	CEh	Timer 2 External Flag. Set by a reload or capture caused by a 1-0 transition on T2EX (P1.1), but only when EXEN2 is set. When T2 interrupt is enabled, this bit will cause the interrupt to be triggered.
5	RCLK	CDh	Timer 2 Receive Clock. When this bit is set, timer 2 will be used to determine the serial port receive baud rate. When clear, timer 1 will be used.
4	TCLK	CCh	Timer 2 Transmit Clock. When this bit is set, timer 2 will be used to determine the serial port transmit baud rate. When clear, timer 1 will be used.
3	EXEN2	CBh	Timer 2 External Enable. When set, a 1-0 transition on T2EX (P1.1) will cause a capture or reload to occur.
2	TR2	CAh	Timer 2 Run. When set, timer 2 will be turned on. Otherwise, it is turned off.
1	C/T2	C9h	Timer 2 Counter/Interval Timer. If clear, timer 2 is an interval counter. If set, timer 2 is incremented by 1-0 transition on T2 (P1.0).
0	CP/RL2	C8h	Timer 2 Capture/Reload. If clear, auto reload occurs on timer 2 overflow, or T2EX 1-0 transition if EXEN2 is set. If set, a capture will occur on a 1-0 transition of T2EX if EXEN2 is set.

8.4.2 Timer 2 in Auto-Reload Mode

The first mode in which timer 2 may be used is auto-reload. The auto-reload mode functions just like timer 0 and timer 1 in auto-reload mode (section 8.2.3.3) except that timer 2 auto-reload mode performs a full 16-

bit reload rather than 8-bit reloads as is the case with the other two timers. When a reload occurs, the values of TH2 and TL2 will be reloaded with the values contained in RCAP2H and RCAP2L, respectively.

To operate timer 2 in auto-reload mode, CP/RL2 (T2CON.0) must be clear. In this mode, timer 2 (TH2/TL2) will be reloaded with the reload value (RCAP2H/RCAP2L) whenever timer 2 overflows; that is to say, whenever timer 2 overflows past FFFFh. An overflow of timer 2 will cause the TF2 bit to be set which will cause an interrupt to be triggered if timer 2 interrupt is enabled. Note that TF2 will not be set on an overflow condition if either RCLK or TCLK (T2CON.5 or T2CON.4) are set.

By also setting EXEN2 (T2CON.3), a reload will also occur whenever a 1-0 transition is detected on T2EX (P1.1). A reload that occurs as a result of such a transition will cause the EXF2 (T2CON.6) flag to be set, triggering a timer 2 interrupt if that interrupt has been enabled.

8.4.3 Timer 2 in Capture Mode

A new mode specific to timer 2 is called "capture mode." As the name implies, this mode captures the value of timer 2 (TH2 and TL2) into the capture SFRs (RCAP2H and RCAP2L). To put timer 2 in capture mode, both CP/RL2 (T2CON.0) and EXEN2 (T2CON.3) must be set.

When configured in capture mode, a capture will occur whenever a 1-0 transition is detected on T2EX (P1.1). At the moment the transition is detected, the current values of TH2 and TL2 will be copied into RCAP2H and RCAP2L, respectively. At the same time, the EXF2 (T2CON.6) bit will be set which will trigger an interrupt if timer 2 interrupt is enabled.

Note that even in capture mode an overflow of timer 2 will result in TF2 being set and an interrupt being triggered.

Capture mode is an efficient way to measure the time between events. At the moment that an event occurs the current value of timer 2 will be copied into RCAP2H/L. However, timer 2 will not stop and an interrupt will be triggered. Thus the interrupt routine may copy the value of RCAP2H/L to a temporary holding variable without having to stop timer 2. When another capture occurs, the interrupt can take the difference between the two values to determine the time that has transpired. Again, the main advantage is that, unlike timer 0 and timer 1, the program doesn't need to stop timer 2 to read its value.

8.4.4 Timer 2 as a Baud Rate Generator

Timer 2 may be used as a baud rate generator for the serial port. This is accomplished by setting either RCLK (T2CON.5) or TCLK (T2CON.4). This will be discussed in the serial communications chapter (chapter 9).

When timer 2 is used as a baud rate generator (either TCLK or RCLK are set), the timer 2 overflow flag (TF2) will not be set.

CHAPTER 9: SERIAL COMMUNICATION

One of the 8052's many powerful features is its integrated Universal Asynchronous Receiver-Transmitter (UART), otherwise known as a serial port. This feature means that a program may very easily transmit and receive data via the serial port with a minimal amount of code. If it were not for the integrated serial port, writing a byte to a serial line would be a rather tedious process requiring turning on and off one of the I/O lines in rapid succession to properly "clock out" each individual bit, including start bits, stop bits, and parity bits (a process known as "bit banging").

Instead, the program simply configures the serial port's operation mode and baud rate. Once configured, all that is necessary to send a byte to the serial port is write it to an SFR. Likewise, receiving a byte from the serial point is a simple matter of reading the same SFR. The 8052 will automatically let the program know when it has finished sending a character and will also let alert the program whenever it has received a byte so that it can be processed. It is not necessary to worry about transmission at the bit level. This saves the developer quite a bit of coding and processing time.

9.1 Serial Control SFR (SCON)

The first step in using the 8052's integrated serial port is to configure it. This is done primarily with the Serial Control (SCON) SFR which configures the communication mode and also contains operational flags that indicate when a byte has been successfully transferred or received.

Serial Control (SCON): SFR Address 98h

Bit	Name	Bit Address	Explanation
7	SM0	9Fh	Serial port mode bit 0
6	SM1	9Eh	Serial port mode bit 1.
5	SM2	9Dh	Multiprocessor Communications Enable (explained later)
4	REN	9Ch	Receiver Enable. Must be set in order to receive characters.
3	TB8	9Bh	Transmit bit 8. The 9th bit to transmit in mode 2 and 3.
2	RB8	9Ah	Receive bit 8. The 9th bit received in mode 2 and 3.
1	TI	99h	Transmit Flag. Set when a byte has been completely transmitted.
0	RI	98h	Receive Flag. Set when a byte has been completely received.

SM0	SM1	Serial Mode	Explanation	Baud Rate
0	0	0	8-bit shift register	Oscillator / 12
0	1	1	8-bit UART	Set by timer 1 or 2*
1	0	2	9-bit UART	Oscillator / 32 *
1	1	3	9-bit UART	Set by timer 1 or 2*

(*) The baud rate indicated is doubled if PCON.7 (SMOD) is set.

The high four bits (bits 4 through 7) of SCON are configuration bits.

Bits **SM0** and **SM1** configure the serial mode to a value between 0 and 3, inclusive. The four modes are defined in the previous table. Selecting the serial mode selects the mode of operation (8-bit/9-bit, UART or shift register) and also determines how the baud rate is calculated.

In modes 0 and 2, the baud rate is fixed based on the oscillator's frequency. In modes 1 and 3, the baud rate is variable and based on how often timer 1 overflows. The four serial modes will be fully explained later in this chapter.

The **SM2** bit is a flag for "multiprocessor communication." Generally, whenever a byte has been received the 8052 will set the "RI" (Receive Interrupt) flag. This lets the program know that a byte has been received and that it needs to be processed. However, when SM2 is set the "RI" flag will only be triggered if the 9th bit received was a "1". That is to say, if SM2 is set and a byte is received whose 9th bit is clear, the RI flag will *not* be set. This can be useful in certain advanced serial applications that allow multiple 8052s (or other hardware) to communicate amongst themselves. For now it is safe to say that this bit should virtually always be cleared so that the flag is set upon reception of any character.

The **REN** bit is "Receiver Enable." This bit must be set to receive data via the serial port. This bit is normally set as part of the initial serial configuration process.

The low four bits (bits 0 through 3) are operational bits. They are used when actually sending and receiving data—they are not used to configure the serial port.

The **TB8** bit is used in modes 2 and 3. In these two modes, a total of nine data bits are transmitted. The first eight data bits are the eight bits of the value written to the serial port while the ninth bit is taken from TB8. If TB8 is set and a value is written to the serial port, the data's bits will be written to the serial line followed by a "set" ninth bit. If TB8 is clear, the ninth bit will be zero.

The **RB8** bit also operates in modes 2 and 3 and functions essentially the same way as TB8, but on the reception side. When a byte is received in these modes, a total of nine bits are received. In this case, the first eight bits received are the data of the serial byte received and the value of the ninth bit received will be placed in RB8.

The **TI** bit means "Transmit Interrupt." When a program writes a value to the serial port, a certain amount of time will pass before the individual bits of the byte are clocked out the serial port. If the program were to write another byte to the serial port before the first byte was completely output, the data being sent would be garbled. Thus, the 8052 lets the program know that it has clocked out the byte by setting the TI bit. When the TI bit is set, the program may assume that the serial port is available to send the next byte.

Finally, the **RI** bit means "Receive Interrupt." It functions similarly to the "TI" bit, but it indicates that a byte has been received. That is to say, whenever the 8052 has received a complete byte it will trigger the RI bit to let the program know that it needs to read the value quickly, before another byte is received.

9.1.1 Serial Mode 0: 8-bit Shift Register, Oscillator-based Baud Rate

In serial mode 0, the RXD and TXD lines act together as an 8-bit shift register and the baud rate is the oscillator frequency divided by 12—so an 11.0592 MHz crystal will generate 921,600 baud. In this mode, the TXD line (P3.1) is used to generate a regular clock signal while RXD (P3.0) is used to clock out or clock in serial data one bit at a time. Mode 0 is half-duplex which means data is either being sent or received—it can't be sent and received simultaneously. In either case, the microcontroller drives the serial clock. Since the clock rate is 1/12th of the crystal frequency—the same as an instruction cycle—it takes eight instruction cycles to clock in or out a single byte.

Data is transmitted in mode 0 by writing the value to the SBUF SFR. At that point the microcontroller will activate the serial clock on TXD and output one bit of the value during each clock cycle, starting with the least significant bit. The state of the output on RXD is changed while the clock is high and is valid while the clock is low. The TI flag is set as soon as the last of the eight bits have been clocked out. The REN bit (SCON.4) must be clear in order to transmit in mode 0.

Data is received in mode 0 by setting the REN bit (SCON.4) and clearing the RI bit. When this condition exists the microcontroller will immediately activate the serial clock on TXD and input one bit during each clock cycle, starting with the least significant bit. As soon as eight bits have been clocked in, the RI bit will be set and the data that was received can be read from SBUF. Subsequent input is triggered by clearing the RI bit. In mode 0, the REN and RI bit are both normally set when the serial mode is configured and RI is then cleared each time the program desires to clock in data from the external device.

This mode may be useful for clocking data out to shift registers such as the 74LS164 or clocking in data from shift registers such as the 74LS165.

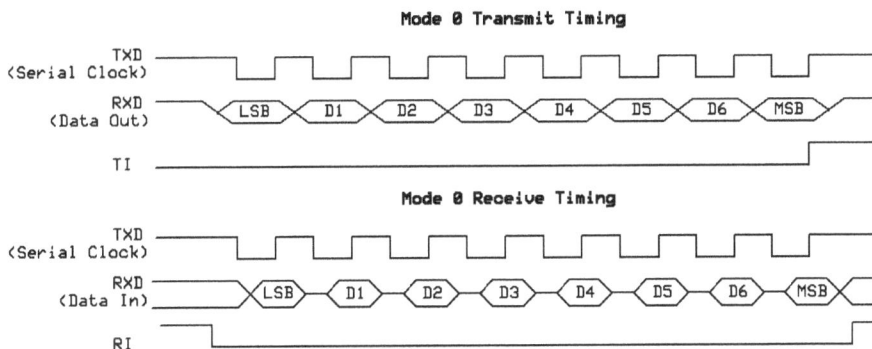

Mode 0 Transmit Timing

TXD (Serial Clock)
RXD (Data Out): LSB D1 D2 D3 D4 D5 D6 MSB
TI

Mode 0 Receive Timing

TXD (Serial Clock)
RXD (Data In): LSB D1 D2 D3 D4 D5 D6 MSB
RI

9.1.2 Serial Mode 1: 8-bit UART, Timer-based Baud Rate

Mode 1 is the most commonly used serial mode since it is full-duplex which allows the serial port to receive data on the RXD (P3.0) line and transmit data on the TXD (P3.1) line simultaneously. This mode is typically used to communicate with a PC via its serial port, external modems, and the vast majority of RS-232-type serial devices. In this mode, the baud rate is determined by the timer 1 overflow frequency which allows the program to configure the baud rate by changing the overflow rate. The baud rate can be doubled by setting the SMOD bit in PCON (PCON.7).

Data is transmitted by writing the value to be sent to the SBUF SFR. Ten bits are then sent at the specified baud rate: a leading start bit (always zero), eight data bits (least significant bit first), and a trailing stop bit (always one). The microcontroller sets the TI bit as soon as the last data bit has been sent or, in other words, as soon as the stop bit is asserted on the TXD line.

Data is received by the serial port by recognizing an incoming start bit on the RXD line—since the stop bit of the previous byte is always one and the start bit of the next byte is always zero, the reception of data is effectively initiated by a high-to-low transition on the RXD line. Once reception has been initiated, eight data bits are clocked in followed by the final stop bit. If the RI flag is clear, the received byte will be placed in SBUF and the final stop bit will be loaded into RB8 (SCON.2). Once this has occurred, the RI flag will be set to indicate that a byte of data has been received. If the RI flag is still set when a byte has been received, the byte that was just received will be discarded resulting in a loss of data.

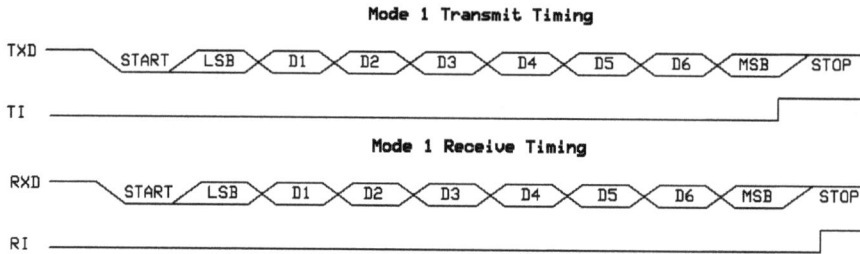

9.1.3 Serial Mode 2: 9-bit UART, Oscillator-based Baud Rate

Mode 2 is full-duplex which means that TXD is used to transmit data at the same time as RXD is used to receive data. Normally, the baud rate is 1/64th of the oscillator frequency such that an 11.0592 MHz oscillator will result in a baud rate of 172,800. However, if SMOD (PCON.7) is set then the baud rate is doubled to 1/32nd of the oscillator frequency—345,600 baud in the case of an 11.0592 MHz oscillator.

Mode 2 transmits and receives eleven bits of data per frame: a leading start bit (always zero), eight data bits (least significant bit first), an extra ninth bit which may be used for parity, multiprocessor communication, or other purposes, and a trailing stop bit (always one).

Sending data is triggered by writing a value to the SBUF SFR. The value of the ninth bit must first be written to TB8 (SCON.3). The data is then clocked out, first the eight bits that were written to SBUF followed by the TB8 bit from SCON. The TI flag is set as soon as the stop bit is placed on TXD.

Reception of data operates the same as it does in serial mode 1, but the ninth data bit is placed in RB8 rather than the stop bit.

Additionally, the received byte will be discarded if SM2 (SCON.5) is set and the final stop bit is zero. This feature aids in multi-processor communication in which one 8052 is the master and others are the slaves. In this case, the master 8052 will send out address bytes to all slaves with the ninth bit in a high state so that all the slaves receive the byte. The slave that recognizes its address in that byte will then switch to normal, non-multiprocessor communication to receive the rest of the bytes which are sent by the master with the stop bit clear. The other slaves, however, will not be notified of those bytes since they will remain in multi-processor mode and will reject all transmissions that end in with a zero stop bit.

The concept of multi-processor communication assumes that multiple 8052 microcontrollers exist in a single circuit design. Such designs are no longer very common so the topic of multi-processor communication will not be explored in further detail.

9.1.4 Serial Mode 3: 9-bit UART, Timer-based Baud Rate

Serial mode 3 operates in the exact same fashion as serial mode 2 except for the fact that the baud rate in this mode is determined by the timer overflow rate exactly as it is in serial mode 1.

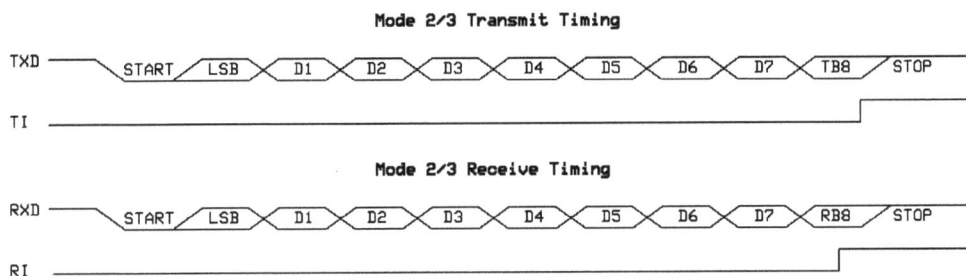

Mode 2/3 Transmit Timing

TXD ──┐ START / LSB ╳ D1 ╳ D2 ╳ D3 ╳ D4 ╳ D5 ╳ D6 ╳ D7 ╳ TB8 / STOP

TI ───┌─────

Mode 2/3 Receive Timing

RXD ──┐ START / LSB ╳ D1 ╳ D2 ╳ D3 ╳ D4 ╳ D5 ╳ D6 ╳ D7 ╳ RB8 / STOP

RI ───┌─────

9.2 Configuring the Serial Port

Configuring the serial port is a matter of selecting the right serial mode and writing the corresponding value to SCON.

The most common serial configuration is serial mode 1 (8-bit UART with baud rate configured by timer 1 or 2) with receive enable. Consulting the SCON table in section 9.1, this means that bits 4 and 6 of SCON should be set—this corresponds to a value of 50h. Thus to configure the serial port for 8-bit UART operation and using timer 1 as the baud rate generator, the following line of code is all that is needed:

```
MOV SCON,#50h        ;8-bit UART, timer 1 or 2 as baud rate generator
```

It is not unusual to see the SCON SFR configured with a value of 52h instead of 50h. By setting SCON to 52h the TI bit is effectively being set. Since the TI bit is set by the microcontroller when a byte is sent successfully, it is used as a flag to let the program know that it is safe to send the next byte. By initializing TI to a "set" state from the beginning, the program may work on the assumption that if TI is set, the next byte may be sent. Likewise, such a program will wait until TI is set before proceeding to send the next character. This is a valid approach in some circumstances, but a drawback is that if serial interrupts are enabled, the serial interrupt will start occurring immediately even though no character has been transmitted.

9.3 Setting the Serial Port Baud Rate

Once the serial mode has been configured using SCON, the program must configure the serial port's baud rate. This only applies to serial modes 1 and 3. In modes 0 and 2, the baud rate is determined based on the oscillator's frequency and no additional configuration is necessary: In mode 0, the baud rate is always the oscillator frequency divided by 12. This means if the crystal's frequency is 11.0592 MHz, the mode 0 baud rate will always be 921,600 baud. In mode 2 the baud rate is always the oscillator frequency divided by 64, so a 11.0592 MHz crystal speed will yield a baud rate of 172,800 (or 345,600 if SMOD is set).

In modes 1 and 3, the baud rate is determined by how frequently timer 1 overflows. The more frequently timer 1 overflows, the higher the baud rate. There are many ways to cause timer 1 to overflow and subsequently affect the baud rate but the most common method is to configure the timer for 8-bit auto-reload mode (timer mode 2) and set a reload value that causes timer 1 to overflow at a frequency appropriate to generate the desired baud rate.

To determine the value that must be placed in TH1 to generate a given baud rate, the following equation may be used (assuming PCON.7 is clear):

```
TH1 = 256 - ((Crystal  Speed / 384) / Baud)
```

If PCON.7 is set then the baud rate is effectively doubled and the equation becomes:

```
TH1 = 256 - ((Crystal Speed / 192) / Baud)
```

For example, to set the baud rate to 19,200 with an 11.0592 MHz crystal, the specific values would be plugged into the first equation:

```
TH1 = 256 - ((Crystal / 384) / Baud)
TH1 = 256 - ((11,059,200 / 384) / 19200)
TH1 = 256 - ((28,800) / 19200)
TH1 = 256 - 1.5 = 254.5
```

According to this equation, to obtain a baud rate of 19,200 with an 11.059 MHz crystal, TH1 would have to be set to 254.5. A value of 254 would result in a baud rate of 14,400 while a value of 255 would result in a baud rate of 28,800 baud. Since a fractional value cannot be stored in TH1, the SMOD (PCON.7) bit must be set which doubles the baud rate and requires that the second equation mentioned above be used.

```
TH1 = 256 - ((Crystal / 192) / Baud)
TH1 = 256 - ((11,059,200 / 192)  / 19200)
TH1 = 256 - ((57600) / 19200)
TH1 = 256 - 3 = 253
```

With the second equation it is possible to calculate an integer value for TH1. Therefore, to obtain 19,200 baud with an 11.0592 MHz crystal, the process is:

1. Configure serial mode to 1 or 3 (for 8-bit or 9-bit serial mode).
2. Configure timer 1 for mode 2 (8-bit auto-reload).
3. Set TH1 to 253 to reflect the correct overflow frequency for 9,600 baud.
4. Set PCON.7 (SMOD) to double the baud rate.

Thus to configure the serial port for 19,200 baud operation with an 11.0592 MHz crystal, the following code could be used.

```
MOV SCON,#50h      ;8-bit UART, timer 1 or 2 as baud rate generator
MOV SMOD,#80h      ;Set SMOD to double baud rate
MOV TMOD,#20h      ;Set timer 1 to mode 2 (8-bit auto-reload)
MOV TH1,#253       ;Set reload value to 253
SETB TR1           ;Turn timer 1 on
```

9.4 Writing to the Serial Port

Once the serial port has been properly configured as explained above, the serial port is ready to be used to send and receive data. Sending and receiving data is even easier than configuring the serial port.

To send a byte to the serial port, the program simply writes the value to the SBUF (99h) SFR. For example, the following instruction would send the letter "A" to the serial port:

```
MOV SBUF,#'A'
```

Upon execution of the above instruction, the 8052 will begin transmitting the character via the serial port. Transmission is not instantaneous—it takes a measurable amount of time to transmit the eight data bits that make up the byte along with its start and stop bits. Since the 8052 does not have a serial output buffer, the program must be sure that a character is completely transmitted before it attempts to transmit the next character.

The 8052 lets the program know when it is done transmitting a byte by setting the TI bit in SCON. When this bit is set the program knows that the previous character has been transmitted and that it may send the next character, if any. Consider the following code segment:

```
CLR TI            ;Be sure the bit is initially clear
MOV SBUF,#'A'     ;Send the letter 'A' to the serial port
JNB TI,$          ;Pause until the TI bit is set.
```

The above three instructions will transmit a character and wait for the TI bit to be set before continuing. The last instruction says "Jump to $ if the TI bit is not set". The "$" character, in most assemblers, means "the same address of the current instruction." Thus the 8052 will pause on the JNB instruction until the TI bit is set by the 8052 upon successful transmission of the character. Once the character has been transmitted the TI bit will be set, the JNB instruction will not be triggered, and program execution will continue with the next instruction which may send the next character to the serial port.

9.5 Reading the Serial Port

Reading data received by the serial port is just as easy as sending data. To read a byte from the serial port, the program waits for the 8052 to set the RI flag in SCON, then reads the value stored in the SBUF (99h) SFR.

For example, if a program needs to wait for a character to be received and subsequently read into the accumulator, the following code segment may be used:

```
JNB RI,$      ;Wait for the 8052 to set the RI flag
MOV A,SBUF    ;Read the character from the serial port
CLR RI        ;Clear the 8052 receive flag
```

The first line of the code waits for the 8052 to set the RI flag. The 8052 sets the RI flag automatically when it receives a character via the serial port, so as long as a character is not received, the bit is not set, and the program repeats the "JNB" instruction continuously. Once a character is received, the RI bit will be set automatically, the above condition fails, and program execution falls through to the "MOV" instruction that reads the character into the accumulator. The last instruction clears the RI bit to indicate that the last byte has been read.

Clearing the RI bit is very important. If the RI bit is not cleared after reading SBUF, the next time the bit is checked by the program it will assume a new character has arrived and will re-read the same value out of SBUF. This will cause the same character to be received over and over by the program until the RI bit is cleared.

9.6 Using Timer 2 for Serial Port Baud Rate

In serial modes 1 and 3, timer 2 may be used to generate the baud rate for transmission and/or reception. This is controlled with the RCLK bit (T2CON.5) and TCLK (T2CON.4). When RCLK is set, timer 2 will be used to generate the receive baud rate. When TCLK is set, timer 2 will be used to generate the transmit baud rate.

Since TCLK and RCLK can be set individually, it is possible to operate the serial port with one baud rate for data reception and another baud rate for data transmission.

When using timer 2 to generate a baud rate, T2CON should be set to either 10h to act as the transmit baud rate generator, 20h to act as the receive baud rate generator, or 30h to generate the baud rate for both transmit and receive.

In this case, the RCAP2L and RCAP2H SFRs must be set to the 16-bit reload value for the desired baud rate. The reload value is calculated as:

```
RCAP2 = 65,536 - (Crystal Speed / 32) / Baud
```

Thus, with an 11.0592 MHz crystal and a desired baud rate of 9600, the calculation would be:

```
RCAP2 = 65,536 - (11059200 / 32) / 9600
RCAP2 = 65,536 - 36 = 65,500 (FFDCh)
```

When timer 2 is used as the baud rate generator, the baud rate cannot be doubled by setting the SMOD bit (PCON.7). The SMOD baud rate doubler only applies to baud rates generated with timer 1.

Thus the code necessary to configure the serial port for 9600 baud with an 11.0592 MHz crystal using timer 2 is:

```
MOV SCON,#50h      ;Serial mode 1
MOV T2CON,#30h     ;Use timer 2 for both send and receive baud rate
MOV RCAP2H,#0FFh   ;Reload value high byte
MOV RCAP2L,#0DCh   ;Reload value low byte
SETB TR2           ;Start timer 2
```

9.7 Serial Communication Sample Program

To illustrate the full concept of serial communication, a very simple serial communication program which does nothing more than echo every character it receives back to the serial port is useful.

```
1        MOV SCON,#50h      ;8-bit UART, timer 1 as baud rate generator
2        MOV TMOD,#20h      ;Set timer 1 to mode 2 (8-bit auto-reload)
3        MOV TH1,#253       ;Set reload value to 253
4        SETB TR1           ;Turn timer 1 on
5  SerialLoop:
6        JNB RI,$           ;Wait for the 8052 to set the RI flag
7        MOV A,SBUF         ;Read the character from the serial port
8        CLR RI             ;Clear the receive flag
9        CLR TI             ;Be sure the bit is initially clear
10       MOV SBUF,A         ;Echo the character back to serial port
11       JNB TI,$           ;Pause until the TI bit is set.
12       SJMP SerialLoop    ;Repeat echo loop
```

Line 1 configures the serial port for normal 8-bit communication and selects timer 1 as the baud rate generator.

Lines 2 through 4 configure and enable timer 1 for mode 2 operation (8-bit auto-reload) and sets the reload value to 253 which generates a baud rate of 9,600 with an 11.0592 MHz crystal.

Line 6 waits for a character to be received by the serial port. Once a character is received, line 7 reads the value into the accumulator and line 8 clears the receive flag.

Line 9 clears the transmit flag before line 10 sends the character to the serial port by writing it back to the SBUF SFR. Line 11 then waits for the character to be completely transmitted. Line 12 returns to the beginning of the loop to wait for the next character.

CHAPTER 10: INTERRUPTS

As the name implies, an interrupt is some event that interrupts normal program execution.

As stated earlier, program flow is always sequential, being altered only by those instructions that expressly cause program flow to deviate in some way. However, interrupts provide a mechanism to "put on hold" the normal program flow, execute a subroutine, and then resume normal program flow as if it had never deviated. This subroutine, called an interrupt handler or **interrupt service routine** (**ISR**), is only executed when a certain event (interrupt) occurs. The event may be one of the timers "overflowing," receiving or transmitting a character via the serial port, or either of two external events. The 8052 may be configured so that when any of these events occur, the main program is temporarily suspended and control is passed to a special section of code which executes some function related to the event that occurred. Once complete, control is returned to the original program. The main program never even knows it was interrupted.

The ability to interrupt normal program execution when certain events occur makes it much easier and much more efficient to handle certain conditions. If it were not for interrupts, a program would have to manually check whether the timers had overflowed, whether another character had been received via the serial port, or if some external event had occurred. Besides making the main program ugly and hard to read, such a situation would make the program inefficient since it would waste valuable instruction cycles checking for events that happen infrequently.

For example, a large 16k program executing many subroutines performing many tasks also needs to toggle the P3.0 port every time timer 0 overflows. The code to do this isn't complicated:

```
    JNB TF0,SKIP_TOGGLE
    CPL P3.0
    CLR TF0
SKIP_TOGGLE: ...
```

Since the TF0 flag is set whenever timer 0 overflows, the above code will toggle P3.0 every time timer 0 overflows. This accomplishes the desired goal but is inefficient.

The JNB instruction consumes two instruction cycles to determine that the flag is not set and to jump over the unnecessary code. In the event that timer 0 overflows, the CPL and CLR instruction require two instruction cycles to execute. To make the math easy, assume that the rest of the code in the program requires 98 instruction cycles. In all, the code consumes 100 instruction cycles (98 instruction cycles plus the 2 that are executed every iteration to determine whether or not timer 0 has overflowed). If the program is using 16-bit timer mode, timer 0 will overflow every 65,536 machine cycles. In that time, the program would have performed 655 JNB tests for a total of 1310 instruction cycles, plus another 2 instruction cycles to perform the code. So to achieve the stated goal the program will have spent 1312 instruction cycles. A full 2.002% of the execution time is being spent just checking when to toggle P3.0. The code is also ugly because the timer check needs to be made after every iteration of the main program loop. Additionally, if the timer overflows while some other code is running there may be a significant delay before P3.0 is toggled in response to the timer overflow.

Interrupts allow the developer to forget about checking for such a condition. The microcontroller itself will check for the condition automatically and, when the condition is met, will jump to a subroutine (the ISR), execute the code, then return. In this case, the subroutine would be nothing more than:

```
CPL P3.0      ;Toggle P3.0
RETI          ;Return from the interrupt
```

Notice that the CLR TF0 command has disappeared. That's because when the 8052 executes the timer 0 ISR, it automatically clears the TF0 flag. Also notice that instead of a normal RET instruction the routine ends with a RETI instruction. The RETI instruction performs the same function as a RET instruction but tells the 8052 that an ISR has finished. ISRs must always end with RETI.

Thus by using interrupts, every 65,536 instruction cycles the ISR will be called resulting in the execution of the CPL instruction and the RETI instruction. Those two instructions together require three instruction cycles. As a result, the same goal has been achieved as in the first example, but it is accomplished in three rather than 1312 instruction cycles. As far as the toggling of P3.0 is concerned, the code is 437 times more efficient! Additionally, the ISR will be called the instant the timer overflows so that P3.0 can be immediately toggled whereas the previous approach would result in a delay until the normal program flow returned to the point of the code that checks for the overflow. The code is also much easier to read and understand because the program doesn't have to check for the timer 0 flag in the main program. The code just sets up the interrupt and forgets about it, secure in the knowledge that the 8052 will execute the ISR whenever it is necessary.

The same idea applies to receiving data via the serial port. One method is to continuously check the status of the RI flag in an endless loop or, alternatively, check the RI flag as part of a larger program loop. However, in the latter case there is a risk of missing characters—what happens if a character is received right after the program checks RI, the rest of the main program loop executes, and before it checks for RI again a second character has already been received? Data will be lost. With interrupts, the 8052 will put the main program "on hold" and call the ISR to handle the reception of the character. This allows the program to avoid a cumbersome check in the main code and, at the same time, avoid loss of data.

10.1 Events that can Trigger Interrupts

The 8052 may be configured so that any (or all) of the following events will cause an interrupt:

1. Timer 0 overflow/reload
2. Timer 1 overflow/reload
3. Reception/transmission of serial data
4. External event 0
5. External event 1
6. Timer 2 overflow/reload

The program needs to be able to distinguish between various interrupts and be able to execute different code depending on which interrupt is triggered. This is accomplished by jumping to a fixed address that will be called when a specific interrupt occurs.

By consulting the following table , it can be seen that whenever timer 0 overflows (i.e., the TF0 bit is set), the main program will be temporarily suspended and control will jump to 000Bh. It is assumed that there is code at address 000Bh that will handle the situation of timer 0 overflowing and subsequently return.

Interrupt	Flag	Interrupt Service Routine Address
External 0	IE0	0003h
Timer 0	TF0	000Bh
External 1	IE1	0013h
Timer 1	TF1	001Bh
Serial	RI/TI	0023h
Timer 2	TF2/EXF2	002Bh

Many derivative chips include additional interrupt sources in addition to the six mentioned above. In such cases, additional interrupt service routine vector entries will continue at 33h, 3Bh, 43h, and so on, with the vector address being increased by eight for each additional interrupt.

10.2 Configuring Interrupts

All interrupts are disabled by default at power-up. This means that even if, for example, the TF0 bit is set, the 8052 will not execute the interrupt. The program must specifically configure the 8052 to enable interrupts and indicate which interrupts it wishes to enable. The program may enable and disable interrupts by modifying the Interrupt Enable SFR (A8h):

Interrupt Enable (IE): SFR Address A8h

Bit	Name	Bit Address	Explanation
7	EA	AFh	Global Interrupt Enable/Disable (0=Disable, 1=Enable)
6	-	AEh	Undefined
5	-	ADh	Undefined
4	ES	ACh	Enable Serial Interrupt
3	ET1	ABh	Enable Timer 1 Interrupt
2	EX1	AAh	Enable External 1 Interrupt
1	ET0	A9h	Enable Timer 0 Interrupt
0	EX0	A8h	Enable External 0 Interrupt

Each of the 8052's interrupts has its own bit in the IE SFR. A given interrupt is enabled by setting the corresponding bit. For example, to enable timer 1 interrupt the following instruction would be executed:

```
MOV IE,#08h
```

or

```
SETB ET1
```

Both of these instructions set bit 3 of IE, thus enabling timer 1 interrupt. Once timer 1 interrupt is enabled, whenever the TF1 bit is set the 8052 will automatically put the main program "on hold" and execute the timer 1 ISR at address 001Bh.

But before any interrupt is truly enabled the program must also set bit 7 of IE. Bit 7, the global interrupt enable/disable, acts as a master switch for all the enabled interrupts simultaneously. If bit 7 is clear then no interrupts will occur even if all the other bits of IE are set. Setting bit 7 will enable all the interrupts that have been selected by setting the other bits in IE. This is useful in program execution if there is time-critical code that needs to execute uninterrupted. In this case, the program may need the code to execute from start to finish without any interrupt getting in the way. To accomplish this, bit 7 of IE may be cleared prior to executing the time-critical code (CLR EA) and then set after it is complete (SETB EA).

So, to summarize what has been stated in this section, to enable timer 1 interrupt the most common approach is to execute the following two instructions:

```
SETB ET1      ;Enable timer 1 Interrupt
SETB EA       ;Enable Global Interrupt flag
```

Thereafter, the timer 1 ISR at 001Bh will automatically be called whenever the TF1 bit is set upon timer 1 overflow.

10.3 Polling Sequence

The 8052 automatically evaluates whether an interrupt should occur after every instruction. When checking for interrupt conditions, it checks them in the following order:

1. External 0 Interrupt
2. Timer 0 Interrupt
3. External 1 Interrupt
4. Timer 1 Interrupt
5. Serial Interrupt
6. Timer 2 Interrupt

This means that if a serial interrupt occurs at the exact same instant that an external 0 interrupt occurs, the external 0 interrupt will be executed first and the serial interrupt will be executed once the external 0 interrupt has completed.

10.4 Interrupt Priorities

The 8052 offers two levels of interrupt priority: high and low. By using interrupt priorities a program may assign higher priority to certain interrupt conditions.

For example, a program may have enabled a timer 1 interrupt that is automatically called every time timer 1 overflows. Additionally, it may have enabled the serial interrupt that is called every time a character is received via the serial port. If, however, the program considers that receiving a character is more important than the timer interrupt, it it may be desirable for the serial interrupt to be able to interrupt a timer 1 interrupt that is already executing. When the serial interrupt is complete, control passes back to timer 1 interrupt and finally back to the main program. This may be accomplished by assigning a high priority to the serial interrupt and a low priority to the timer 1 interrupt.

Interrupt priorities are controlled by the Interrupt Priority SFR (B8h). The IP SFR has the following format:

Interrupt Priority (IP): SFR Address B8h

Bit	Name	Bit Address	Explanation
7	-	BFh	Undefined
6	-	BEh	Undefined
5	-	BDh	Undefined
4	PS	BCh	Serial Interrupt Priority (0=Low priority, 1=High Priority)
3	PT1	BBh	Timer 1 Interrupt Priority (0=Low priority, 1=High Priority)
2	PX1	BAh	External 1 Interrupt Priority (0=Low priority, 1=High Priority)
1	PT0	B9h	Timer 0 Interrupt Priority (0=Low priority, 1=High Priority)
0	PX0	B8h	External 0 Interrupt Priority (0=Low priority, 1=High Priority)

When considering interrupt priorities, the following rules apply:

1. Nothing can interrupt a high-priority interrupt—not even another high priority interrupt.
2. A high-priority interrupt may interrupt a low-priority interrupt.
3. A low-priority interrupt may only occur if no other interrupt is already executing.
4. If two interrupts occur at the same time, the interrupt with higher priority will execute first. If both interrupts are of the same priority, the interrupt that is serviced first by polling sequence will be executed first.

Some derivative chips include a secondary interrupt priority register that, when combined with the standard IP SFR, yields four interrupt priorities rather than two.

10.5 Interrupt Triggering

When an interrupt is triggered, the following actions are taken automatically by the microcontroller:

1. The current program counter is saved on the stack, low-byte first, high-byte second.
2. Interrupts of the same and lower priority are blocked.
3. In the case of timer 0/1 and external interrupts, the corresponding interrupt flag is cleared.
4. Program execution transfers to the corresponding ISR vector address.
5. The ISR is executed.

Take special note of the third step: If the interrupt being handled is a timer 0, timer 1, or external interrupt, the microcontroller automatically clears the interrupt flag before passing control to the interrupt service routine. This means it is not necessary to clear the flag bit in program code. It also means the interrupt code should not depend on the bit being set.

10.6 Exiting Interrupt

An interrupt ends when the program executes the RETI (Return from Interrupt) instruction. When the RETI instruction is executed the following actions are taken by the microcontroller:

1. Two bytes are popped off the stack into the program counter to restore normal program execution at the location the program was running when it was interrupted.
2. Interrupt status is restored to its pre-interrupt state.

10.7 Specific Interrupt Types

In general terms, each interrupt works in basically the same way. In practice, there are a few specific points to keep in mind for each type of interrupt.

10.7.1 Timer 0 and Timer 1 Interrupts

Timer 0 and timer 1 interrupts are good examples of "simple" interrupts. These interrupts are triggered when the corresponding TFx flag is set by a timer overflow. Thus an overflow of timer 0 will set TF0 which will cause a timer 0 interrupt to be triggered, while an overflow of timer 1 will set the TF1 bit which will cause a timer 1 interrupt to be triggered.

In both cases the bit that triggered the interrupt (TF0 or TF1) is cleared automatically by the 8052 *before* the ISR is called. This means that a timer 0 or timer 1 ISR that attempts to check TF0 or TF1 will always find the corresponding bit cleared.

10.7.2 External 0 and External 1 Interrupts

External 0 and external 1 interrupts are triggered by certain conditions on the INT0 and INT1 pins, respectively. The TCON SFR was previously mentioned in section 8.2.4 but only the upper four bits were discussed. The lower four bits of TCON relate to external interrupts and are as follows:

Timer Control (TCON): SFR Address 88h

Bit	Name	Bit Address	Explanation
3	IE1	8Bh	External 1 Interrupt Flag. Flag that triggers an external 1 interrupt.
2	IT1	8Ah	External 1 Interrupt Type. When set external 1 interrupt is triggered on 1-0 transition of the INT1 pin. When clear external 1 interrupt is trigged on a low-level condition on INT1.
1	IE0	89h	External 0 Interrupt Flag. Flag that triggers an external 0 interrupt.
0	IT0	88h	External 0 Interrupt Type. When set external 0 interrupt is triggered on 1-0 transition of the INT0 pin. When clear external 0 interrupt is trigged on a low-level condition on INT0.

An external 0 interrupt will be triggered when the IE0 bit in TCON is set while an external 1 interrupt will be triggered when the IE1 bit is set. These flags will be set when certain conditions are triggered by activity on either INT0 (for external 0 interrupt) or INT1 (for external 1 interrupt). The condition that triggers the flag and, subsequently, the interrupt depends on IT0 (for external 0 interrupt) and IT1 (for external 1 interrupt).

If ITx is clear, the corresponding IEx flag will continuously be set as long as INTx remains low. In this case, the IEx flag cannot be cleared by software since it will be immediately set again by the 8052 as long as INTx remains low. Normally an interrupt that is activated by a low-level on INTx will execute code that causes the external source to terminate the interrupt condition. If the low-level condition on INTx does not terminate while the ISR is executing, the same interrupt will be triggered again as soon as the ISR is complete.

If ITx is set, the corresponding IEx flag will be set on a 1-0 transition of INTx. In this case the 8052 will automatically detect a change in state on INTx from high on one instruction cycle to low at the subsequent instruction cycle. The IEx flag will be set and the external interrupt will be triggered. When the ISR is triggered, the IEx flag will automatically be cleared before the ISR is called.

10.7.3 Serial Interrupts

Serial Interrupts are slightly different than the previously mentioned interrupts. This is due to the fact that there are two interrupt flags: RI and TI. If *either* flag is set, a serial interrupt is triggered. As mentioned in the Serial Communications chapter (chapter 9), the RI bit is set when a byte is received by the serial port and the TI bit is set when a byte has been transmitted.

This means that when a serial interrupt triggers the serial ISR, it may have been triggered because the RI flag was set or because the TI flag was set—or because both flags were set. Thus, the serial ISR must check the status of both of these flags to determine what action is appropriate. Also, since the 8052 does not automatically clear the RI and TI flags, the ISR must clear these bits somewhere in the ISR itself.

A brief code example is in order:

```
SerialInterrupt:
    JNB RI,CHECK_TI     ;If RI flag is not set, jump to check TI
    MOV A,SBUF          ;The RI bit *was* set so read SBUF
    CLR RI              ;Clear the RI bit after it's been processed
CHECK_TI:
    JNB TI,EXIT_INT     ;If TI flag not set, jump to the exit point
    CLR TI              ;Clear TI bit before sending next character
    MOV SBUF,#'A'       ;Send another character to the serial port
EXIT_INT:
    RETI                ;Exit interrupt handler
```

The code above checks the status of both interrupts flags. If both flags are set, both sections of code will be executed. Also note that each section of code clears its corresponding interrupt flag. If the program omits the step of clearing the flags, the serial interrupt will be executed over and over until the bits are cleared. For this reason it is very important to always clear the interrupt flags in a serial interrupt.

10.7.4 Timer 2 Interrupts

The timer 2 interrupt, like the serial interrupt, may be triggered by two different flags: the TF2 bit which indicates that timer 2 has overflowed or by the EXF2 bit which indicates that a capture or reload was caused by a 1-0 transition on the T2EX pin. These bits must be cleared by the ISR to prevent the timer 2 interrupt from executing repeatedly as these bits are not cleared automatically by the 8052.

10.8 Register Protection

One very important rule applies to all interrupt service routines: interrupts must leave the processor in the same state as it was in when the interrupt began. The idea behind interrupts is that the main program not be aware that they are executing in the "background." Consider the following code:

```
CLR C          ;Clear carry
MOV A,#25h     ;Load the accumulator with 25h
ADDC A,#10h    ;Add 10h, with carry
```

After the above three instructions are executed the accumulator will contain a value of 35h. But what would happen if an interrupt occurred right after the MOV instruction? During this interrupt, consider what would happen if the the carry bit were set and the value of the accumulator were changed to 40h. When the interrupt finished and control was passed back to the main program, the ADDC would add 10h to 40h, and additionally add an additional 01h because the carry bit is set. In this case, the accumulator will contain the value 51h at the end of execution.

In this example the main program has seemingly calculated the wrong answer. How can 25h + 10h yield 51h as a result? A developer unfamiliar with interrupts would be convinced that the microcontroller was damaged in some way and was having problems with mathematical calculations.

What has happened, in reality, is that the interrupt did not protect the registers it used. Restated: *An interrupt must leave the processor in the same state as it was in when the interrupt initiated.*

This means that if the ISR uses the accumulator, it must ensure that the value of the accumulator is the same at the end of the interrupt as it was at the beginning. This is generally accomplished with a PUSH and POP sequence at the beginning and end of each interrupt handler.

For example:

```
INTERRUPT_HANDLER:
    PUSH ACC       ;Push the initial value of accumulator onto stack
    PUSH PSW       ;Push the initial value of PSW SFR onto stack
    MOV A,#0FFh    ;Use accumulator & PSW for whatever you want
    ADD A,#02h     ;Use accumulator & PSW for whatever you want
    POP PSW        ;Restore the initial value of the PSW from the stack
    POP ACC        ;Restore initial value of the accumulator from stack
```

The guts of the interrupt are the MOV and ADD instructions which modify the accumulator (the MOV instruction) and also modify the value of the carry bit (the ADD instruction will cause the carry bit to be set in PSW). Since an ISR must guarantee that the registers remain unchanged, the routine pushes the original values onto the stack using the PUSH instruction. It is then free to use the registers it protected in any way it likes. Once the interrupt has finished its task, it POPs the original values back into the registers. When the interrupt exits, the main program will never know the difference because the registers are exactly the same as they were before the interrupt executed.

In general, an ISR must protect the following registers if they are used within the routine:

1. Program Status Word SFR (PSW)
2. Data Pointer SFRs (DPH/DPL)
3. Accumulator (ACC)
4. "B" Register (B)
5. "R" Registers (R0-R7)

PSW consists of many individual bits that are set by various 8052 instructions so it is generally a good idea to always protect PSW by pushing it onto and popping it off the stack at the beginning and end of all interrupts unless the developer is very familiar with the instructions being used in the ISR and is absolutely certain that PSW will not be affected.

Note also that most assemblers will not allow the instruction:

```
PUSH R0        ;Error - Invalid instruction!
```

This is due to the fact that, depending on which register bank is selected, R0 may refer to either internal RAM address 00h, 08h, 10h, or 18h. R0, in and of itself, is not a valid memory address that the PUSH and POP instructions can use. Thus, if the ISR is using any "R" registers it will generally have to do an additional step by first moving the "R" register to the accumulator and then pushing the accumulator on the stack. For example, instead of PUSH R0 the ISR would execute:

```
MOV A,R0       ;Copies the contents of R0 to the accumulator
PUSH ACC       ;Pushes the accumulator (which holds R0) onto the stack
```

Of course, the ISR must first protect the accumulator itself by pushing the accumulator onto the stack *before* it starts using it to push "R" registers onto the stack. Subsequently, when the ISR is done restoring the "R" registers at the end of the routine, it must restore the accumulator itself with one additional POP ACC instruction.

```
POP ACC        ;Pops the last value on stack into accumulator
MOV R0,A       ;Copies the contents of the accumulator into R0
```

10.9 Locating Large ISRs in Memory

Reviewing the table in section 10.1, it is quite noticeable that each ISR only has eight bytes of memory allocated to it. External 0 interrupt is at 0003H while timer 0 interrupt is at 000Bh, and so on. An obvious problem exists if a given ISR consists of more than eight bytes of program code since it would extend into the next ISR address.

In most programs, the code at the ISR address is nothing more than an LJMP which jumps to the real ISR code elsewhere in memory. An LJMP requires only three bytes of memory so normally if a program uses, for example, external 0 interrupt, the program would place an LJMP instruction at 0003H which jumps to the ISR code elsewhere in memory. The ISR address vectors become nothing more than a "jump table" which jumps to the real ISR code. In this way, ISRs may exceed eight bytes of memory without overwriting the ISR addresses of the other interrupts.

10.10 Common Problems with Interrupts

Interrupts are a very powerful tool available to the 8052 developer but, when used incorrectly, can be a source of a huge number of debugging hours. Errors in interrupt routines are often very difficult to diagnose and correct.

If a program using interrupts crashes or does not seem to be performing as expected, always review the following interrupt-related issues:

1. **Register Protection**: Make sure all registers are protected as explained above. Forgetting to protect a register that is used in an interrupt will often produce very strange results. In the example above it was shown how a failure to protect registers caused the main program to apparently calculate that 25h + 10h = 51h. If problems with registers changing values unexpectedly or operations producing "incorrect" values are observed, it is very likely that the program isn't protecting registers properly.

2. **Forgetting to restore protected values**: Another common error is to push registers onto the stack to protect them and subsequently forget to pop them off the stack before exiting the interrupt. For example, a program may push ACC, B, and PSW onto the stack in order to protect them and subsequently pop only ACC and PSW off the stack before exiting. In this case, since the program didn't restore the value of "B" an extra value remains on the stack. When the RETI instruction is executed the 8052 will use that value as the return address instead of the correct value. In this case the program will almost certainly crash. Always make sure the program pops the same number of values off the stack as were pushed onto it.

3. **Using RET instead of RETI**: Remember that interrupts are always terminated with the RETI instruction. It is easy to inadvertently use the RET instruction instead. The RET instruction, however, will not end the interrupt. Usually, using RET instead of RETI will cause the illusion of the main program running normally but the interrupt will only be executed once. If it appears that the interrupt mysteriously stops executing, verify that the interrupt is terminating with RETI.

10.11 Functional Interrupt Example

The concept of interrupts is demonstrated in section 22.3 in which a timer interrupt is used to develop a software-based real-time clock.

ASSEMBLY LANGUAGE

CHAPTER 11: 8052 ASSEMBLY LANGUAGE

Assembly language is a low-level, pseudo-English representation of the microcontroller's machine language. Each assembly language instruction has a one-to-one relation with the microcontroller's machine-level instructions.

High-level languages, such as 'C', Basic, Visual Basic, Java, etc. are one or more steps above assembly language so that no significant knowledge of the underlying architecture is necessary. It is possible (and common) for a developer to program a Visual Basic application in Windows without knowing much of anything about the Windows API, much less the underlying architecture of the Intel Pentium. Further, a developer who has written code in 'C' for Unix won't have significant problems adapting to writing code in 'C' for Windows or a microcontroller such as an 8052—while there are some variations, the C compiler itself takes care of most of the processor-specific issues.

Assembly language, on the other hand, is very processor-specific. A prior knowledge of assembly language with any given processor will be helpful when beginning to code in the assembly language of another processor but the two assembly languages may be extremely different. Different architectures have different instruction sets and different forms of addressing. In fact, only general concepts may "port" from one processor to another.

The low-level nature of assembly language programming requires an understanding of the underlying architecture of the processor for which one is developing. This is why this book explained the 8052's architecture fully before attempting to introduce the reader to assembly language programming. Many aspects of assembly language would be completely confusing without a prior knowledge of the architecture.

In this chapter the reader will be introduced to 8052 assembly language, concepts, and programming style.

11.1 Syntax

Each line of an assembly language program consists of the following syntax, each field of which is optional. However, when used, the elements of the line must appear in the following order:

- **Label** – A programmer-chosen symbol that names the instruction's address in memory. The label, if present, must be terminated with a colon immediately following the label.
- **Instruction** – An assembly language instruction which, when assembled, will perform some specific function when executed by the microcontroller. The instruction is a pseudo-English "mnemonic" which relates directly to one machine language instruction.
- **Comment** – The developer may include a comment on each line for in-line documentation. These comments are ignored by the assembler but may make it easier to understand the code. A comment, if used, must be preceded by a semicolon.

In summary, then, a typical 8052 assembly language line might appear as:

```
MYLABEL: MOV A,#25h ;This is just a sample comment
```

In this example the label is MYLABEL. This means that if other instructions in the program need to make reference to this instruction, they may do so by referring to "MYLABEL" rather than the memory address of the instruction.

The 8052 assembly language instruction in this line is MOV A,#25h. This is the actual instruction that the assembler will analyze and assemble into the two bytes 74h 25h. The first number, 74h, is the 8052 machine language instruction (opcode) "MOV A,#*dataValue*"—which means "move the value of the next byte in code memory into the accumulator." Since this instruction loads the accumulator with the value 25h, the byte following the opcode is 25h. There is a one-to-one relationship between the assembly language instruction and the machine language code that is generated by the assembler.

Finally, the example above includes the optional comment ;This is just a sample comment. The comment must always start with a semicolon. The semicolon tells the assembler that the rest of the line is a human-readable comment that should be ignored by the assembler.

Since all fields are optional, the following are also alternatives to the above syntax:

```
Label without instruction and comment:      LABEL:
Line with label and instruction:            LABEL: MOV A,#25h
Line with instruction and comment:                 MOV A,#25h ;This is a comment
Line with label and a comment:              LABEL: ;This is a comment
Line with just a comment:                          ;This is a comment
```

All of the above permutations are completely valid. It is up to the developer which components of the assembly language syntax will be used. But, when used, they must follow the above syntax and in the correct order.

It does not matter in which column each field begins. That is, a label may start at the beginning of the line or after any number of blank spaces or tabs. Likewise, an instruction may start in any column of the line as long as it follows any label that may also be on that line.

The case of assembly language instructions and pre-defined symbols—such as SFRs—is normally irrelevant. Each of the following examples is functionally identical:

```
MOV A,#10h    ;These three
mov a,#10H    ;instructions are
Mov A,#10h    ;all the same
```

Labels, however, are normally case sensitive. For example, the following would generate an assembly error:

```
MyLabel: LJMP MYLABEL      ;Error! Label is "MyLabel", not "MYLABEL"
```

Labels may generally include any uppercase or lowercase alphabetic character, numbers, the underscore character, and the dollar sign. Labels must never start with a numeric value, however, nor may they contain spaces.

```
Label1:              ;Valid
Label_1:             ;Valid
1_Label:             ;INVALID! Labels cannot start with a digit
My Label:            ;INVALID! Labels cannot contains spaces
```

11.2 Number Bases

Most assemblers are capable of accepting numeric data in a variety of number bases. Commonly supported are decimal, hexadecimal, binary, and octal.

Decimal: To express a decimal number in assembly language, simply enter the number as normal.

Hexadecimal: To express a hexadecimal number, enter the number as a hexadecimal value and terminate the number with the suffix "h". For example, the hexadecimal number 45 would be expressed as 45h. If the hexadecimal number begins with an alphabetic character (A, B, C, D, E, or F), the number must be preceded with a leading zero. For example, the hex number E4 would be written as 0E4h. The leading zero allows the assembler to differentiate the hex number from a symbol since a symbol can never start with a number.

Binary: To express a binary number, enter the binary number itself followed by a trailing "B" to indicate binary. For example, the binary number 100010 would be expressed as 100010B.

Octal: To express an octal number, enter the octal number itself followed by a trailing "Q" to indicate octal. For example, the octal number 177 would be expressed as 177Q.

All of the following instructions load the accumulator with 30 (decimal):

```
MOV A,#30
MOV A,#11110B
MOV A,#1EH
MOV A,#36Q
```

11.3 Expressions

Mathematical expressions may be used in assembly language instructions anywhere a numeric value may be used. For example, both of the following are valid assembly language instructions:

```
MOV A,#20h + 34h       ;Equivalent to #54h
MOV 35h + 2h,#10101B   ;Equivalent to MOV 37h,#10101B
```

11.4 Operator Precedence

Mathematical operators within an expression are subject to the following order of precedence. Operators at the same "level" are evaluated left-to-right.

1 (Highest Precedence)	( )
2	HIGH LOW NOT
3	* / MOD SHL SHR
4	+ -
5	AND
6	OR XOR
7 (Lowest)	< <= => > =

If there is any doubt about operator precedence, it is useful to use parenthesis to force the order of operation that is desired. Even if it isn't strictly necessary it is often easier to read mathematical expressions when parenthesis have been added.

11.5 Characters and Character Strings

Characters and character strings are enclosed in quotes and are converted to their numeric equivalent at assemble-time. For example, the following two instructions are the same:

```
MOV A,#'C'
MOV A,#43H
```

These two instructions are the same because the assembler will see the 'C' sequence, convert the character contained in quotes to its ASCII equivalent (43H), and use that value.

Strings of characters are sometimes enclosed in single-quotes and sometimes enclosed in double-quotes. Pinnacle 52 uses double-quotes to indicate a string of characters and a single-quote to indicate a single character. Thus:

```
MOV A,#'C'        ;Single character - ok
MOV A,#"STRING"   ;String - ERROR! Can't load a string into accumulator
```

Strings are invalid in the above context although there are other special assembler directives that do allow strings. Be sure to check the manual for the assembler being used to determine whether character strings should be placed within single-quotes or double-quotes.

11.6 Assembly Language Directives

Some commands in an assembly language program are not actual 8052 instructions but rather what are known as assembler **directives**. These are not converted to 8052 machine language and downloaded to the microcontroller, but rather are used by the assembler at assemble-time to influence the assembly process. In short, they direct the assembler to take some specific action.

11.6.1 Setting the Address of Program Assembly (ORG)

Unless otherwise specified, an 8052 assembly language program will start to be assembled at program memory address 0000h which is also the address at which an 8052 microcontroller will start executing code when it is powered-up. It is often necessary to instruct the assembler to place code at different places in code memory, especially when dealing with interrupt service routines (ISRs).

The **ORG** directive is handled by the assembler at assemble-time and instructs it to change where the instructions which follow are to be stored in code memory. For example:

```
ORG 0000h              ;Tell assembler to continue assembly at 0000h
LJMP MAIN              ;Code memory 0000h: jumps to "MAIN"

ORG 000Bh              ;Tell assembler to continue assembly at 000Bh
LJMP EXT0ISR           ;Code memory 000Bh: jumps to "EXT0ISR"

ORG 0030h              ;Tell assembler to continue assembly at 0030h
MAIN: NOP              ;Code memory 0030h: the "MAIN" routine starts here

ORG 0050h              ;Tell assembler to continue assembly at 0050h
EXT0ISR: RETI          ;Code memory 0050h: the "EXT0ISR" routine starts here
```

The first line (ORG 0000h) tells the assembler that the next instruction(s) (LJMP MAIN) should be assembled starting at code memory address 0000h. The third line (ORG 000Bh) tells the assembler that the next instruction (LJMP EXT0ISR) should be assembled starting at code memory address 000Bh. The fifth line (ORG 0030h) tells the assembler that the next instruction (NOP) should be assembled starting at code memory address 0030h. Finally, the seventh line (ORG 0050h) tells the assembler that the next instruction (RETI) should be assembler starting at code memory address 0050h.

This allows the developer to, among other things, locate specific code in the interrupt service routine vector table that exists at 0003h, 000Bh, 0013h, 001Bh, 0023h and 002Bh. The main code is then placed at an address that is safely out of the code memory space occupied by the ISR vector table.

11.6.2 Establishing Symbol Equates (EQU)

The assembler recognizes certain symbols as having specific values. For example, if the assembler encounters the instruction MOV SBUF,#25, it knows that SBUF refers to SFR address 99h so it internally replaces SBUF with 99h and assembles the instruction MOV 99h,#25. This is clearly helpful to the developer since it makes it unnecessary to memorize the address associated with the SFR name.

It is often useful for the developer to establish other symbols that have fixed values throughout the program so they may be referred to by name. If a yellow LED is attached to P1.0 it could be flashed with:

```
CLR P1.0
SETB P1.0
```

There are two potential problems with this: 1) It is not clear that P1.0 is attached to a LED so the code doesn't really help anyone understand what it's doing. 2) If it is later decided that the LED should be placed on another port pin, the developer will have to go through the code and replace all references to P1.0 with the new value.

Assemblers allow the developer to alleviate this problem by defining **equates**. An equate allows a user-defined symbol name to be associated with a value. For example:

```
YELLOW_LED EQU P1.0
CLR YELLOW_LED
SETB YELLOW_LED
```

In this case, the symbol YELLOW_LED has been defined to be equal to P1.0 (90h). This allows the developer to refer to P1.0 more descriptively as YELLOW_LED throughout the program. If the design ever changes so that the yellow LED is on a different I/O line, only the EQU line needs to be updated—the rest of the program remains unchanged.

It is a very good idea to use equates when possible as it makes the program easier to read and maintain.

11.6.3 Inserting 8-bit Data In Program (DB)

It is often necessary to embed raw data in code memory. This may consist of numeric data, character strings, or other data that is needed by a program at run-time but which is not, itself, executable code. This is accomplished with the **DB** directive.

```
DB "HELLO WORLD", 13, 0
```

The above code will be converted to the following sequence of bytes:

```
48h 45h 4Ch 4Ch 4Fh 20h 57h 4Fh 52h 4Ch 44h 0Dh 00h
```

In this case, the DB directive will cause each of the characters within the quotes to be converted to its ASCII code equivalent. Any parameters outside the quotes—and separated by commas—will also be converted to data within the code. The value 13 in this example is the ASCII code for a carriage-return and is represented by 0Dh (in hexadecimal) in the converted sequence. The final 0 is converted to 00h and is the NULL character which signifies the end of the string.

The DB directive only inserts 8-bit data. If values exceeding FFh are included in a DB directive, only the least significant byte will be inserted into the code. For example:

```
DB 1234h, 4567h
```

Since the two values in the DB directive are 16-bit values, only the low eight bits will be inserted into code:

```
34h 67h
```

11.6.4 Inserting 16-bit Data In Program (DW)

If 16-bit data needs to be included in a program, the data word (DW) directive should be used instead of the DB directive. DW functions exactly the same as DB but generates 16-bit values. Data is generated high-byte first, low-byte second. Any 8-bit value is converted to a 16-bit value and its high byte will be padded with zeros.

Consider the example:

```
DW 1234h, 56h, 8912h
```

74

This will produce the following data in code memory:

```
12h 34h 00h 56h 89h 12h
```

Note that the 16-bit values are converted as two bytes with the high byte first. In the case of the 8-bit value (56h), it is treated as the value 0056h and inserted in the code accordingly.

11.7 Changing Program Flow (LJMP, SJMP, AJMP)

LJMP, SJMP and **AJMP** are used as a "GOTO" in assembly language. They cause program execution to continue at the address or label they specify. For example:

```
LJMP  LABEL3        ;Program execution is transferred to LABEL3
SJMP  LABEL4        ;Program execution is transferred to LABEL4
AJMP  LABEL7        ;Program execution is transferred to LABEL7
```

The differences between LJMP, SJMP, and AJMP are:

1. **LJMP** requires 3 bytes of program memory and can jump to any address in the program.
2. **SJMP** requires 2 bytes of program memory but can only jump to an address within the 128 bytes preceding or the 127 bytes following the address that follows the SJMP instruction.
3. **AJMP** requires 2 bytes of program memory but can only jump to an address in the same 2k block of memory.

These instructions perform the same task but differ in what addresses they can jump to and how many bytes of program memory they require.

LJMP will always work. You can always use LJMP to jump to any address in your program.

SJMP requires two bytes of memory but has the restriction that it can only jump to an instruction or label that is within 128 bytes before or 127 bytes after the address that follows the instruction. This is useful if the program is branching to an address that is very close to the jump itself. One byte of memory will be saved by using SJMP instead of LJMP.

AJMP also requires two bytes of memory but has the restriction that it can only jump to an instruction or label that is in the same 2k block of program memory. For example, if the AJMP instruction is at address 0200h it can only jump to addresses between 0000h and 07FFh. It could not jump to 800h, for example.

Some optimizing assemblers allow the use of "JUMP" or "JMP". While there isn't a JUMP instruction in the 8052 instruction set, the optimizing assembler will automatically replace the JUMP with the most memory-efficient instruction.

11.8 Subroutines (LCALL, ACALL, RET)

8052 assembly language permits the use of subroutines. A subroutine is a section of code that is called by a program, executes a task, and then returns to the instruction immediately following that of the instruction that made the call.

LCALL and **ACALL** are both used to call a subroutine. LCALL requires three bytes of program memory and can call any subroutine anywhere in memory; ACALL requires two bytes of program memory and can only call a subroutine within the same 2k block of program memory.

Both call instructions will save the current address on the stack and jump to the specified address or label. The subroutine at that address will perform a task and then return to the original instruction by executing the **RET** instruction.

For example, consider the following code:

```
            LCALL SUBROUTINE1    ;Call the SUBROUTINE1 subroutine
            LCALL SUBROUTINE2    ;Call the SUBROUTINE2 subroutine
                     .
                     .
 SUBROUTINE1: {subroutine code}  ;Insert subroutine code here
            RET                  ;Return from subroutine

 SUBROUTINE2: {subroutine code}  ;Insert subroutine code here
            RET                  ;Return from subroutine
```

The code above starts by calling SUBROUTINE1. Execution will transfer to SUBROUTINE1 and execute whatever code is found there. When the MCU encounters the RET instruction it will automatically return to the next instruction, which is LCALL SUBROUTINE2. SUBROUTINE2 will then be called, execute its code, and return to the main program when it encounters the RET instruction.

It is very important that all subroutines end with the RET instruction. Failing to do so will lead to unpredictable results.

Subroutines may call other subroutines. For example, in the code above SUBROUTINE1 could include an instruction that calls SUBROUTINE2. SUBROUTINE2 would then execute and return to SUBROUTINE1 which would then return to the instruction that called it. However, keep in mind that every LCALL or ACALL that is executed expands the stack by two bytes. If the stack starts at internal RAM address 30h and 10 nested calls are made from within subroutines to other subroutines the stack will expand by 20 bytes to 44h.

Recursive subroutines (subroutines that call themselves) are a very popular method of solving some common programming problems. However, unless you know for certain that the subroutine will call itself only a certain number of times it is generally not possible to use subroutine recursion in 8052 assembly language. Due to the small amount of internal RAM a recursive subroutine could quickly cause the stack to fill all of internal RAM.

11.9 Register Assignment (MOV)

One of the most commonly used 8052 assembly language instructions is the **MOV** instruction. 57 of the 254 opcodes are MOV instructions which is due to the fact that there are many ways that data can be moved between the various registers using various addressing modes.

The MOV instruction is used to move data from one register to another—or to simply assign a value to a register—and has the following general syntax:

MOV *DestinationRegister, SourceValue*

"*DestinationRegister*" always indicates the register or address in which "*SourceValue*" will be stored whereas *SourceValue* indicates the register from which the value will be taken, or the value itself if it is preceded by a pound sign (#).

For example:

```
MOV A,25h      ;Moves contents of internal RAM address 25h to accumulator
MOV 25h,A      ;Move contents of accumulator into internal RAM address 25h
MOV 80h,A      ;Move the contents of the accumulator to P0 SFR (80h)
MOV A,#25h     ;Moves the value 25h into the accumulator
```

The first parameter is the register, internal RAM address, or SFR address that a value is being moved to. Another way of looking at it is that the first parameter is the register that is going to be assigned a new value.

The second parameter tells the 8052 where to get the new value. Normally, the value of the second parameter indicates the internal RAM or SFR address from which the value should be obtained. However, if the second parameter is preceded by a pound sign (#), the register will be assigned the *value* of the number that follows the pound sign (as is demonstrated in the last example above).

As already mentioned, the MOV instruction is one of the most common and vital instructions that an 8052 assembly language programmer will use. The prospective assembly language programmer must fully master the MOV instruction. This may seem simple but it requires knowing all of the permutations of the MOV instruction as well as when to use them. This knowledge comes with time and experience and by reviewing the "8052 Instruction Set Overview" (Appendix A).

It is important that all types of MOV instructions be understood so that the programmer knows what types of MOV instructions are available as well as what kinds of MOV instructions are *not* available.

Careful inspection of the MOV instructions in the instruction set reference will reveal that there is no "MOV from 'R' register to 'R' register." That is to say, the following instruction is invalid:

```
MOV R2,R1      ;INVALID!!
```

This is a logical type of operation for a programmer to want to implement but such an instruction does not exist. Instead, the above must be coded as:

```
MOV A,R1       ;Move R1 to accumulator
MOV R2,A       ;Move accumulator to R2
```

Another combination that is not supported is "MOV indirectly from internal RAM to another indirect RAM address". Again, the following instruction is invalid:

```
MOV @R0,@R1    ;INVALID!!
```

Instead this could be programmed as:

```
MOV A,@R1      ;Move contents of IRAM pointed to by R1 to accumulator
MOV @R0,A      ;Move accumulator to internal RAM address pointed to by R0
```

Also note that only R0 and R1 can be used for indirect addressing.

When a situation arises that requires a type of MOV instruction that doesn't exist, it is generally helpful to use the accumulator. If a given MOV instruction doesn't exist, it can usually be accomplished by using two MOV instructions that both use the accumulator as a transfer or temporary register.

With this knowledge of the MOV instruction simple memory assignment tasks can be performed.

1. Clear the contents of internal RAM address FFh:

```
MOV A,#00h     ;Move the value 00h to the accumulator (accumulator=00h)
MOV R0,#0FFh   ;Move the value FFh to R0 (R0=0FFh)
MOV @R0,A      ;Move accumulator to @R0, thus clearing contents of FFh
```

2. Clear the contents of internal RAM address FFh (more efficient):

```
MOV R0,#0FFh   ;Move the value FFh to R0 (R0=0FFh)
MOV @R0,#00h   ;Move 00h to @R0 (FFh), thus clearing contents of FFh
```

3. Clear the contents of all bit memory (internal RAM addresses 20h through 2Fh):

```
MOV 20h,#00h   ;Clear internal RAM address 20h
MOV 21h,#00h   ;Clear internal RAM address 21h
MOV 22h,#00h   ;Clear internal RAM address 22h
MOV 23h,#00h   ;Clear internal RAM address 23h
MOV 24h,#00h   ;Clear internal RAM address 24h
MOV 25h,#00h   ;Clear internal RAM address 25h
MOV 26h,#00h   ;Clear internal RAM address 26h
MOV 27h,#00h   ;Clear internal RAM address 27h
MOV 28h,#00h   ;Clear internal RAM address 28h
MOV 29h,#00h   ;Clear internal RAM address 29h
MOV 2Ah,#00h   ;Clear internal RAM address 2Ah
MOV 2Bh,#00h   ;Clear internal RAM address 2Bh
MOV 2Ch,#00h   ;Clear internal RAM address 2Ch
MOV 2Dh,#00h   ;Clear internal RAM address 2Dh
MOV 2Eh,#00h   ;Clear internal RAM address 2Eh
MOV 2Fh,#00h   ;Clear internal RAM address 2Fh
```

11.10 Incrementing and Decrementing Registers (INC/ DEC)

Two instructions, **INC** and **DEC**, can be used to increment or decrement the value of a register, internal RAM, or SFR by one. These instructions are rather self-explanatory.

The INC instruction will add 1 to the current value of the specified register; if the current value is 255, it will "overflow" back to 0. For example, if the accumulator holds the value 240 and the INC A instruction is executed, the accumulator will be incremented to 241.

```
INC A        ;Increment the accumulator by 1
INC R1       ;Increment R1 by 1
INC 40h      ;Increment internal RAM address 40h by 1
```

The DEC instruction will subtract 1 from the current value of the specified register; if the current value is 0, it will "underflow" back to 255. For example, if the accumulator holds the value 240 and the DEC A instruction is executed, the accumulator will be decremented to 239.

```
DEC A        ;Decrement the accumulator by 1
DEC R1       ;Decrement R1 by 1
DEC 40h      ;Decrement internal RAM address 40h by 1
```

Those that have programmed in assembly language under other architectures may be used to the INC and DEC instructions setting an overflow or underflow flag when the register overflows from 255 to 0 or underflows from 0 back to 255. This is not the case in 8052 assembly language. Neither of these instructions affect any flags whatsoever.

11.11 Program Loops (DJNZ)

Many operations are conducted within finite loops. That is, a given code segment is executed repeatedly until a given condition is met.

A common type of loop is a simple "counter loop" in which a code segment is executed a certain number of times and finishes. This is accomplished easily in assembly language with the **DJNZ** instruction. DJNZ means "decrement, jump if resulting value not zero". Consider the following code:

```
        MOV R0,#08h  ;Set number of loop cycles to 8
LOOP:   INC A        ;Increment acc (or do whatever the loop needs to do)
        DJNZ R0,LOOP ;Decrement R0, loop back to LOOP if R0 is not 0
        DEC A        ;This is the first instruction after the loop
```

This is a very simple counter loop. The first line initializes R0 to 8 which will be the number of times the loop will be executed.

To execute a loop 256 times, set the initial value of the counter variable to 00h.

The second line, "LOOP", is the actual body of the loop. This could contain any instruction (or multiple instructions) that need to be executed repeatedly. In this example, the only thing the loop does is increment the accumulator with the INC A instruction.

The key is the last line with the DJNZ instruction. This instruction reads "Decrement the R0 Register, and if it's not now zero, jump back to LOOP". This instruction will decrement the R0 register. It will then check to see if the new value is zero and, if it isn't, will go back to LOOP. The first time this loop executes, R0 will be decremented from 08 to 07, then 07 to 06, and so on until it decrements from 01 to 00. At that point the test in the DJNZ instruction will fail since the accumulator is zero; this will cause the program to *not* go back to LOOP and, thus, it will then continue executing at the DEC instruction.

DJNZ is one of the most common ways to perform programming loops that execute a specific number of times. The number of times the loop will be executed depends on the initial value of the "R" register that is used by the DJNZ instruction.

11.11.1 DJNZ With More Than 256 Repetitions

Since the "R" registers that are available with the DJNZ instruction can only handle 8-bit numbers in the range of 0-255, DJNZ can repeat a loop a maximum of 256 times.

If it is necessary to repeat a loop more than 256 times, DJNZ can be "nested." Consider the following code:

```
          MOV R1,#03h  ;Set number of R1 loops to 3
TOP:      MOV R0,#00h  ;Set number of primary loop cycles to 256
LOOP:     INC A        ;Increment acc (or do whatever the loop needs to do)
          DJNZ R0,LOOP ;Decrement R0, loop back to LOOP if R0 is not 0
          DJNZ R1,TOP  ;Decrement R1, loop back to TOP if R1 is not 0
          DEC A        ;Decrement accumulator (or what comes after loop)
```

This code is the same as the previous example except the "TOP" label is added to the line that initializes R0 and there are two additional lines that refer to R1. In this example R1 is initially set to 3. The code then executes as before—counting R0 down from 256 to 0. When R0 reaches 0, the loop will have been executed 256 times and execution will proceed with the DJNZ R1, TOP instruction. At that point, R1 will be decremented from 3 to 2 and, since R1 is not zero, it will jump back to TOP which will reset the R0 loop for another 256 iterations. This will happen again as R1 is decremented from 2 to 1 and then from 1 to 0. When R1 reaches 0 execution will proceed with the DEC A instruction.

In this fashion, the loop will be executed a total of 768 times (3 * 256). Virtually any number of repetitions can be achieved by adjusting the initialization values of the two registers. For example, 398 repetitions could be obtained by setting R1 initially to 2 and R0 initially to 199 (199 * 2 = 398).

This technique can be used to execute a maximum of 65,536 repetitions of the loop. If even more repetitions are necessary, the above technique can be expanded to three, four, or even more "R" registers.

11.12 Setting, Clearing, and Moving Bits (SETB/CLR/CPL/MOV)

One very powerful feature of the 8052 architecture is its ability to manipulate individual bits on a bit-by-bit basis. As mentioned in section 2.3.1.3, there are 128 numbered bits (00h through 7Fh) that may be used by the user's program as bit variables. Additionally, bits 80h through FFh allow access to SFRs that are divisible by 8 on a bit-by-bit basis. The two basic instructions to manipulate bits are **SETB** and **CLR** while a third instruction, **CPL**, is also often used.

The SETB instruction will set the specified bit which means the bit will then have a value of "1" or "on".

For example:

```
SETB 20h      ;Sets user bit 20h (sets bit 0 of IRAM address 24h to 1)
SETB 80h      ;Sets bit 0 of SFR 80h (P0) to 1
SETB P0.0     ;Exactly the same as the previous instruction
SETB C        ;Sets the carry bit to 1
SETB TR1      ;Sets the TR1 bit to 1 (turns on timer 1)
```

The first example, SETB 20h, sets user bit 20h. Since all bits between 00h and 7Fh are user bits, the address 20h automatically corresponds to a user-defined bit. Since these 128 user bits reside in internal RAM at the addresses of 20h through 2Fh, bit 20h is the 32nd user-defined bit. Each byte of internal RAM, by definition, holds 8 individual bits so bit 20h would be the lowest bit of internal RAM address 24h.

It is very important to understand that bit memory is a part of internal RAM. In the case of SETB 20h it was calculated that bit 20h is actually the low bit of internal RAM address 24h. This is because bits 00h-07h are internal RAM address 20h, bits 08h-0Fh are internal RAM address 21h, bits 10h-17h are internal RAM address 22h, bits 18h-1Fh are internal RAM address 23h, and bits 20h-27h are internal RAM address 24h.

The second example, SETB 80h, is similar to SETB 20h. SETB 80h sets bit 80h. However, remember that bits 80h-FFh correspond to individual bits of SFRs, not internal RAM. Thus SETB 80h will set bit 0 of SFR 80h which is the P0 SFR.

The next instruction, SETB P0.0, is identical to SETB 80h. The only difference is that the bit is now being referenced by name rather than number. This will make an assembly language program more readable. The assembler will automatically convert "P0.0" to 80h when the program is assembled. It is also recommended that user bits utilized by the program be named using the EQU directive for clarity (see section 11.6.2).

The next example, SETB C, is a special case. This instruction sets the carry bit which is a very important bit that is used for many purposes. It is also special in that there is an opcode that means "SETB C". While other SETB instructions require two bytes of code memory, SETB C only requires one.

Finally, the SETB TR1 example shows a typical use of SETB to set an individual bit of an SFR. In this case, TR1 is TCON.6 (bit 6 of TCON SFR, SFR address 88h). Since TCON's SFR address is 88h, it is divisible by eight and, thus, addressable on a bit-by-bit basis. The assembler will convert the bit name TR1 to its bit address 8Eh when the program is assembled.

The CLR instruction functions in the same manner but clears the specified bit. For example:

```
CLR 20h       ;Clears user bit 20h to 0
CLR P0.0      ;Sets bit 0 of P0 to 0
CLR TR1       ;Clears TR1 bit to 0 (stops timer 1)
```

These two instructions, CLR and SETB, are the two fundamental instructions used to manipulate individual bits in 8052 assembly language.

A third bit instruction, CPL, complements the value of the given bit. The instruction syntax is exactly the same as SETB and CLR, but CPL will "flip the bit." If the bit was clear, CPL will set it; likewise, if the bit was set, CPL will be clear it.

An additional instruction, `CLR A`, exists which is used to clear the contents of the accumulator. This is the only CLR instruction that clears an entire SFR rather than just a single bit. The `CLR A` instruction is the equivalent of `MOV A,#00h`. The advantage of using `CLR A` is that it requires only one byte of program memory whereas the `MOV A,#00h` solution requires two bytes. An additional instruction, `CPL A`, also exists. This instruction will flip (complement) each bit in the accumulator. Thus if the accumulator holds 179 (10110011 binary), it will hold 76 (01001100 binary) after the `CPL A` instruction has executed.

Finally, the MOV instruction can be used to move bit values between any given bit—user or SFR bits—and the carry bit. The instructions `MOV C,`*`bit`* and `MOV `*`bit`*`,C` allow these bit movements to occur. They function like the MOV instruction described earlier—copying the value of the second bit to the first bit.

Consider the following examples:

```
MOV C,P0.0    ;Move the value of the P0.0 line to the carry bit
MOV C,30h     :Move the value of user bit 30h to the carry bit
MOV 25h,C     ;Move the carry bit to user bit 25h
```

These MOV instructions, which allow bits to be moved through the carry flag, allow for more advanced bit operations without the need for workarounds that would otherwise be required to move bit values.

The MOV instruction, when used with bits, can only move values to and from the carry bit. There is no instruction that allows the program to copy directly from one bit to another bit when neither is the carry bit. Thus it is often necessary to use the carry bit as a temporary bit register when moving bits.

11.13 Bit-Based Decisions & Branching (JB, JBC, JNB, JC, JNC)

It is often useful, especially in microcontroller applications, to execute different code based on whether or not a given bit is set or cleared. The 8052 instruction set offers five instructions that do precisely that.

JB means "jump if bit set". The 8052 will check the specified bit and, if it is set, will jump to the specified address or label.

JBC means "jump if bit set, and clear bit". This instruction is identical to JB except that the bit will be cleared if it was set. That is to say, if the specified bit is set, the 8052 will jump to the specified address or label and also clear the bit. This can save the programmer the use of an extra CLR instruction.

JNB means "jump if bit not set". This instruction is the opposite of JB—it tests the specified bit and will jump to the specified label or address if the bit was *not* set.

JC means "jump if carry set." This is the same as the JB instruction but it only tests the carry bit. Since many operations and decisions are based on whether or not the carry flag is set, this additional instruction was included in the instruction set to test for this common condition. Thus instead of using the instruction `JB C,`label, which requires three bytes of code memory, the `JC label` takes two.

JNC means "jump if carry bit not set." This is the opposite of JC. This instruction tests the carry bit and will jump to the specified label or address if the carry bit is clear.

Some examples of these instructions are:

```
JB  40h,LABEL1      ;Jumps to LABEL1 if user bit 40h is set
JBC 45h,LABEL2      ;Jumps to LABEL2 if user bit 45h set, then clears it
JNB 50h,LABEL3      ;Jumps to LABEL3 if user bit 50h is clear
JC  LABEL4          ;Jumps to LABEL4 if the carry bit is set
JNC LABEL5          ;Jumps to LABEL5 if the carry bit is clear
```

These instructions are very common and very useful. Virtually all 8052 assembly language programs of any complexity will use them—especially the JC and JNC instructions.

11.14 Value Comparison (CJNE)

CJNE (Compare, Jump if Not Equal) is a very important instruction. It is used to compare the value of a register to another value and branch to another instruction based on whether or not the values are the same. This is a very common way of building a typical switch…case decision structure or an IF…THEN…ELSE structure.

The CJNE instruction compares the values of the first two parameters of the instruction and jumps to the address contained in the third parameter if the first two parameters are *not* equal.

```
CJNE A,#24h,NOT24    ;Jumps to the label NOT24 if accumulator isn't 24h
CJNE A,40h,NOT40     ;Jumps to the label NOT40 if accumulator is
                     ;different than the value at internal RAM address 40h
CJNE R2,#36h,NOT36   ;Jumps to the label NOT36 if R2 isn't 36h
CJNE @R1,#25h,NOT25  ;Jumps to the label NOT25 if the internal RAM
                     ;address pointed to by R1 does not contain 25h
```

The MCU will compare the first parameter to the second parameter. If they are different, it will jump to the label provided; if the two values are the same then execution will continue with instruction following the CJNE instruction. This can allow the programming of extensive condition evaluations.

For example, to call the PROC_A subroutine if the accumulator is equal to 30h, call the CHECK_LCD subroutine if the accumulator equals 42h, and call the DEBOUNCE_KEY subroutine if the accumulator equals 50h, the following code could be used:

```
        CJNE A,#30h,CHECK2  ;If A is not 30h, jump to CHECK2 label
        LCALL PROC_A        ;If A is 30h, call the PROC_A subroutine
        SJMP CONT           ;When we get back, we jump to CONTINUE label
CHECK2: CJNE A,#42h,CHECK3  ;If A is not 42h, jump to CHECK3 label
        LCALL CHECK_LCD     ;If A is 42h, call the CHECK_LCD subroutine
        SJMP CONT           ;When we get back, we jump to CONTINUE label
CHECK3: CJNE A,#50h,CONT    ;If A is not 50h, we jump to CONTINUE label
        LCALL DEBOUNCE_KEY  ;If A=50h, call DEBOUNCE_KEY subroutine
CONT:{Code continues here}  ;The rest of the program continues here
```

The first line compares the accumulator with 30h. If the accumulator is not 30h it jumps to CHECK2 where the next comparison will be made. If the accumulator is 30h, however, program execution continues with the next instruction which calls the PROC_A subroutine. When the subroutine returns the SJMP instruction causes the program to jump ahead to the CONTINUE label—thus bypassing the rest of the checks.

The code at label CHECK2 works in the same fashion as the first check. It first compares the accumulator with 42h and then either branches to CHECK3 or calls the CHECK_LCD subroutine and jumps to CONTINUE. The code at CHECK3 does a final check for the value of 50h. This time there is no SJMP instruction following the call to DEBOUNCE_KEY since the next instruction is CONTINUE.

Code structures similar to the one shown above are very common in 8052 assembly language programs to execute certain code or subroutines based on the value of some register.

11.15 Less Than and Greater Than Comparison (CJNE)

Often it is necessary not to check whether a register is or isn't a certain value but rather to determine whether a register is less than or greater than another register or value. As it turns out, the CJNE instruction—in combination with the carry flag—accomplishes this.

When the CJNE instruction is executed, not only does it compare *parameter1* to *parameter2* and branch if they aren't equal, it also sets or clears the carry bit based on which parameter is greater or less than the other.

1. If parameter1 < parameter2 the carry bit will be set (set to 1).
2. If parameter1 >= parameter2 the carry bit will be cleared (cleared to 0).

This aspect of CJNE is how an assembly language program can perform a greater than/less than comparison. For example, if the accumulator holds some number and the program wants to know if it is less than or greater than 40h, the following code could be used:

```
                CJNE A,#40h,CHECK_LESS    ;If A is not 40h, check if  < or > 40h
                LJMP A_IS_EQUAL           ;If A is 40h, jump to A_IS_EQUAL code
CHECK_LESS:     JC A_IS_LESS              ;If carry is set, A is less than 40h
A_IS_GREATER:              {Code}         ;Otherwise, A is greater than 40h
```

The code first compares the accumulator to 40h. If they are the same, the program will fall through to the next line and jump to A_IS_EQUAL since it is already established that they are equal. If they aren't the same, execution will continue at CHECK_LESS. If the carry bit is set it means that the accumulator was less than the second parameter (40h) it will jump to the label A_IS_LESS which will handle the "less than" condition. If the carry bit wasn't set, execution will fall through to A_IS_GREATER at which point the developer would insert code to handle the "greater than" condition.

Keep in mind that CJNE will clear the carry bit if parameter1 is greater than *or equal* to *parameter2*. This means it is very important to check to see if the values are equal before using the carry bit to determine less than/greater than. Otherwise, the program may branch to the "greater than" condition when, in fact, the two parameters are equal.

84

11.16 Zero and Non-Zero Decisions (JZ/JNZ)

Sometimes it is useful to be able to simply determine if the accumulator holds a zero or not. This could be done with a CJNE instruction, but since these types of tests are so common the 8052 instruction set provides two instructions for this purpose: **JZ** and **JNZ**.

JZ will jump to the given address or label if the accumulator is zero. The instruction means "jump if zero."

JNZ will jump to the given address or label if the accumulator is *not* zero. The instruction means "jump if not zero".

For example:

```
JZ  ACCUM_ZERO      ;Jump to ACCUM_ZERO if the accumulator = 0
JNZ NOT_ZERO        ;Jump to NOT_ZERO if the accumulator is not 0
```

Using JZ and/or JNZ is much easier and faster than using CJNE if all that the program is interested in is testing for a zero/non-zero value in the accumulator.

Other non-8052 architectures have a "zero flag" that is affected by certain instructions and the zero-test instruction tests that flag, not the accumulator. The 8052, however, has no zero flag; JZ and JNZ both test the value of the accumulator, not the status of any flag.

11.17 Performing Additions (ADD, ADDC)

The **ADD** and **ADDC** instructions provide a way to perform 8-bit addition. All addition involves adding some number or register to the accumulator and leaving the result in the accumulator. The original value in the accumulator is always overwritten with the result of the addition.

```
ADD  A,#25h    ;Add 25h to whatever value is in the accumulator
ADD  A,40h     ;Add contents of internal RAM address 40h to accumulator
ADD  A,R4      ;Add the contents of R4 to the accumulator
ADDC A,#22h    ;Add 22h to the accumulator, plus carry bit
```

The ADD and ADDC instructions are identical except that ADD will only add the accumulator and the specified value or register whereas ADDC will also add the carry bit. The difference between the two, and the use of both, can be seen in the next example.

The following code assumes that a 16-bit number is in internal RAM address 30h (high byte) and address 31h (low byte). The code will add 1045h to the number and leave the result in addresses 32h (high byte) and 33h (low byte).

```
MOV  A,31h     ;Move value from IRAM address 31h (low byte) to accumulator
ADD  A,#45h    ;Add 45h to the accumulator (45h is low byte of 1045h)
MOV  33h,A     ;Move the result from accumulator to IRAM address 33h
MOV  A,30h     ;Move value from IRAM address 30h (hi byte) to accumulator
ADDC A,#10h    ;Add 10h to the accumulator (10h is the high byte of 1045h)
MOV  32h,A     ;Move result from accumulator to internal RAM address 32h
```

This code first loads the accumulator with the low byte of the original number from internal RAM address 31h. It then adds 45h to it. Since the ADD instruction is used, it doesn't matter what the carry bit holds. The result is moved to 33h and the high byte of the original address is moved to the accumulator from address 30h. The ADDC instruction then adds 10h to it, plus any carry that might have occurred in the first ADD step. The final answer is stored in internal RAM address 32h which holds the high byte of the result.

Both ADD and ADDC will set the carry flag (C) if an addition of unsigned integers results in an overflow that can't be held in the accumulator, or will clear the carry flag if the addition did not result in an overflow. For example, if the accumulator holds the value F0h and the value 20h is added to it, the accumulator will hold the result of 10h and the carry bit will be set. The fact that the carry bit is set can be subsequently be used with the ADDC to add the carry into the next addition instruction.

The auxiliary carry (AC) flag is set if there is a carry from bit 3 and cleared otherwise. For example, if the accumulator holds the value 2Eh and the value 05h is added to it, the accumulator will then equal 33h and the AC bit will be set since the low nibble overflowed from Eh to 3h.

The overflow (OV) flag is set if there is a carry out of bit 7 but not out of bit 6, or out of bit 6 but not out of bit 7. This is used in the addition of signed numbers to indicate that a negative number was produced as a result of the addition of two positive numbers, or that a positive number was produced by the addition of two negative numbers. For example, adding 20h to 70h (32 + 112, two positive numbers) would produce the value 90h (144). However, if the accumulator is being treated as a signed number, the value 90h would represent the number –10h. The fact that the OV bit was set means that the value in the accumulator shouldn't really be interpreted as a negative number.

Many non-8052 architectures only have a single type of ADD instruction—one that always includes the carry bit in the addition. The 8052 has two different types of ADD instructions is to avoid the need to start every addition calculation with a CLR C instruction. Using the ADD instruction is the same as using the CLR C instruction followed by the ADDC instruction.

11.18 Performing Subtractions (SUBB)

The **SUBB** instruction provides a way to perform 8-bit subtraction. All subtraction involves subtracting some number or register from the accumulator and leaving the result in the accumulator. The original value in the accumulator is always overwritten with the result of the subtraction.

```
SUBB A,#25h   ;Subtract 25h from whatever value is in the accumulator
SUBB A,40h    ;Subtract contents of IRAM address 40h from the accumulator
SUBB A,R4     ;Subtract the contents of R4 from the accumulator
```

The SUBB instruction *always* includes the carry bit in the subtract operation. This means if the accumulator holds the value 38h and the carry bit is set, subtracting 06h will result in 31h (38h – 6h – carry bit).

Since SUBB always includes the carry bit in its operation, the carry bit must always be cleared (CLR C) before executing the first SUBB in a subtraction operation so that the prior status of the carry flag does not affect the calculation.

SUBB sets and clears the carry, auxiliary carry, and overflow bits in much the same way as the ADD and ADDC instructions.

SUBB will set the carry flag (C) if the number (plus carry) being subtracted from the accumulator is larger than the value in the accumulator. In other words, the carry will be set if a borrow is needed for bit 7. Otherwise the carry bit will be cleared.

The auxiliary carry (AC) flag will be set if a borrow is needed for bit 3, otherwise it is cleared.

The overflow (OV) flag will be set if there is a borrow into bit 7 but not into bit 6, or into bit 6 but not into bit 7. This is used when subtracting signed integers. If subtracting a negative value from a positive value produces a negative number, OV will be set. Likewise, if subtracting a positive number from a negative number produces a positive number, the OV flag will also be set.

The 8052 uses what is known as 2's complement representation for representing signed numbers used with the ADD, ADCC, and SUBB instructions. Using this representation, any value between 0 and 127 (highest bit clear) represents that number while values between 128 and 255 (highest bit set) represent a negative number. In this system, the value 255 represents -1, the value 254 represents -2, etc. Therefore, if the accumulator holds the value 00h and the value 10 is subtracted, the accumulator will then hold the value 245 which is the 2's complement representation of -10. Likewise, if 20 is subtracted from 245 (-10 as a signed number), the result of 225 will represent -30 if treated as a signed value.

11.19 Performing Multiplication (MUL)

In addition to addition and subtraction, the 8052 also offers the **MUL AB** instruction to multiply two 8-bit values. Unlike addition and subtraction, the MUL AB instruction always multiplies the contents of the accumulator by the contents of the "B" register (SFR F0h). The result overwrites both the accumulator and B, placing the low-byte of the result in the accumulator and the high-byte of the result in B.

For example, to multiply 20h by 75h, the following code could be used:

```
MOV A,#20h    ;Load accumulator with 20h
MOV B,#75h    ;Load B with 75h
MUL AB        ;Multiply A by B
```

The result of 20h x 75h is 0EA0h, thus after the MUL instruction the accumulator would hold the low-byte of the answer (A0h) and B would hold the high-byte of the answer (0Eh). The original values of the accumulator and B are overwritten.

If the result is greater than 255 the Overflow (OV) bit will be set, otherwise it will be cleared. The carry bit is always cleared and the auxiliary carry (AC) bit is unaffected.

A program may multiply any two 8-bit values using MUL AB and obtain a result that will fit in the 16-bits available for the result in A and B. This is because the largest possible multiplication would be FFh X FFh which would result in FE01h which comfortably fits into the 16-bit space. It is not possible to overflow a 16-bit result space with two 8-bit multipliers.

11.20 Performing Division (DIV)

The last of the basic mathematics functions offered by the 8052 is the **DIV AB** instruction. This instruction, as the name implies, divides the accumulator by the value held in the B register. Like the MUL instruction, this instruction *always* uses the accumulator and B registers. The integer (whole-number) portion of the answer is placed in the accumulator and any remainder is placed in the B register. The original values of the accumulator and B are overwritten.

For example, to divide F3h by 13h the following code could be used:

```
MOV A,#0F3h  ;Load accumulator with F3h
MOV B,#13h   ;Load B with 13h
DIV AB       ;Divide A by B
```

The result of F3h / 13h is 0Ch with remainder 0Fh, thus after this DIV instruction the accumulator will hold the value 0Ch and B will hold the value 0Fh.

The carry bit and the overflow bits are both cleared by DIV, unless a division by zero is attempted in which case the overflow bit is set. In the case of division by zero, the result in the accumulator and B after the instruction are undefined.

NOTE: While the MUL instruction takes two 8-bit values and multiplies them into a 16-bit value, the DIV instruction takes two 8-bit values and divides it into an 8-bit value and a remainder. The 8052 does not provide an instruction that will divide a 16-bit number. The topic of 16-bit division is covered in chapter 12.

CODE LIBRARY: Source code that includes 16-bit and 32-bit division may be found in the Code Library at **http://www.8052.com/codelib**.

11.21 Shifting Bits (RR, RRC, RL, RLC)

The 8052 offers four instructions which are used to shift the bits in the accumulator to the left or right by one bit: **RR A, RRC A, RL A, RLC A**. There are two instructions that shift bits to the right, RR A and RRC A, and two that shift bits to the left, RL A and RLC A. The RRC and RLC instructions are different in that they rotate bits through the carry bit whereas RR and RL don't involve the carry bit.

```
RR A   ;Rotate accumulator one bit to right, bit 0 is rotated into bit 7
RRC A  ;Rotate accumulator to right, bit 0 is rotated into carry,
       ;carry into bit 7
RL A   ;Rotate accumulator one bit to left, bit 7 is rotated into bit 0
RLC A  ;Rotate the accumulator to the left, bit 7 is rotated into
       ;carry, carry into bit 0
```

The following illustrations show how each of the instructions manipulates the eight bits of the accumulator and the carry bit.

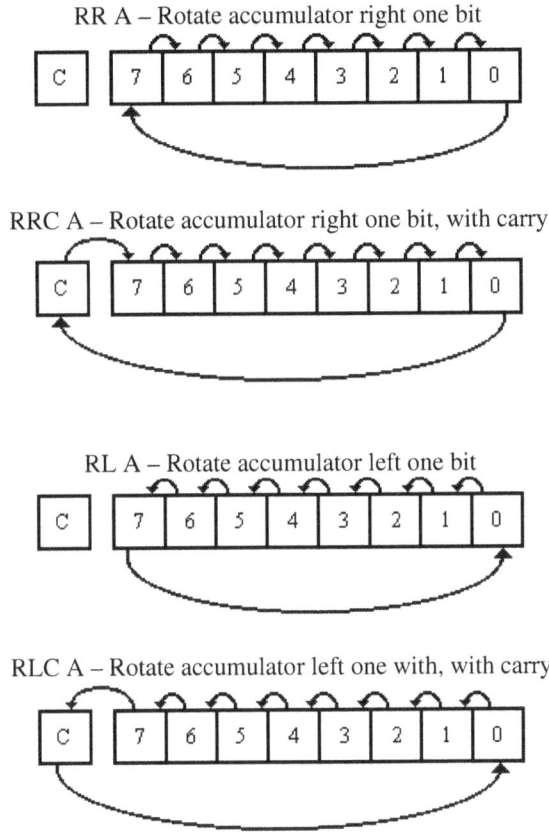

RR A – Rotate accumulator right one bit

| C | | 7 | 6 | 5 | 4 | 3 | 2 | 1 | 0 |

RRC A – Rotate accumulator right one bit, with carry

| C | | 7 | 6 | 5 | 4 | 3 | 2 | 1 | 0 |

RL A – Rotate accumulator left one bit

| C | | 7 | 6 | 5 | 4 | 3 | 2 | 1 | 0 |

RLC A – Rotate accumulator left one with, with carry

| C | | 7 | 6 | 5 | 4 | 3 | 2 | 1 | 0 |

Using the shift instructions is, obviously, useful for bit manipulations. They can also be used to quickly multiply or divide by multiples of two.

For example, to multiply the accumulator by two, two approaches could be used:

```
MOV B,#02h    ;Load B with 2
MUL AB        ;Multiply accumulator by B (2), leaving low
              ;byte in accumulator
```

Or the RLC instruction could be used:

```
CLR C         ;Make sure carry bit is initially clear
RLC A         ;Rotate left, multiplying by two
```

This may look like the same amount of work but it isn't to the microcontroller. The first approach requires

4 bytes of program memory and takes 6 instruction cycles whereas the second approach requires only 2 bytes of program memory and 2 instruction cycles. The RLC approach requires half as much memory and is three times as fast.

11.22 Bit-wise Logical Instructions (ANL, ORL, XRL)

The 8052 instruction set offers three instructions to perform the three most common types of bit-level logic which are "Logical And" (**ANL**), "Logical OR" (**ORL**), and "Logical Exclusive OR" (**XRL**). The instructions are capable of operating on the accumulator or an internal RAM address.

Some examples of these instructions are:

```
ANL A,#35h    ;Performs logical AND between accumulator and 35h,
              ;result in accumulator
ORL 20h,A     ;Performs logical OR between internal RAM 20h and
              ;accumulator, result in internal RAM 20h
XRL 25h,#15h  ;Performs logical Exclusive OR between IRAM 25h and 15h
```

ANL (Logical AND) looks at each bit of *parameter1* and compares it to the same bit in *parameter2*. If the bit is set in both parameters the bit remains set—otherwise the bit is cleared. The result is left in *parameter1*.

ORL (Logical OR) looks at each bit of *parameter1* and compares it to the same bit in *parameter2*. If the bit is set in either parameter the bit remains set—otherwise the bit is cleared. The result is left in *parameter1*.

XRL (Logical Exclusive OR) looks at each bit of *parameter1* and compares it to the same bit in *parameter2*. If the bit is set in *one* of the two parameters, the bit is set—otherwise the bit is cleared. This means if the bit is set in both parameters it will be cleared. If it is set in one of the two parameters it will remain set. If it is clear in both parameters it will remain clear. The result is left in *parameter1*.

The following tables show the result of each of these logical instructions:

ANL	0	1
0	0	0
1	0	1

ORL	0	1
0	0	1
1	1	1

XRL	0	1
0	0	1
1	1	0

Most of the logical bit-wise instructions affect entire 8-bit memory registers. However, the following instructions are available to perform logical operations on the carry bit. The result of these instructions is always left in the carry bit and the other bit is left unchanged.

ANL C,*bit*: This instruction will perform a logical AND between the carry flag and the specified *bit*. If both bits are set the carry bit will remain set. Otherwise the carry bit is cleared.

ANL C,/*bit*: This instruction performs a logical AND between the carry flag and the *complement* of the specified *bit*. That means if the *bit* is set the carry bit will be ANDed as if it were clear. If the specified *bit* is clear it will be ANDed with the carry bit as if it were set.

ORL C,*bit*: This instruction will perform a logical OR between the carry flag and the specified *bit*. If either the carry bit or *bit* is set the carry bit will be set. If neither bit is set the carry bit will be cleared.

ORL C,/*bit*: This instruction performs a logical OR between the carry flag and the *complement* of the specified *bit.* That means if the *bit* is set the carry bit will be ORed as if it were clear. If the specified *bit* is clear it will be ORed with the carry bit as if it were set.

There is no bit-wise XRL instruction. Only the ANL and ORL logical instructions are supported as bit-wise instructions.

11.23 Exchanging Register Values (XCH)

Often it is convenient to swap the value of the accumulator with the value of another SFR or internal RAM address. The **XCH** instruction allows this to be accomplished quickly and without using additional temporary holding registers.

XCH will take the value of the accumulator and write it to the specified SFR or internal RAM address while at the same time writing the original value of that SFR or internal RAM address to the accumulator.

For example:

```
MOV A,#25h    ;Accumulator now holds 25h
MOV 60h,#45h  ;Internal RAM 60h now holds 45h
XCH A,60h     ;Accumulator now holds 45, IRAM 60h now holds 25h
```

11.24 Swapping Accumulator Nibbles (SWAP)

In some cases it can be useful to swap the nibbles of the accumulator. A nibble is four bits so there are two nibbles in the accumulator. The "high" nibble consists of bits 4 through 7 while the "low" nibble consists of bits 0 through 3.

The **SWAP A** instruction will swap the two nibbles of the accumulator. For example, if the accumulator holds the value 56h, the SWAP instruction will convert it to 65h. Likewise, F7h will be converted into 7Fh.

The SWAP A instruction is identical to executing four RL A instructions or four RR A instructions.

11.25 Exchanging Nibbles Between Accumulator and Internal RAM (XCHD)

The **XCHD** instruction swaps the low nibble of the accumulator with the low nibble of the register or internal RAM address specified in the instruction.

For example, if R0 holds 87h and the accumulator holds 24h, then the XCHD R0 instruction will result in the accumulator holding 27h and R0 holding 84h. The low nibbles of the two were simply exchanged.

11.26 Adjusting Accumulator for BCD Addition (DA)

DA A is a very useful instruction if the program is engaged in Binary Coded Decimal (BCD) encoded addition.

BCD is a form of expressing two decimal digits in a single 8-bit byte. Any 8-bit value can be expressed in a single byte as a number between 00 and FF. Obviously, it is possible to express all normal decimal numbers between 0 and 99 in hexadecimal format so that, printed as hexadecimal, they appear to be two decimal digits.

For example, the decimal digits "00" would be represented in BCD as, not surprisingly, 00h. The decimal digits "09" would be represented in BCD as 09h. The decimal digits "10", however, would be represented in BCD as 10h—note that 10h is actually 16 (decimal). That's because in BCD the hex values A, B, C, D, E, and F are not used, thus explaining the jump from 09h to 10h.

But what happens if a program needs to add two BCD numbers together, say 38 and 25? In normal decimal math 38 + 25 = 63. Doing the same addition on BCD-encoded values should have the same result.

But 38 encoded as BCD is 38h and 25 encoded as BCD is 25h. 38h + 25h = 5Dh (93 decimal). Obviously the result no longer looks like a decimal value—and that's not surprising since BCD doesn't use the values A, B, C, D, E, and F.

DA A automatically "adjusts" the accumulator after the addition of two BCD values to compensate for this problem. In the above example, executing DA A when the accumulator holds 5Dh will result in the accumulator being adjusted to 63h, thus "righting" the rather strange addition.

The details of how DA A works, and why, are not extremely important at this point and would tend to confuse things rather than clarify them. If a program will be implementing BCD addition, the developer should investigate this instruction further. For the majority of readers that will not be doing BCD addition, this instruction can be safely ignored.

11.27 Using the Stack (PUSH/POP)

The 8052 stack, as in any processor, is an area of memory that can be used to store information temporarily, including the return address for returning from subroutines that are called by ACALL or LCALL. The 8052 automatically utilizes the stack when it executes an ACALL or LCALL as well as when it executes the RET instruction. The stack is also used automatically when an interrupt service routine is triggered by an interrupt and when the program returns from the ISR with the RETI instruction.

The stack can also be used by the program for storing values temporarily by using the **PUSH** and **POP** instructions. The PUSH instruction will "push" a value onto the stack and the POP instruction will "pop" the last value off the stack. A program may save a value temporarily by PUSHing it onto the stack and it may restore the value by POPping it.

> The stack operates on a Last In-First Out (LIFO) basis. This means if the values 4, 5, and 6 are PUSHed onto the stack (in that order), POPping them one at a time will return 6, 5,

and then 4. The value most recently added to the stack is the first value that will come off when a POP instruction is executed.

An example using the PUSH and POP instruction is:

```
MOV A,#35h    ;Load the accumulator with the value 35h
PUSH ACC      ;Push accumulator onto stack, accumulator still holds 35h
ADD A,#40h    ;Add 40h to the accumulator, accumulator now holds 75h
POP ACC       ;Pop the accumulator from stack, accumulator holds
              ;35h again
```

The above code, though functionally useless, illustrates how to use PUSH and POP.

The code starts by assigning 35h to the accumulator. It then PUSHes it onto the stack. It then adds 40h to the accumulator for the sole purpose of changing its value to something else. At this point the accumulator holds 75h. Finally, the program POPs the accumulator off the stack. Since the last value pushed onto the stack was 35h, the POP restores the value of the accumulator to 35h.

When PUSHing or POPping the accumulator, it must be referred to as ACC since that is the SFR name of the accumulator. The instructions PUSH A and POP A are both invalid and will result in an assemble-time error with most, if not all, 8052 assemblers.

When using PUSH, the SFR or internal RAM address that follows the PUSH instruction is the value that will be PUSHed onto the stack. For example, PUSH ACC will push the value of the accumulator onto the stack. PUSH 60h will push the value of internal RAM address 60h onto the stack.

Likewise, the internal RAM address or SFR that follows a POP instruction indicates where the value should be stored when it is POPped from the stack. For example, POP ACC will pop the next value off the stack and move it into the accumulator. POP 40h will pop the next value off the stack and into internal RAM address 40h.

The stack itself resides in internal RAM and is managed by the stack pointer (SP) SFR. SP will always point to the internal RAM address from which the next POP instruction should obtain the data.

1. **POP** will return the value of internal RAM pointed to by SP, then decrement SP by 1.
2. **PUSH** will increment SP by 1, then store the value in internal RAM at the address then pointed to by SP.

SP is initialized to 07h when an 8052 is first powered-up. That means the stack will begin at 08h and start growing from there. If 16 values are pushed onto the stack, for example, the stack will occupy addresses 08h through 17h.

Using the stack can be both useful and powerful, but it can also be dangerous when incorrectly used. Remember that the stack is also used by the 8052 to determine the return addresses of subroutines and interrupts. If the stack is modified incorrectly, it is very easy to cause a program to crash or to behave in very unexpected ways.

When using the stack, all but advanced developers users should observe the following recommendations:

1. When using the stack from within a subroutine or interrupt service routine, be sure to have one POP instruction for every PUSH instruction. If the number of POPs and PUSHes aren't the same, the program will probably end up crashing since the return address on the stack will be incorrect.

2. When using PUSH, be sure to always POP that value off the stack—even if the code is not within a subroutine.

3. Be sure not to jump over the section of code that POPs a value off the stack. A common error is to PUSH a value onto the stack and then execute a conditional instruction that jumps over the instruction that POPs that value off. This results in an unbalanced stack and will probably end up crashing the program. Remember, not only must there be a POP instruction for every PUSH, but a POP instruction must be *executed* for every PUSH that is executed. Make sure the program doesn't jump over the POP instructions.

4. Always be sure to use the RET instruction to return from subroutines and RETI instruction to return from interrupt service routines.

5. As a general rule, only modify SP at the very beginning of the program. Once the program starts using the stack or making calls to subroutines, SP should *not* be modified.

6. Make sure the stack has enough room. For example, the stack will start by default at address 08h. If a program has a variable at internal RAM address 20h then the stack has only 24 bytes available to it, from 08h through 1Fh. If the stack is 24 bytes long and another value is pushed onto the stack or another subroutine is called, the variable at 20h will be overwritten.

7. Remember that the stack is Last In-First Out, which means that values must be POPped off the stack in the reverse order they were PUSHed onto the stack.

Keep in mind, too, that the 8052 can only use internal RAM for its stack. Even if an 8052 system has 64k of external RAM, it can only use its 256 bytes of internal RAM for the stack. This means the stack should be used very sparingly.

11.28 Setting the Data Pointer DPTR (MOV DPTR)

The next few instructions use the data pointer (DPTR), the 8052's only 16-bit register. DPTR is used to point to a RAM or ROM address when used with these instructions.

As described in section 3.5, DPTR is made up of two individual SFRs: DPH and DPL which hold the high and low bytes, respectively, of the 16-bit data pointer. But when DPTR is used to access memory, the 8052 will treat DPTR as a single register.

The **MOV DPTR** instruction is used to set DPTR to a specific value. This instruction lets the program set both DPH and DPL in a single instruction. Nevertheless, the program may still modify DPTR by accessing DPH and DPL directly as illustrated in the following examples:

```
MOV DPTR,#1234h     ;Sets DPTR to 1234h
MOV DPTR,#0F123h    ;Sets DPTR to F123h
MOV DPH,#40h        ;Sets DPTR high-byte to 40h (DPTR now 4023h)
MOV DPL,#56h        ;Sets DPTR low-byte to 56h (DPTR now 4056h)
```

The first two instructions set DPTR to 1234h and then to F123h. The next example sets DPH to 40h while DPTR's low-byte is unchanged. Since the low byte is still 23h from the previous instruction, changing DPH to 40h will result in DPTR being equal to 4023h. Finally, the low-byte is changed to 56h leaving the high-byte unchanged. Since the high-byte was set to 40h in the previous instruction, setting the low-byte to 56h will leave the DPTR with a value of 4056h.

In other words, MOV DPTR,#1567h is the same as MOV DPH,#15h and MOV DPL,#67h. The advantage to using MOV DPTR is that it uses only 3 bytes of memory and 2 instruction cycles whereas the other method requires 6 bytes of memory and 4 instruction cycles.

11.29 Reading External RAM/Data Memory (MOVX)

The 8052 generally has 128 or 256 bytes of internal RAM which are accessed with the MOV instruction as described earlier. However, many projects will require more than 256 bytes of RAM. The 8052 has the ability to address up to 64k of external RAM which takes the form of additional, off-chip ICs. Many derivatives also provide additional on-chip RAM that is accessed with the MOVX instruction as if it were actually located off-chip.

The **MOVX** instruction is used to read from and write to external RAM. The MOVX instruction has four forms:

1. **MOVX A,@DPTR**: Reads external RAM address DPTR into the accumulator.
2. **MOVX A,@Ri**: Reads external RAM address pointed to by R0 or R1 into the accumulator.
3. **MOVX @DPTR,A**: Sets external RAM address DPTR to the value of the accumulator
4. **MOVX @Ri,A**: Sets the external RAM address held in R0 or R1 to the value of the accumulator

The first two forms move data from external RAM into the accumulator whereas the last two forms move data from the accumulator into external RAM.

MOVX with DPTR: When using the forms of MOVX that use DPTR, DPTR will be used as a 16-bit memory address. The 8052 will automatically communicate with the off-chip RAM and obtain the value of that memory address and store it in the accumulator (MOVX A,@DPTR) or will write the accumulator to the off-chip RAM (MOVX @DPTR,A).

For example, to add 5 to the value contained in external RAM address 2356h, the following code could be used:

```
MOV DPTR,#2356h      ;Set DPTR to 2356h
MOVX A,@DPTR         ;Read external RAM address 2356h into accumulator
ADD A,#05h           ;Add 5 to the accumulator
MOVX @DPTR,A         ;Write new value of accumulator back to
                     ;external RAM 2356h
```

MOVX with @R0 or @R1: When using the forms of MOVX that use @R0 or @R1, R0 or R1 will be used to determine the address of external RAM to access. Since both R0 and R1 are 8-bit registers, these forms of MOVX can only access external RAM addresses 0000h through 00FFh unless special action is taken to control the high-byte of the address on P2.

11.30 Reading Code Memory/Tables (MOVC)

It is often useful to be able to read code memory itself from within a program. This allows for the placement of data or tables in code memory to be read at run-time by the program. This is accomplished with the **MOVC** instruction.

The MOVC instruction comes in two forms: `MOVC A,@A+DPTR` and `MOVC A,@A+PC`. Both instructions move a byte of code memory into the accumulator. The code memory address from which the byte is read depends on which of the two forms is used.

`MOVC A,@A+DPTR` will read the byte from the code memory address calculated by adding the current value of the accumulator to that of DPTR. For example, if DPTR holds the value 1234h and the accumulator holds the value 10h, the instruction would copy the value of code memory address 1244h into the accumulator. This can be thought of as an "absolute" read since the byte will always be read from the address pointed to by the accumulator plus DPTR. DPTR is initialized to point to the first byte of the table and the accumulator is used as an offset into the table.

For example, if there is a table of values that resides at 2000h in code memory, the program may need a subroutine that obtains a value from the table based on the value of the accumulator. This could be coded as:

```
          MOV A,#04h          ;Set accumulator to offset into the
                              ;table we want to read
          LCALL SUB           ;Call subroutine to read 4th byte of the table
             ...
SUB:      MOV DPTR,#2000h     ;Set DPTR to the beginning of the value table
          MOVC A,@A+DPTR      ;Read the 5th byte from the table
          RET                 ;Return from the subroutine
```

`MOVC A,@A+PC` will read the byte from the code memory address calculated by adding the current value of the accumulator to the program counter; that is, the address of the currently executing instruction. This can be thought of as a "relative" read since the address of code memory from which the byte will be read depends on where the MOVC instruction is found in memory. This form of MOVC is used when the data to be read immediately follows the code that is to read it.

For example, if instead of the table being located at code memory 2000h it were located right after the routine that read it, the subroutine would be changed to:

```
SUB:      INC A                   ;Increment Acc to account for RET instruction
          MOVC A,@A+PC            ;Get the data from the table
          RET                     ;Return from subroutine
          DB 01h,02h,03h,04h,05h  ;The actual data table
```

Note that in the above example the accumulator is first incremented by 1. This is because the value of the program counter will be that of the instruction immediately following the MOVC instruction—in this case, the RET instruction. The subroutine doesn't want to get the value of the RET opcode but rather wants the data that follows RET. Since the RET instruction requires one byte of code memory the program needs to increment the accumulator by 1 to skip over the RET instruction.

The value that the accumulator must be incremented by is the number of bytes between the MOVC instruction and the first data of the table being read. For example, if the RET instruction above were replaced with an LJMP instruction, which is 3 bytes long, the INC A instruction would be replaced with ADD A,#03h to increment the accumulator by 3.

11.31 Using Jump Tables (JMP @A+DPTR)

A frequently used method for quickly branching to many different areas in a program is jump tables. For example, if the program had to branch to different subroutines based on the value of the accumulator, it could be accomplished with the CJNE instruction as was previously explained:

```
                CJNE A,#00h,CHECK1   ;If it's not zero, jump to CHECK1
                AJMP SUB0            ;Go to SUB0 subroutine
        CHECK1: CJNE A,#01h,CHECK2   ;If it's not 1, jump to CHECK2
                AJMP SUB1            ;Go to SUB1 subroutine
        CHECK2:
```

The above code will work but each additional possible value will increase the size of the program by 5 bytes—3 bytes for the CJNE instruction and 2 bytes for the AJMP instruction.

A more efficient way is to create a "jump table" by using the JMP @A+DPTR instruction. Like the MOVC @A+DPTR, this instruction will calculate an address by summing the accumulator and DPTR and then jump to that address. So if DPTR holds 2000h and the accumulator holds 14h the JMP instruction will jump to 2014h.

Consider the following code:

```
            RL A                    ;Rotate accumulator left, multiply by 2
            MOV DPTR,#JUMP_TABLE     ;Load DPTR with address of jump table
            JMP  @A+DPTR             ;Jump to the corresponding address
        JUMP_TABLE:
            AJMP SUB0                ;Jump table entry to SUB0
            AJMP SUB1
```

This code first takes the value of the accumulator and multiplies it by 2 by shifting the accumulator to the left by one bit. Since each AJMP entry in JUMP_TABLE is two bytes long, the accumulator must first be multiplied by two.

The code then loads the DPTR with the address of the JUMP_TABLE and proceeds to JMP to the address of the accumulator *plus* DPTR. Since the program already knows that it is to jump to the offset indicated by the accumulator, no additional checks are necessary. It jumps directly into the table that jumps to the appropriate subroutine. Each additional entry in the jump table will require only 2 additional bytes (2 bytes for each AJMP instruction).

It's almost always a good idea to use a jump table if a sequence will have 2 or more choices based on a zero-based index. A jump table with just two entries, like the above example, will save 1 byte of memory over using the CJNE approach and will save 3 bytes of memory for each additional entry.

An additional benefit of jump tables is that the execution time is *always the same*. Using a sequence of CJNE instructions means that the code will execute faster if the value being searched for is at the beginning of the CJNE sequence and will execute slower if it is at the end of the CJNE sequence. A jump table always takes the same amount of time to execute and, thus, can be very helpful in time-critical code which must take a specific amount of time to execute.

CHAPTER 12: 16-BIT MATHEMATICS WITH THE 8052

The 8052 is an 8-bit microcontroller. This basically means that each machine language opcode in its instruction set consists of a single 8-bit value. This permits a maximum of 256 instruction codes (of which 255 are actually used in the 8052 instruction set).

The 8052 also works almost exclusively with 8-bit values. The accumulator is an 8-bit value, as is each register in the register banks, the stack pointer, and each of the many Special Function Registers (SFRs) that exist in the architecture. In reality, the only values that the 8052 handles that are truly 16-bit values are the program counter, which internally indicates the next instruction to be executed, and the data pointer (DPTR), which the program may utilize to access external RAM as well as directly access code memory. Other than these two registers, the 8052 works exclusively with 8-bit values.

For example, the ADD instruction will add two 8-bit values to produce a third 8-bit value. The SUBB instruction subtracts an 8-bit value from another 8-bit value and produces a third 8-bit value. The MUL instruction will multiply two 8-bit values and produce a 16-bit value.

It could be said that the MUL instruction is a 16-bit math instruction since it produces a 16-bit answer. However, its inputs are only 8-bit. The result is 16-bits out of necessity since any multiplication with two operands greater than the number 16 will produce a 16-bit result. Thus for the MUL operation to have any value at all it *must* produce a 16-bit result.

As already discussed, the 8052 provides a number of instructions aimed at performing mathematical calculations. Unfortunately, they all work with 8-bit input values even though it is often necessary to work with values that simply cannot be expressed in 8-bits.

This chapter will discuss techniques that allow the developer to work with 16-bit values in the 8052's 8-bit architecture. While only 16-bit operations will be discussed, the same techniques can be extended to any number of bits (24-bit, 32-bit, 64-bit, etc.). It's just a matter of expanding the code to support the additional bytes. The concepts remain the same.

12.1 How did we learn math in primary school?

Before jumping into multi-byte mathematics in assembly language, a quick review of how mathematics is taught in schools is useful. For example, to perform the calculation 156 + 248, schools teach the process:

100's	10's	1's
1	5	6
+ 2	4	8
= 4	0	4

How is the above calculated? The digits in the 1's column are added, 6 + 8 = 14. Since 14 can't fit in a single column, the 4 is left in the 1's column and the 1 is carried into the 10's column. The next step is to add 5 + 4 = 9, add the 1 that was carried, and get 10. Again, 10 does not fit in a single column. So the 0 is left in the 10's column and the 1 is carried into the 100's column. Finally, the digits in the 100's column are added, 1 + 2 = 3, plus the 1 that was carried, results in a final answer of 4 for the 100's column. The final answer, therefore, is 404.

It is important to remember how these operations were taught in school because the exact same process is going to be used in multi-byte mathematics.

12.2 16-bit Addition

16-bit addition is the addition of two 16-bit values. The addition of any two 16-bit values will result in a value that is, at most, 17 bits long. Why is this so? The largest value that can fit in 16-bits is 256 * 256 - 1 = 65,535. If 65,535 is added to 65,535 the result is 131,070. This value fits in 17 bits. Adding two 16-bit values results in a 24-bit value. Of course, seven of the highest eight bits will never be used—but the entire answer will be contained in 3 bytes. Also keep in mind that this process deals with unsigned integers.

Another option, instead of using 3 full bytes for the answer, is to use 2 bytes (16-bits) for the answer and the carry bit (C) to hold the 17th bit. This is perfectly acceptable and probably even preferred. The more advanced programmer will understand and recognize this option and be able to make use of it. However, since this is an *introduction* to 16-bit mathematics it is our goal that the answer produced by the routines be in a form that is easy for the reader to utilize once calculated. This is most likely best achieved by leaving the answer fully expressed as three 8-bit values.

Consider the addition of the following two decimal values: 6724 + 8923. The answer is 15,647. How does a program go about adding these values with the 8052? The first step is to always work with hexadecimal values: convert the two values to hexadecimal. In this example, this results in the following equivalent hexadecimal addition: 1A44 + 22DB.

How are add these two numbers added? The exact same method that is taught in primary school applies:

	256's	1's
	1A	44
+	22	DB
=	3D	1F

The columns no longer represent 1's, 10's, and 100's. Instead there are just two columns: the 1's column and the 256's column. In familiar computer terms, this is the low byte (the 1's column) and the high byte (the 256's column). The addition process, however, is exactly the same.

First add the values in the 1's column (low byte): 44 + DB = 11F. Only a 2-digit hexadecimal value can fit in a single column so 1F is left in the low-byte column and the 1 is carried to the high-byte column. Next, the high bytes are added 1A + 22 = 3C, plus the 1 that was carried from the low-byte column. The resulting value is 3D.

Thus the completed answer is 3D1F. Converting 3D1F back to decimal results in the answer 15,647. This matches with the original addition that was calculated in decimal and so the process is shown to work.

The only remaining challenge is to code the above process in 8052 assembly language. The following table will be used to explain how the addition is to be accomplished:

	65536's	256's	1's
		R6	R7
+		R4	R5
=	R1	R2	R3

Since the program is adding 16-bit values, each value requires two 8-bit registers. The first value to added will be held in R6 and R7 (the high byte in R6 and the low byte in R7) while the second value to add will be held in R4 and R5 (the high byte in R4 and the low byte in R5). The result of the addition will be left in R1, R2, and R3.

Remember that the sum of two 16-bit values is a 17-bit value. In this case a 24-bit space (R1, R2, and R3) is being used for the answer even though no more than 1 bit of R1 will ever be used.

Reviewing the steps involved in adding the values above:

1. Add the low bytes R7 and R5, leave the answer in R3.
2. Add the high bytes R6 and R4, adding any carry from step 1, and leave the answer in R2.
3. Put any carry from step 2 in the final byte R1.

Converting the above steps to assembly language results in the following:

Step 1: Add the low bytes R7 and R5, leave the answer in R3.
```
MOV A,R7      ;Move the low-byte into the accumulator
ADD A,R5      ;Add the second low-byte to the accumulator
MOV R3,A      ;Move the answer to the low-byte of the result
```

Step 2: Add the high bytes R6 and R4, adding any carry from step 1, and leave the answer in R2.
```
MOV A,R6      ;Move the high-byte into the accumulator
ADDC A,R4     ;Add second high-byte to the accumulator, plus carry.
MOV R2,A      ;Move the answer to the high-byte of the result
```

Step 3: Put any carry from step 2 in the final byte, R1.
```
MOV A,#00h    ;By default, the highest byte will be zero.
ADDC A,#00h   ;Add zero, but this will add one if there was a
              ;carry from step 2.
MOV R1,A      ;Move the answer to the highest byte of  the result
```

That's it! Combining the code from the three steps results in the following functional subroutine:

```
ADD16_16:
        ;Step 1 of the process
        MOV A,R7      ;Move the low-byte into the accumulator
        ADD A,R5      ;Add the second low-byte to the accumulator
        MOV R3,A      ;Move the answer to the low-byte of the result
        ;Step 2 of the process
        MOV A,R6      ;Move the high-byte into the accumulator
        ADDC A,R4     ;Add second high-byte to the accumulator, plus carry.
```

```
       MOV R2,A      ;Move the answer to the high-byte of the result
       ;Step 3 of the process
       MOV A,#00h    ;By default, the highest byte will be zero.
       ADDC A,#00h   ;Add zero, but this will add one if there was
                     ;a carry from step 2.
       MOV R1,A      ;Move the answer to the highest byte of  the result
       RET           ;Return - answer now resides in R1, R2, and R3.
```

To call the ADD16_16 subroutine, the program would first load the values into R6/R7 and R4/R5 and then call the subroutine. This could be coded as:

```
       ;Load the first value into R6 and R7
       MOV R6,#1Ah
       MOV R7,#44h

       ;Load the second value into R4 and R5
       MOV R4,#22h
       MOV R5,#0DBh

       ;Call the 16-bit addition routine
       LCALL ADD16_16
```

12.3 16-bit Subtraction

16-bit subtraction is the subtraction of one 16-bit value from another. A subtraction of this nature results in another 16-bit value. Why? The number 65,535 is a 16-bit value. If the value 1 is subtracted, the answer will be 65,534 which is also a 16-bit value. Thus any 16-bit subtraction will result in another 16-bit value.

Consider the subtraction of the following two decimal values: 8923 - 6905. The answer is 2018. How does a program go about subtracting these values with the 8052? As was the case with addition, the first step is to convert the expression to hexadecimal. The above decimal subtraction is equivalent to the following hexadecimal subtraction: 22DB - 1AF9.

Again, going back to the way subtraction is taught:

	256's	1's
	22	DB
-	1A	F9
=	07	E2

First, the second value in the 1's column (low byte) is subtracted from the first value in the 1's column: DB - F9. Since F9 is greater than DB, a "borrow" from the 256's column is required. Thus, in reality, the subtraction being performed is 1DB - F9 = E2. The value E2 is what is left in the 1's column.

Now the subtraction of high bytes must be performed: 22 - 1A. But remember that a 1 was "borrowed" from the 256's column so an additional 1 must be subtracted. Thus the subtraction for the 256's column becomes 22 - 1A - 1 = 7, which is the value that is left in the 256's column.

The final answer is 07E2. If this is converted back to decimal, the value is 2018 which coincides with the answer that was originally obtained in decimal. Once again the process is confirmed.

102

Again a a small table will be used to help convert the above process to 8052 assembly language:

	256's	1's
	R6	R7
-	R4	R5
=	R2	R3

Since two 16-bit values are being subtracted, each value requires two 8-bit registers. The value to be subtracted from will be held in R6 and R7 (the high byte in R6 and the low byte in R7) while the value to be subtracted will be held in R4 and R5 (the high byte in R4 and the low byte in R5). The answer will b e left in in R2 and R3.

Reviewing the steps involved in the above subtraction:

1. Subtract the low byte R5 from R7, leave the answer in R3.
2. Subtract the high byte R4 from R6, less any borrow, and leave the answer in R2.

Converting the above process into assembly language results in the following:

Step 1: Subtract the low byte R5 from R7, leave the answer in R3.
```
MOV A,R7          ;Move the low-byte into the accumulator
CLR C             ;Always clear carry before first subtraction
SUBB A,R5         ;Subtract the second low-byte from the accumulator
MOV R3,A          ;Move the answer to the low-byte of the result
```

Step 2: Subtract the high byte R4 from R6, less any borrow, and leave the answer in R2.
```
MOV A,R6          ;Move the high-byte into the accumulator
SUBB A,R4         ;Subtract the second high-byte from the accumulator
MOV R2,A          ;Move the answer to the low-byte of the result
```

The SUBB instruction always subtracts the second value in the instruction from the first, less any borrow. While there are two versions of the ADD instruction (ADD and ADDC), one of which ignores the carry bit, there is no such distinction with the SUBB instruction. This means before the first subtraction is performed the carry bit must always be cleared. Otherwise, if the carry bit happens to be set the result of the first subtraction will be incorrect.

Combining the code from the two steps above the following functional subroutine is obtained:

```
SUBB16_16:
  ;Step 1 of the process
  MOV A,R7          ;Move the low-byte into the accumulator
  CLR C             ;Always clear carry before first subtraction
  SUBB A,R5         ;Subtract the second low-byte from the accumulator
  MOV R3,A          ;Move the answer to the low-byte of the result
  ;Step 2 of the process
  MOV A,R6          ;Move the high-byte into the accumulator
  SUBB A,R4         ;Subtract the second high-byte from the accumulator
  MOV R2,A          ;Move the answer to the low-byte of the result
  RET               ;Return - answer now resides in R2, and R3.
```

To call the SUBB16_16 subroutine, the program would first load the values into R6/R7 and R4/R5 and then call the subroutine. This could be coded as:

```
;Load the first value into R6 and R7
MOV R6,#22h
MOV R7,#0DBh

;Load the first value into R6 and R7
MOV R4,#1Ah
MOV R5,#0F9h

;Call the 16-bit subtraction routine
LCALL SUBB16_16
```

12.4 16-bit Multiplication

16-bit multiplication is the multiplication of two 16-bit values from another. Such a multiplication results in a 32-bit value.

Any multiplication results in an answer that has a size equal to the sum of the number of bits in each of the multiplicands. For example, multiplying an 8-bit value by a 16-bit value results in a 24-bit value (8 + 16). A 16-bit value multiplied by another 16-bit value results in a 32-bit value (16 + 16), etc.

For the sake of example, the program will multiply 25,136 by 17,198. The answer is 432,288,928. As with both addition and subtraction, the expression will first be converted to hexadecimal: 6230h x 432Eh.

Once again, the following table arranges the numbers in columns, although now the grid becomes more complicated. The first two lines represent the original two values. The next 4 lines represent intermediate calculations obtained by multiplying each byte of the original values. The final line indicates the final answer obtained by summing the columns in the four intermediate lines.

	Byte 4	Byte 3	Byte 2	Byte 1
			62	30
X			43	2E
	====	====	====	====
			08	A0
		11	9C	
		0C	0A	
	19	A6		
	====	====	====	====
	19	C4	34	A0

Going back to the technique taught in school, first 2Eh is multiplied by 30h (byte 1 of both numbers) and the result is placed directly below. Then 2Eh is multiplied by 62h (byte 1 of the bottom number by byte 2 of the upper number). This result is positioned such that the right-most byte of the answer is left in byte 2 of the intermediate answer. Next 43h is multiplied by 30h (byte 2 of the bottom number by byte 1 of the top number) and the result is positioned such that the right-most byte of the answer is left in byte 2 of the intermediate result. Finally, 43h is multiplied by 62h (byte 2 of both numbers) and the result is positioned such that right-most column ends up in byte 3. At that point each column is added, with appropriate carries, to arrive at the final answer.

The process in assembly language will be identical. Using the now-familiar grid to help conceptualize the process:

	Byte 4	Byte 3	Byte 2	Byte 1
			R6	R7
*	__	__	R4	R5
=	R0	R1	R2	R3

The first number will be contained in R6 and R7 while the second number will be held in R4 and R5. The result of the multiplication will be placed in R0, R1, R2 and R3. The process will be the following:

1. Multiply R5 by R7, leaving the 16-bit result in R2 and R3.
2. Multiply R5 by R6, adding the 16-bit result to R1 and R2.
3. Multiply R4 by R7, adding the 16-bit result to R1 and R2.
4. Multiply R4 by R6, adding the 16-bit result to R0 and R1.

This process will now be converted to assembly language, step by step.

Step 1. Multiply R5 by R7, leaving the 16-bit result in R2 and R3.
```
MOV A,R5      ;Move the R5 into the accumulator
MOV B,R7      ;Move R7 into B
MUL AB        ;Multiply the two values
MOV R2,B      ;Move B (the high-byte) into R2
MOV R3,A      ;Move A (the low-byte) into R3
```

Step 2. Multiply R5 by R6, adding the 16-bit result to R1 and R2.
```
MOV A,R5      ;Move R5 back into the accumulator
MOV B,R6      ;Move R6 into B
MUL AB        ;Multiply the two values
ADD A,R2      ;Add the low-byte into the value already in R2
MOV R2,A      ;Move the resulting value back into R2
MOV A,B       ;Move the high-byte into the accumulator
ADDC A,#00h   ;Add zero (plus the carry, if any)
MOV R1,A      ;Move the resulting answer into R1
MOV A,#00h    ;Load the accumulator with zero
ADDC A,#00h   ;Add zero (plus the carry, if any)
MOV R0,A      ;Move the resulting answer to R0.
```

Step 3. Multiply R4 by R7, adding the 16-bit result to R1 and R2.
```
MOV A,R4      ;Move R4 into the accumulator
MOV B,R7      ;Move R7 into B
MUL AB        ;Multiply the two values
ADD A,R2      ;Add the low-byte into the value already in R2
MOV R2,A      ;Move the resulting value back into R2
MOV A,B       ;Move the high-byte into the accumulator
ADDC A,R1     ;Add the current value of R1 (plus any carry)
MOV R1,A      ;Move the resulting answer into R1.
MOV A,#00h    ;Load the accumulator with zero
```

```
        ADDC A,R0      ;Add the current value of R0 (plus any carry)
        MOV R0,A       ;Move the resulting answer to R1.
```

Step 4. Multiply R4 by R6, adding the 16-bit result to R0 and R1.

```
        MOV A,R4       ;Move R4 back into the accumulator
        MOV B,R6       ;Move R6 into B
        MUL AB         ;Multiply the two values
        ADD A,R1       ;Add the low-byte into the value already in R1
        MOV R1,A       ;Move the resulting value back into R1
        MOV A,B        ;Move the high-byte into the accumulator
        ADDC A,R0      ;Add it to the value already in R0 (plus any carry)
        MOV R0,A       ;Move the resulting answer back to R0
```

Combining the code from the four steps above results in the following functional subroutine:

```
MUL16_16:
        ;Multiply R5 by R7
        MOV A,R5       ;Move the R5 into the accumulator
        MOV B,R7       ;Move R7 into B
        MUL AB         ;Multiply the two values
        MOV R2,B       ;Move B (the high-byte) into R2
        MOV R3,A       ;Move A (the low-byte) into R3
        ;Multiply R5 by R6
        MOV A,R5       ;Move R5 back into the accumulator
        MOV B,R6       ;Move R6 into B
        MUL AB         ;Multiply the two values
        ADD A,R2       ;Add the low-byte into the value already in R2
        MOV R2,A       ;Move the resulting value back into R2
        MOV A,B        ;Move the high-byte into the accumulator
        ADDC A,#00h    ;Add zero (plus the carry, if any)
        MOV R1,A       ;Move the resulting answer into R1
        MOV A,#00h     ;Load the accumulator with  zero
        ADDC A,#00h    ;Add zero (plus the carry, if any)
        MOV R0,A       ;Move the resulting answer to R0.
        ;Multiply R4 by R7
        MOV A,R4       ;Move R4 into the accumulator
        MOV B,R7       ;Move R7 into B
        MUL AB         ;Multiply the two values
        ADD A,R2       ;Add the low-byte into the value already in R2
        MOV R2,A       ;Move the resulting value back into R2
        MOV A,B        ;Move the high-byte into the accumulator
        ADDC A,R1      ;Add the current value of R1 (plus any carry)
        MOV R1,A       ;Move the resulting answer into R1.
        MOV A,#00h     ;Load the accumulator with zero
        ADDC A,R0      ;Add the current value of R0 (plus any carry)
        MOV R0,A       ;Move the resulting answer to R1.
        ;Multiply R4 by R6
        MOV A,R4       ;Move R4 back into the accumulator
        MOV B,R6       ;Move R6 into B
        MUL AB         ;Multiply the two values
        ADD A,R1       ;Add the low-byte into the value already in R1
        MOV R1,A       ;Move the resulting value back into R1
        MOV A,B        ;Move the high-byte into the accumulator
        ADDC A,R0      ;Add it to the value already in R0 (plus any carry)
        MOV R0,A       ;Move the resulting answer back to R0
        ;Return - answer is now in R0, R1, R2, and R3
        RET
```

To call the routine to multiply the two values used in the example above, the following code could be used:

```
;Load the first value into R6 and R7
MOV R6,#62h
MOV R7,#30h

;Load the second value into R6 and R7
MOV R4,#43h
MOV R5,#2Eh

;Call the 16-bit subtraction routine
LCALL MUL16_16
```

12.5 16-bit Division

16-bit division is the division of one 16-bit value by another 16-bit value resulting in a 16-bit quotient and a 16-bit remainder.

> The number of bits in the quotient and the remainder can never be larger than the number of bits in the original dividend. For example, in the division of a 16-bit value by a 2-bit value both the quotient and the remainder must be able to handle a 16-bit result. If a 24-bit value is being divided by a 16-bit value the quotient and remainder must both be able to handle a 24-bit result.

In this case, the program will divide 179 by 8. These are actually two 8-bit numbers and the DIV instruction could be used; but these two numbers will be used for the sake of simplicity knowing that the same approach will work equally well with 16-bit values—it would just take longer to walk through the steps.

```
              2      2
 8  |  1      7      9
       1      6
              1      9
              1      6
                     3
```

In this example 179 is divided by 8 by using the following steps:

1. The program attempts to divide 8 into 1, but it doesn't fit.
2. The program attempts to divide 8 into 17 and gets 2 (first digit of the answer).
3. The program multiplies 8 by 2 (the result obtained in step #2) to get 16.
4. The program subtracts 16 from 17 and gets 1.
5. The program attempts to divide 8 by the 1 obtained in step #4, but it doesn't fit.
6. The program brings down the 9 from the dividend to arrive at the number 19.
7. The program attempts to divide 8 into 19 and gets 2 (second digit of the answer).
8. The program subtracts 16 from 19 (the number divided into in step #7) and gets 3.
9. The value 3 left over in step #8 is the remainder of the division, so the answer is 22 with remainder 3.

This can be accomplished in code in the following manner:

1. Make the divisor the same length as the dividend. In this example the dividend was 3 digits long (179) so the divisor must be made 3 digits long. Thus two zeros are added and results in a divisor of 800.
2. 179 can't be divided by 800 so the divisor is "shifted" to the right one digit. The 800 divisor becomes 80.
3. 179 can be divided by 80 twice so the first digit of the answer is 2.
4. The program multiplies 2 (the result of step #3) by 80 (the current divisor) and gets 160.
5. The program subtracts 160 (result of step #4) from the dividend and ends up 19.
6. The divisor is "shifted" to the right one digit. The 80 divisor becomes 8.
7. 19 (the result from step #5) can be divided by the divisor, 8, twice so the next digit of the answer is 2.
8. The program multiplies 2 (the result of step #7) by 8 (the current divisor) and gets 16.
9. The program subtracts 16 (the result of step #8) from the current dividend, 19, for an answer of 3.
10. There are no more digits to shift to the right so the "left-over" answer from step #9 is the remainder.

This same approach can be used for 16-bit values if we treat those values as binary values for the sake of the division process. Just keep in mind the following points:

1. Rather than determining the "largest" bit in the dividend and adjusting the divisor to that same bit position, the program just shifts the divisor as far to the left as possible. That is, 179 decimal is 0000000010110011 as a 16-bit binary value and 8 decimal is 0000000000001000. While the program could determine that the highest bit used in 179 is bit 7 and shift the divisor until its highest bit is also in bit 7 this turns out to be needlessly complicated. It is easier just to shift the value 8 all the way to the left so it becomes 1000000000000000.
2. When doing long division as described above it is necessary to determine at each step whether the divisor fits into the dividend. The program is basically asking "Is divisor < dividend?" In code, all that needs to be done is attempt to do a 16-bit subtraction of divisor from dividend. If this produces a carry (carry bit set) then it means the divisor didn't fit into dividend and a "0" goes into that position of the answer. If the subtraction doesn't produce a carry (carry bit clear) then it means the divisor does fit into the dividend and a "1" goes into that position of the answer. Since this is binary math the answer is either 0 or 1.

The code used to accomplish each of these steps is:

Step 1: Shift Divisor as far left as possible

```
  MOV B,#00h;Clear B.  B will count the number of left-shifted bits
  CLR C              ;Clear carry bit initially
  MOV R2,#00h        ;Clear R2 initially
  MOV R3,#00h        ;Clear R3 initially
Div1:
  INC B       ;Increment counter for each left shift, minimum 1 shift
  MOV A,R2    ;Move the current divisor low byte into the accumulator
  RLC A       ;Shift low-byte left, high bit goes to carry
  MOV R2,A    ;Save the updated divisor low-byte
```

```
MOV A,R3    ;Move the current divisor high byte into the accumulator
RLC A       ;Shift high-byte left, rotate carry from low-byte into low bit
MOV R3,A    ;Save the updated divisor high-byte
JNC div1    ;Repeat until carry flag is set from high-byte
```

In the example, once the above code is executed the registers will be as follows (including the carry bit 'C'):

```
C/R1/R0 0 00000000 10110011      (Dividend)
C/R3/R2 1 00000000 00000000      (Divisor)
```

Step 2: Perform the actual division loop

At this point the program can do the actual division. As the logic is working in binary mode there is no need for a real division—it's just a comparison as described above. The comparison appears in the code as a 16-bit subtraction.

The first six instructions shift the current 16-bit divisor one bit to the right. A safe (backup) copy of the dividend is then made in internal RAM 06h and 07h. The next section of code subtracts the current shifted divisor from the dividend. If the subtraction occurred without error then it knows that the division was successful and inserts a "1" into the result byte at that position; otherwise it knows the subtraction failed so it inserts a "0."

```
Div2:          ;Perform division cycle
  ;Shift current divisor one bit to the right
  MOV A,R3     ;Move high-byte of divisor into accumulator
  RRC A        ;Rotate high-byte of divisor right and into carry
  MOV R3,A     ;Save updated value of high-byte of divisor
  MOV A,R2     ;Move low-byte of divisor into accumulator
  RRC A        ;Rotate low-byte of divisor right, w/ carry from high-byte
  MOV R2,A     ;Save updated value of low-byte of divisor
  CLR C        ;Clear carry, don't need it, clear it for SUBB operations

  ;Make a safe copy of the dividend so that if our subtraction
  ;shows that the divisor didn't fit into the dividend we can
  ;restore the dividend to its current state.

  MOV 07h,R1 ;Make a safe copy of the dividend high-byte
  MOV 06h,R0 ;Make a safe copy of the dividend low-byte

  ;Subtract divisor from dividend, overwriting the current dividend
  ;This subtraction will tell us whether or not the divisor fits in
  ;the value of the dividend.

  MOV A,R0     ;Move low-byte of dividend into accumulator
  SUBB A,R2    ;Dividend-shifted divisor=result bit (only 0 or 1)
  MOV R0,A     ;Save updated dividend
  MOV A,R1     ;Move high-byte of dividend into accumulator
  SUBB A,R3    ;Subtract high-byte of divisor (16-bit subtraction)
  MOV R1,A     ;Save updated high-byte back in high-byte of divisor

  ;At this point if the carry flag is not set then it means that the
  ;subtraction occurred without a carry, meaning the divisor did fit in
  ;the dividend.  If the carry flag is set then it means the subtraction
  ;create a negative number, meaning the divisor did not fit in the
  ;dividend.
```

```
        JNC div3    ;If carry flag is NOT set, result is 1

          ;If we get here, means the carry bit was set which means the
          ;subtraction created a negative number which means the divisor didn't
          ;fit in the dividend.  Since we have overwritten the dividend with a
          ;now-invalid negative number, we restore the value of the dividend by
          ;restoring the safe values stored prior to initiating the subtraction

        MOV R1,07h ;Result is 0, save copy of divisor to undo subtraction
        MOV R0,06h

    div3:
          ;If the carry bit is set then it means the divisor didn't fit in the
          ;dividend, so the answer for this position in the result is "0".
          ;If the carry bit is clear, the divisor did fit in the dividend so the
          ;answer for this position in the result is "1".  The bit in the
          ;answer is the opposite of the carry bit; thus the carry bit is
          ;inverted and inserted it into the lowest bit of the result
          ;(R4 and R5).

        CPL C        ;Invert carry
        MOV A,R4     ;Get current low byte of the temporary result
        RLC A        ;Shift carry flag into low-bit of temporary result
        MOV R4,A     ;Update low byte of the temporary result by shifting left
        MOV A,R5     ;Get current high byte of the temporary result
        RLC A        ;Shift high byte of the temporary result left
        MOV R5,A     ;Update the high byte of the temporary result
        DJNZ B,div2  ;Repeat division loop until "B" is zero
```

Step 3. Final Clean-up.

Since the working result is stored in R4 and R5 the routine finishes by moving the answer into R2 and R3 for consistency with the other 16-bit mathematical functions in this chapter.

```
        MOV R3,05h    ;Move result to R3/R2
        MOV R2,04h    ;Move result to R3/R2
```

To use the routine the main program will call it with R0/R1 as the low/high byte of the dividend and R2/R3 as the low/high byte of the divisor. The routine leaves the quotient in R2 and R3 and leaves the remainder in R0 and R1.

110

8052 HARDWARE & SINGLE BOARD COMPUTER

CHAPTER 13: 8052 MICROCONTROLLER PIN FUNCTIONS

While there are many packages in which 8052 and derivatives are produced, including surface-mount, plastic leaded chip carrier (PLCC), etc. this chapter will explain the traditional 8052 40-pin DIP pinout. Many 8052 derivatives are also pin-compatible with the original 8052 and, thus, with the pinout described in this chapter.

Even when the pinout is different, the names and functions of these pins are almost always found on 8052-compatible parts—the only aspect which varies is on which pin the given function may be found. In all cases, the part's datasheet should be consulted to confirm the function of each pin.

```
        T2 / P1.0 | 1      40 | VCC
      T2EX / P1.1 | 2      39 | P0.0 / AD0
            P1.2  | 3      38 | P0.1 / AD1
            P1.3  | 4      37 | P0.2 / AD2
            P1.4  | 5      36 | P0.3 / AD3
            P1.5  | 6      35 | P0.4 / AD4
            P1.6  | 7      34 | P0.5 / AD5
            P1.7  | 8      33 | P0.6 / AD6
            RST   | 9      32 | P0.7 / AD7
       RXD / P3.0 | 10     31 | -EA
       TXD / P3.1 | 11     30 | ALE
      -INT0 / P3.2 | 12    29 | -PSEN
      -INT1 / P3.3 | 13    28 | P2.7 / A15
        T0 / P3.4 | 14     27 | P2.6 / A14
        T1 / P3.5 | 15     26 | P2.5 / A13
       -WR / P3.6 | 16     25 | P2.4 / A12
       -RD / P3.7 | 17     24 | P2.3 / A11
            XTAL2 | 18     23 | P2.2 / A10
            XTAL1 | 19     22 | P2.1 / A9
            GND   | 20     21 | P2.0 / A8
```

13.1 I/O Ports (P0, P1, P2, P3)

Of the 40 pins on the typical 8052, 32 of them are dedicated to I/O lines that have a one-to-one relation with SFRs P0, P1, P2, and P3. The developer may raise and lower these lines by writing 1s or 0s to the corresponding bits in the SFRs. Likewise, the current state of these lines may be read by reading the corresponding bits of the SFRs, as long as the respective pins are in their high state.

All of the ports have internal pull-up resistors except for port 0.

13.1.1 Port 0

Port 0 is dual-function in that, in some designs, its I/O lines are available to the developer to access external devices while in other designs it is used to access external memory. If the circuit requires external RAM or ROM, the microcontroller will automatically use port 0 to clock in/out the 8-bit data word as well as the low eight bits of the address in response to a MOVX instruction—in this case, port 0 I/O lines may be used for other functions as long as external RAM isn't being accessed at the same time. If the circuit requires external code memory, the microcontroller will automatically use the port 0 I/O lines to access each instruction that is to be executed. In this case, port 0 cannot be utilized for other purposes since the states of the I/O lines are constantly being modified to access the external code memory.

Note that there are no pull-up resistors on port 0 so it may be necessary to include pull-up resistors on these lines depending on the characteristics of the parts that will be connected to port 0.

13.1.2 Port 1

Port 1 consists of eight I/O lines that are for the exclusive purpose of interfacing to external parts. Unlike port 0, most derivatives do not use port 1 for any functions themselves. Port 1 is commonly used to interface to external hardware such as LCDs, keypads, and other devices. With 8052 derivatives, two lines of port 1 are optionally used as previously explained for extended timer 2 functions (see section 8.4). These two lines are not assigned special functions on 8051s since 8051s don't have a timer 2. Even on 8052s that have timer 2, these lines can be used for any purpose if the program doesn't need the special features of timer 2.

P1.0 (T2): If C/T2 (T2CON.1) is set, timer 2 will be incremented whenever there is a 1-0 transition on this line.

P1.1 (T2EX): If timer 2 is in auto-reload mode and EXEN2 (T2CON.3) is set, a 1-0 transition on this line will cause timer 2 to be reloaded with the auto-reload value. This will also cause the EXF2 (T2CON.6) flag to be set which will cause an interrupt, if so enabled.

13.1.3 Port 2

Like port 0, port 2 is dual-function. In some circuit designs it is available for accessing devices while in others it is used to address external RAM or external code memory. When the MOVX @DPTR instruction is executed, port 2 is used to output the high byte of the memory address that is to be accessed. In these cases, port 2 may be used to access other devices as long as the devices are not being accessed at the same time as a MOVX instruction is using port 2 to address external RAM. If the circuit requires external code memory, the microcontroller will automatically use the port 2 I/O lines to access each instruction in code memory. In this case port 2 cannot be utilized for other purposes since the states of the I/O lines are constantly being modified to access the external code memory.

Port 2 is only modified automatically by the microcontroller when accessing external code memory or when the MOVX @DPTR instruction is used. Port 2 is not affected by the MOVX @Ri instructions.

13.1.4 Port 3

Port 3 consists entirely of dual-function I/O lines. While the developer may use any of these lines for specific purposes by reading/writing to the P3 SFR, each pin has a predefined function that the microcontroller handles automatically when configured to do so and/or when necessary.

P3.0 (RXD): The UART/serial port uses P3.0 as the "receive line." In circuit designs that will be using the microcontroller's integrated serial port, this is the line into which serial data will be clocked. Note that when interfacing an 8052 to an RS-232 port that it may not be connected directly to the RS-232 pin; rather, it must pass through a part such as the MAX232 to obtain the correct voltage levels. This pin is available for any use the developer may assign it if the circuit doesn't need the integrated serial port.

P3.1 (TXD): The UART/serial port uses P3.1 as the "transmit line." In designs that will be using the microcontroller's integrated serial port, this is the line that the microcontroller will clock out all data which is written to the SBUF SFR. Note that when interfacing an 8052 to an RS-232 port that it may not be connected directly to the RS-232 pin; rather it is necessary to pass it through a part such as the MAX232 to obtain the correct voltage levels. This pin is available for any use the developer may assign it if the design has no need to transmit data via the integrated serial port.

P3.2 (-INT0): When so configured, this line is used to trigger an "external 0 interrupt." This may either be low-level triggered or may be triggered on a 1-0 transition. These features are covered extensively in the "Interrupts" chapter (chapter 10). This pin is available for any use the developer may assign it if the circuit does not need to trigger an external 0 interrupt.

P3.3 (-INT1): When so configured, this line is used to trigger an "external 1 interrupt." This may either be low-level triggered or may be triggered on a 1-0 transition. These features are covered extensively in the "Interrupts" chapter (chapter 10). This pin is available for any use the developer may assign it if the circuit does not need to trigger an external 1 interrupt.

P3.4 (T0): When so configured, this line is used as the clock source for timer 0. Timer 0 will be incremented either every instruction cycle that T0 is high or every time there is a 1-0 transition on this line, depending on how the timer is configured. These features are covered extensively in the "Timers" chapter (chapter 8). This pin is available for any use the developer may assign it if the circuit does not to control timer 0 externally.

P3.5 (T1): When so configured, this line is used as the clock source for timer 1. Timer 1 will be incremented either every instruction cycle that T1 is high or every time there is a 1-0 transition on this line, depending on how the timer is configured. These features are covered extensively in the "Timers" chapter (chapter 8). This pin is available for any use the developer may assign it if the circuit does not to control timer 1 externally.

P3.6 (-WR): This is the external memory write strobe line. This line will be brought low by the microcontroller whenever a MOVX instruction writes to external RAM. This line should be connected to the RAM's write enable (-WE) line. This pin is available for any use the developer may desire if the circuit does not write to external RAM using MOVX.

P3.7 (-RD): This is the external memory read strobe line. This line will be brought low by the microcontroller whenever a MOVX instruction reads from external RAM. This line should be connected to the RAM's output enable (-OE) line. This pin is available for any use the developer may desire it if the circuit does not read from external RAM using MOVX.

13.2 Oscillator Inputs (XTAL1, XTAL2)

The 8052 is typically driven by a crystal connected to pins 18 (XTAL2) and 19 (XTAL1). Two small capacitors must be connected from XTAL1 and XTAL2 to ground. A typical value for these capacitors is approximately 33pF but depends on the crystal. Common crystal frequencies are 11.0592MHz as well as 12MHz, although many newer derivatives are capable of accepting frequencies as high as 40MHz.

While a crystal is the normal clock source, other clock sources may be used. A TTL clock source may also be attached to XTAL1 and XTAL2 to provide the microcontroller's clock.

13.3 Reset Line (RST)

Pin 9 is the master reset line for the microcontroller. When this pin is brought high for two instruction cycles, the microcontroller will be reset. SFRs, including the I/O ports, are restored to their default conditions and the program counter will be reset to 0000h. Keep in mind that internal RAM is *not* affected by a reset. The microcontroller will begin executing code at 0000h when this pin returns to a low state.

The reset line is often connected to a reset button/switch that the user may press to reset the circuit. It is also common to connect the reset line to a watchdog or supervisor IC. The latter is highly recommended for commercial and professional designs since traditional resistor-capacitor networks attached to the reset line, while often sufficient for students or hobbyists, are not terribly reliable and do not protect the microcontroller from attempting to operate in out-of-tolerance voltage conditions.

13.4 Address Latch Enable (ALE)

The ALE at pin 30 is output-only and is controlled entirely by the microcontroller to allow it to multiplex the low-byte of a memory address and the 8-bit data itself on port 0. This is necessary because, while the high-byte of the memory address is sent on port 2, port 0 is used to send *both* the low-byte of the memory address and the data itself. This is accomplished by placing the low-byte of the address on port 0, asserting ALE high to latch the low-byte of the address into a latch IC (such as the 74LS573), and then placing the eight data bits on port 0. In this way the 8052 is able to output a 16-bit address and an 8-bit data word with 16 I/O lines instead of 24.

The ALE line is used in this fashion both for accessing external RAM with MOVX @DPTR as well as for accessing instructions in external code memory. When a program is executed from external code memory, ALE will pulse at a rate of 1/6th that of the oscillator frequency. Thus if the oscillator is operating at 11.0592MHz, ALE will pulse at a rate of 1,843,200 times per second. The only exception is when the MOVX instruction is executed one ALE pulse is skipped in lieu of a pulse on –WR or –RD.

13.5 Program Store Enable (-PSEN)

The Program Store Enable (PSEN) line at pin 29 is brought low automatically by the microcontroller whenever it accesses external code memory. This line should be attached to the Output Enable (-OE) pin of the EPROM that contains the external code memory. PSEN will not be asserted—and will remain in a high state—if the program is being executed from internal code memory.

13.6 External Access (-EA)

The External Access (-EA) line at pin 31 is used when the part is first powered-up to determine whether the program will be executed from external code memory or from internal code memory. If EA is tied high (connected to +5V), the microcontroller will execute the program it finds in internal/on-chip code memory. If EA is tied low (to ground) then it will attempt to execute the program it finds in the attached external code memory EPROM. Of course, the EPROM must be properly connected for the microcontroller to be able to access the program in external code memory.

EA must be tied low for any microcontroller that does not have internal code memory.

Even when EA is tied high, indicating that the microcontroller should execute from internal code memory, the microcontroller will normally attempt to execute from external code memory if the program counter references an address not available for that chip. For example, if the derivative being used has 4k of internal code-memory and EA is tied high, the derivative will start executing the program it finds on-chip. However, if the on-chip program attempts to execute code above 0FFFh (i.e. exceeding 4k) then the derivative will normally attempt to execute that code at that address from external code memory. Thus it is possible to have a "split" design where some of the code is found on-chip and the rest is found off-chip. This can potentially be very useful from a security standpoint where proprietary and trade secret code could be programmed into the microcontroller itself and protected with security bits while less sensitive code could be stored on an external EPROM. The microcontroller could also be programmed with underlying "solid" firmware that will never change and use a socketed EPROM to carry sections of code that may be subject to upgrades or changes in the future.

CHAPTER 14: AN 8052 SINGLE BOARD COMPUTER (SBC)

Having reviewed the function of each pin of the 8052 in chapter 13, it is now possible to use that knowledge to build a **Single Board Computer** (**SBC**). An SBC is a simple computer that is built on a single circuit board. It is a full computer in the sense that it encompasses all the functions generally attributed to a full computer: input, output, processing capabilities, and the ability to store a program to be executed.

There are many 8052-based SBCs that may be purchased ready-to-use off the Internet. Low-end SBCs may cost as little as US$50 while higher-end SBCs, with very specific features, may cost hundreds of dollars. The user must review the features of each SBC on the market and analyze its expandability and capability to address his or her needs and purchase the SBC that most closely matches the desired feature set. By purchasing a pre-made SBC, the developer may focus on coding the firmware or designing the product-specific circuitry rather than "reinventing the wheel" of constructing an 8052-based system with supporting RAM, capacitors, resistors, etc.

An understanding of the underlying structure of an 8052-based SBC is, however, instrumental in the ability to develop any future 8052 design or project. Most features found in an 8052 SBC will also be found in even the simplest 8052-based products. By understanding the hardware implementation of the various components of a typical SBC, the reader will be prepared to apply and improve those same concepts in real-world functioning devices.

In this chapter, the 8052.com SBC will be presented in schematic form and the function and purpose of each section of the schematic will be completely explained. The rest of this book presents example programs that interact with the components of the SBC. If the reader prefers to purchase another pre-made SBC, the examples in this book may still work as long as the external parts (keypad, LCD, and other external peripherals) are connected on the purchased SBC to the same I/O pins as they are on the 8052.com SBC.

> If you are already comfortable understanding a schematic, feel free to skip the rest of this chapter. The remainder of this chapter will go through each section of the schematic and explain it thoroughly for the benefit of those that may have had limited exposure to schematics and electronics in the past or have questions about specific sections of the hardware.

14.1 Features of the 8052.com SBC

The SBC presented in this chapter and used throughout the rest of this book was designed specifically to complement the interfacing topics that will be discussed in the subsequent chapters. The SBC includes an LCD display, a 4x4 keypad matrix, a serial port for communication to and from a PC or other serial device, a real-time clock for accurate time-keeping, and an external serial EEPROM for the storage of whatever data the user may wish to save in a non-volatile fashion. The LCD may be connected either in "memory-mapped" or "direct connect" mode to illustrate two common approaches to communicating with the LCD. Likewise, the real-time clock and the serial EEPROM were selected to illustrate the I²C and SPI protocols that are common for inter-chip communication. Finally, the SBC was designed so that the user may utilize virtually *any* 40-pin 8052-compatible microcontroller. However, if the user selects an Atmel AT89S8252 or Dallas DS89C420 microcontroller, an added advantage will be the ability to download programs directly to the SBC using in-system programming. Other 8052-compatible microcontrollers may be used but the firmware will have to be downloaded to an EPROM and inserted manually into the EPROM socket of the SBC.

14.2 SBC Schematic

Electrical circuits are described and represented by their schematic drawings. Schematics graphically show how each pin of each component of the circuit (IC chips, capacitors, resistors, etc.) must be connected to duplicate the circuit correctly. Any electrical circuit can be reproduced given the schematic.

The schematic for the SBC discussed in the rest of this book is presented on the following five pages. Each IC and functional area of the circuit is then discussed in the following sections of this chapter. The reader may choose to simply view the schematic and read and understand its purpose, may choose to buy the individual parts and build his or her own SBC that will function exactly the same as the SBC described in this chapter, or may purchase the SBC online. A fully built and functioning 8052.com SBC is pictured below.

A fully built and functioning version of this SBC is available at **http://www.8052.com/sbc**. The SBC may be purchased such that all that is necessary is to plug it in and start using it to test code and experiment with the 8052. Alternatively, the SBC may be purchased as a kit which includes the PCB and all the parts necessary to build it—the reader must then solder each part onto the PCB. Finally, the bare PCB is available in which case the reader must acquire all the necessary parts elsewhere and solder them onto the PCB. Of course the reader may simply build the SBC based on the schematic without the pre-designed PCB. This will, however, require a lot of wire soldering and there is a large opportunity for errors since it takes just one mis-wired connection to keep the SBC from functioning properly.

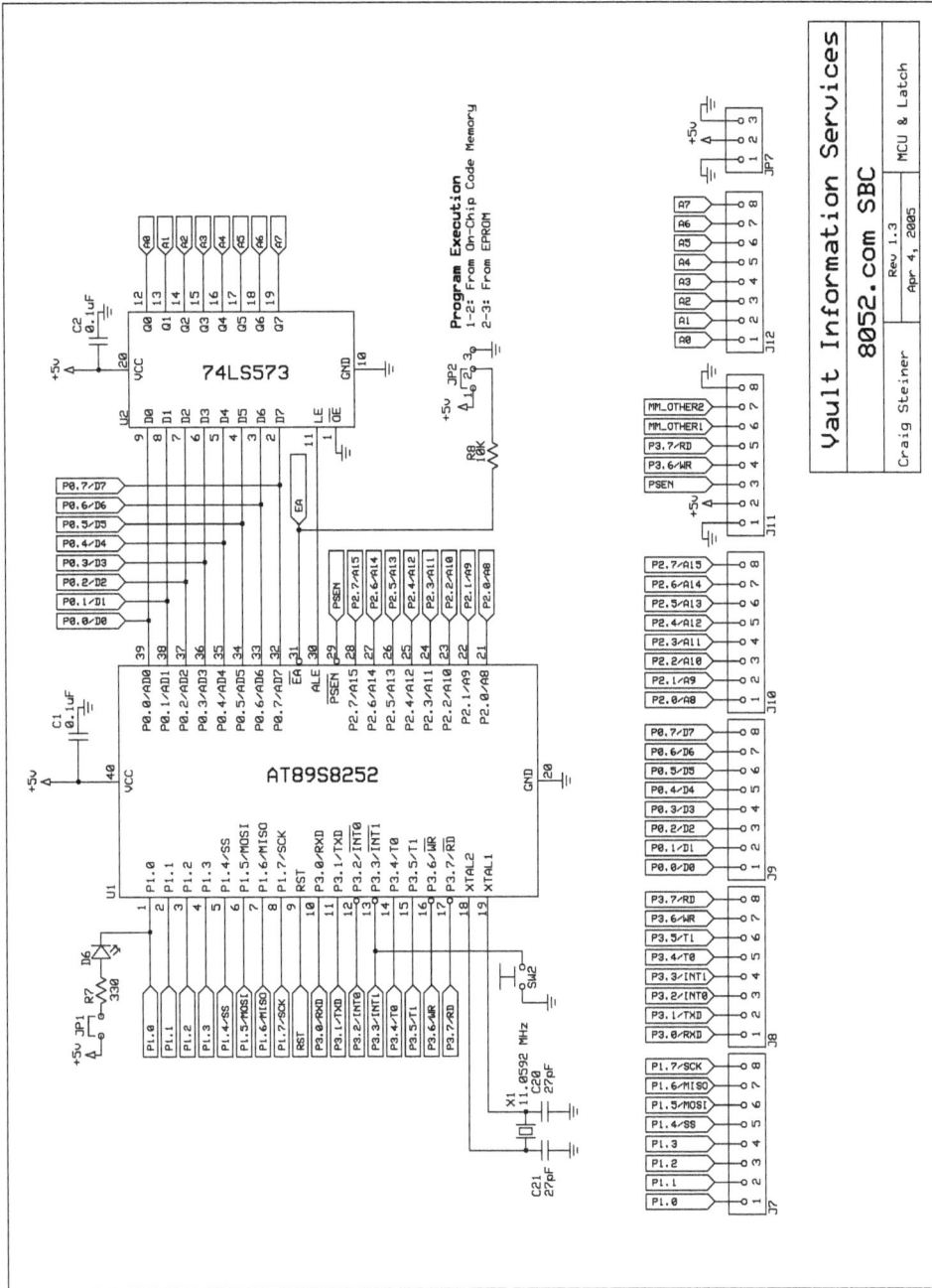

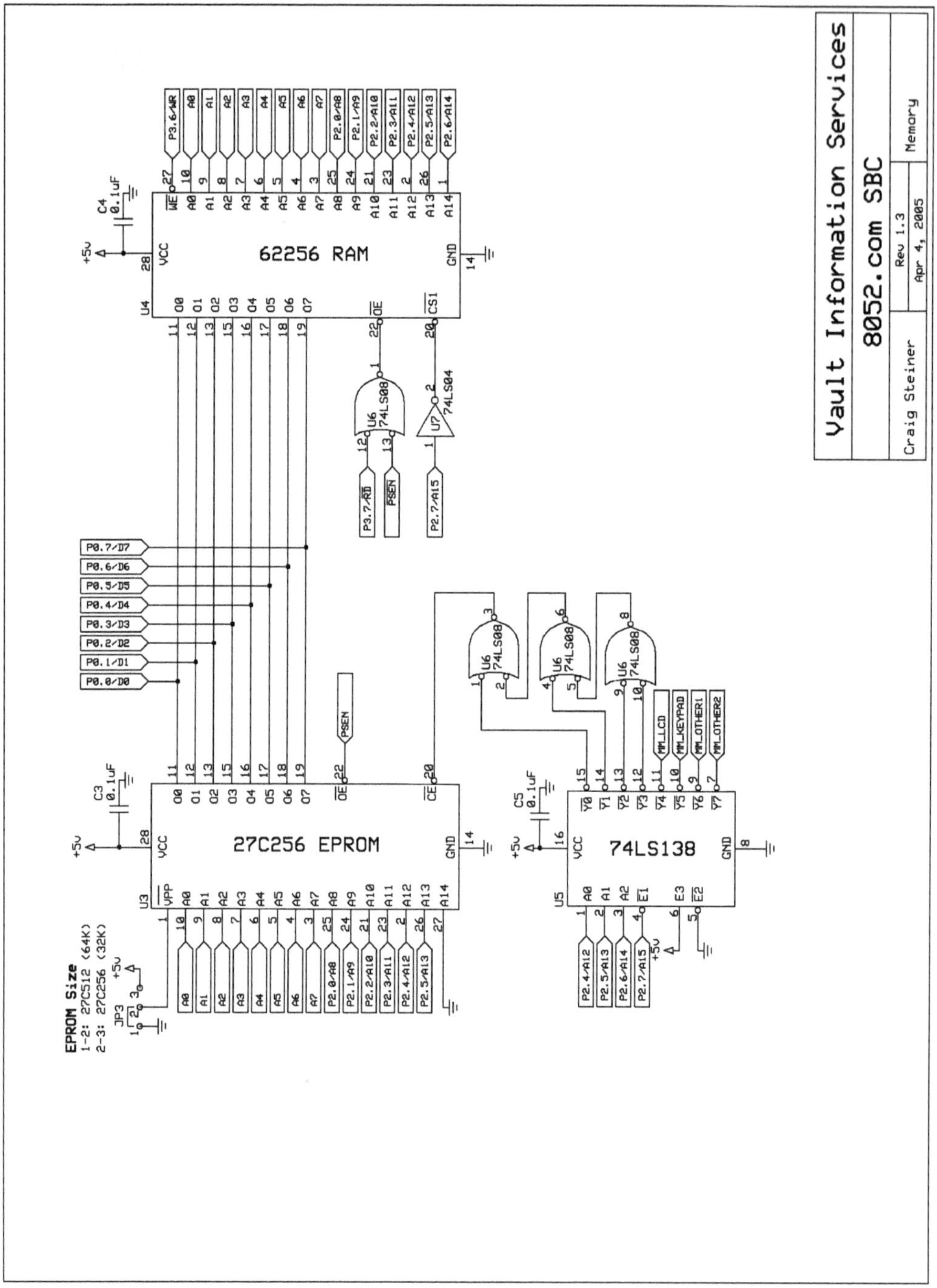
Vault Information Services
8052.com SBC
Rev 1.3
Craig Steiner
Apr 4, 2005
Memory
62256 RAM
27C256 EPROM
74LS138
EPROM Size
1-2: 27C512 (64K)
2-3: 27C256 (32K)

Memory Mapped LCD Enable Logic

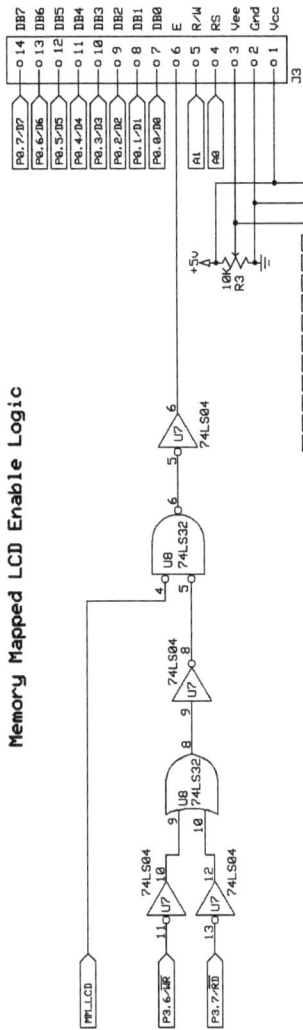

Memory Mapped Keypad Logic

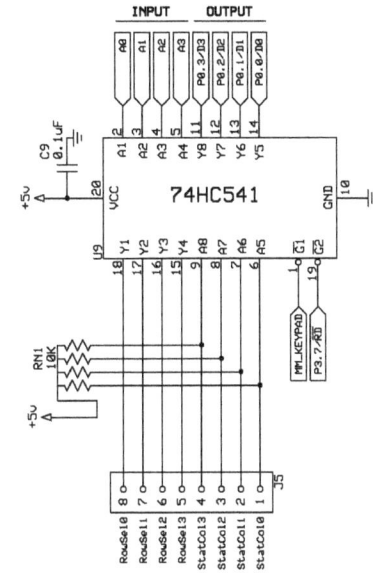

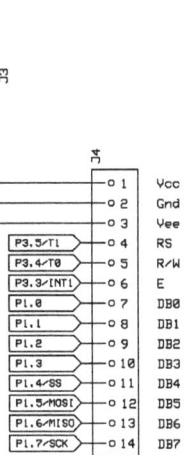

J3

P0.7/D7	14	DB7
P0.6/D6	13	DB6
P0.5/D5	12	DB5
P0.4/D4	11	DB4
P0.3/D3	10	DB3
P0.2/D2	9	DB2
P0.1/D1	8	DB1
P0.0/D0	7	DB0
	6	E
	5	R/W
A1	4	RS
A0	3	Vee
	2	Gnd
	1	Vcc

J4

	1	Vcc
	2	Gnd
	3	Vee
P3.5/T1	4	RS
P3.4/T0	5	R/W
P3.3/INT1	6	E
P1.0	7	DB0
P1.1	8	DB1
P1.2	9	DB2
P1.3	10	DB3
P1.4/SS	11	DB4
P1.5/MOSI	12	DB5
P1.6/MISO	13	DB6
P1.7/SCK	14	DB7

+5v
10K
R3

74LS04 U7 6
74LS32 U8
74LS04 U7 8
74LS32 U8
74LS04 U7 10
74LS04 U7 12

MM.LCD
P3.6/WR
P3.7/RD

74HC541

INPUT
A0
A1
A2
A3

OUTPUT
P0.3/D3
P0.2/D2
P0.1/D1
P0.0/D0

C9
0.1uF
+5v
VCC
GND

U9
Y1 18
Y2 17
Y3 16
Y4 15
A8 9
A7 8
A6 7
A5 6
Y8
Y7
Y6
Y5
G1 1
G2 19

MM.KEYPAD
P3.7/RD

RN1
10K
+5v

J5

RowSel0	8	
RowSel1	7	
RowSel2	6	
RowSel3	5	
StatCol3	4	
StatCol2	3	
StatCol1	2	
StatCol0	1	

Vault Information Services
8052.com SBC
Rev 1.3   Peripherals
Craig Steiner   Apr 4, 2005

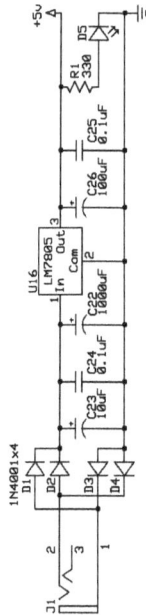

Reset Circuit

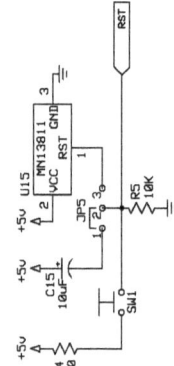

RST

MN13811 U15
VCC GND
RST
JP5
R5 10K
+5U
R4 100
C15 10uF
SW1

Power

1N4001x4
D1 D2 D3 D4
J1
C23 1uF
C24 0.1uF
C22 100uF
U16 LM7805
In Com Out
C26 100uF
C25 0.1uF
R1 330
D5
+5U

AT25010A SPI Serial EEPROM

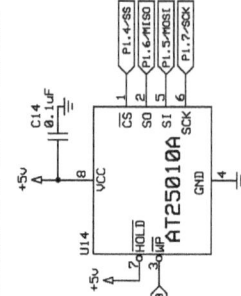

P1.4-SS
P1.6-MISO
P1.5-MOSI
P1.7-SCK

C14 0.1uF
U14
AT25010A
VCC
CS SO SI SCK
HOLD WP
GND
+5U
P3.2-INT0

DS1307 I2C Serial RTC

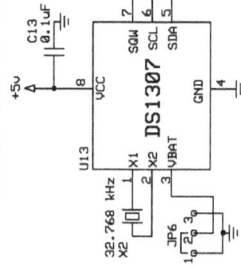

P1.2
P1.1

C13 0.1uF
U13
DS1307
VCC
SQW SCL SDA
X1 X2 VBAT
GND
+5U
32.768 kHz
X2
JP6

RS232/TTL Conversion

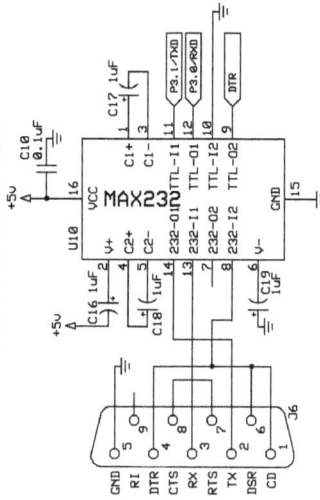

C17 1uF
P3.1-TXD
P3.0-RXD
DTR

C10 0.1uF
U10
MAX232
VCC
V+ C1+ C1-
C2+ C2-
232-O1 TTL-11
232-I1 TTL-O1
232-I2 TTL-I2
232-O2 TTL-O2
V- GND
+5U
C16 1uF
C18 1uF
C19 1uF

J6
GND RI DTR CTS RX RTS TX DSR CD
5 9 4 8 3 7 2 6 1

Vault Information Services
8052.com SBC
Craig Steiner
Rev 1.3
Apr 4, 2005
Power, Misc.

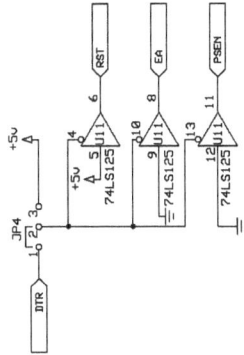

Dallas 89C420
Serial-Based ISP
DTR low sets RST high,
EA low, and PSEN low

Other Connections Not Shown

IC	Ground	+5v with bypass
U6	7	14
U7	3,7	14
U8	1,2,7,12,13	14
U11	1,2,7,9,12	14
U12	7	14

Vault Information Services		
8052.com SBC		
	Rev 1.3	ISP
Craig Steiner	Apr 4, 2005	

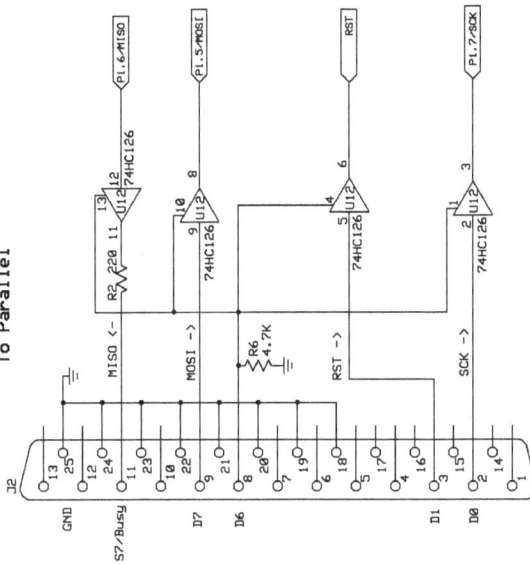

Atmel 89S8252
SPI/ICP Programming
To Parallel

14.2.1 Microcontroller: Atmel AT89S8252, Dallas 89C420, or 8052 (U1)

The 8052-compatible microcontroller (U1 in the schematic) drives the SBC. The SBC is designed for an Atmel AT89S8252 or Dallas DS89C420, but this part could be replaced with an original Intel 8031, 8032 or any other 8052 pin-compatible microcontroller.

The advantage of the AT89S8252 is that it includes two data pointers instead of one, includes an SPI interface for communicating with certain SPI-compatible ICs, has 8k of on-chip flash code memory and also includes 2k of on-chip EEPROM memory. The on-chip flash code memory means that programs can be downloaded directly to the microcontroller with no need for the 27C64 EPROM IC. The 2k of EEPROM memory means that the part can store up to 2k of data that will be retained even after a complete power-down or power failure. This is extremely useful for storing program parameters, configuration, or logging operation history for subsequent downloading.

The Dallas DS89C420, like the AT89S8252, has two data pointers, but it also offers 16k of on-chip flash code memory, 1k of on-chip extended RAM (accessed with MOVX), and has a second serial port for advanced applications that may need to communicate with two devices serially at the same time. Additionally, the DS89C420 is a 1-clock device which means it executes one instruction cycle for each oscillator cycle. This means that a DS89C420 using the same 11.0592 MHz crystal as an AT89S8252 or a traditional 8052 will execute a full twelve times faster.

The individual pins of both the AT89S8252 and the DS89C420 are exactly the same as the 8052. A full description of each pin can be found in the chapter 13. For the rest of this chapter, all references to the AT89S8252 will be just as accurate when applied to any alternative 8052-compatible microcontroller used in the circuit.

The SBC was also designed to support the Dallas 89C420 microcontroller. Specifically, the SBC is capable of downloading programs to the 89C420 via the serial port using the 89C420's ISP protocol.

14.2.1.1 LED on P1.0 (JP1, D6)

Jumper JP1 is connected between pin 1 (P1.0) of the microcontroller and a yellow LED (D6). When this jumper is closed (i.e. a jumper connects the two pins), the yellow LED will be connected to P1.0. In this case, the LED will turn on whenever P1.0 is brought low and will turn off whenever P1.0 is brought high.

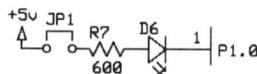

R7 is a current-limiting resistor to prevent the microcontroller pin from sinking more current than it can handle. The following equation may be used to determine the correct value for the resistor:

Resistor Value = (power supply voltage – 1.9) / max amps microcontroller can sink

The value of the power supply voltage is reduced by 1.9 which is the typical voltage drop for this LED—the exact value depends on the LED being used. In the case of the AT89S8252, the maximum current on P1.0 is 10mA but the total current on all eight P1 pins may not exceed 15mA. Assuming that P1 isn't going to sink much current other than the LED, it may be concluded that about 5mA (half the maximum for the pin and one third the maximum for the entire port) is an acceptable value. The power supply voltage is 5V so the equation is:

$$620 \text{ Ohms} = (5.0 - 1.9) / 0.005$$

Thus the value of the resistor R7 should be about 620 Ohms. The exact value is not critical and a value of 600 is used in the SBC. The higher the resistance, the lower the current and the dimmer the LED will be. The lower the resistance, the higher the current and the brighter the LED, but one risks exceeding the microcontroller's maximum current specification for the given pin.

> P1.0 is also used on the SBC to communicate with the LCD when it is connected to the direct connect LCD port (J4). In this case, the LED may flash quickly whenever data is written to the LCD if JP1 is enabled.

14.2.1.2 Push Button on P3.3 (SW2)

A momentary push-button is connected to P3.3. When the button is not pressed, the port will be unaffected by this button. When the button is pressed, P3.3 will be connected to ground which will cause the line to go low for as long as the button is pressed. The microcontroller program may detect this by noting when P3.3 goes to zero or an interrupt may be enabled to automatically trigger an interrupt service routine.

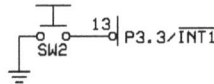

> P3.3 is also used on the SBC to communicate with the LCD when it is connected to the direct connect LCD port (J4). In this case, pressing this button will interfere with any communication that the SBC attempts with the LCD while the button remains pressed.

14.2.1.3 External Access Jumper (JP2)

Jumper JP2 is a two-position jumper. The jumper may either be connected between pin 1 and pin 2 to connect EA to +5V or between pin 2 and pin 3 to connect EA to ground. The position of this jumper selects whether the microcontroller executes code from its on-chip code memory or from external EPROM. When JP2 is connected between pin 1 and pin 2 (+5V), the microcontroller will execute code from its on-chip memory. When JP2 is connected between pin 2 and pin 3 (ground), the microcontroller will execute code from external EPROM (U3).

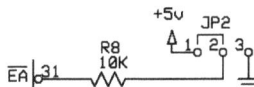

A 10k resistor (R8) is placed between the the EA pin of the microcontroller and JP2. This is due to the fact that the DS89C420's in-system programming protocol requires the EA pin be driven low when ISP is in progress. If EA were connected directly to JP2 and JP2 were connected to +5V, it would be impossible for the ISP circuit to drive EA low and might create a short-circuit condition. By inserting a 10k resistor between EA and JP2, the jumper can control whether EA is pulled high or low under normal conditions but the ISP circuit will be able to drive EA low when necessary.

The on-chip code memory option only applies if the microcontroller being used has on-chip code memory and has been loaded with a program. Microcontrollers such as the AT89S8252 and DS89C420 have on-chip code memory which means the programs can be loaded directly onto the microcontroller with no need for external EPROM. If the microcontroller doesn't have on-chip code memory or the program is stored in external EPROM, JP2 should be configured in the 2-3 position.

14.2.1.4 Crystal (X1)

The microcontroller is normally driven by a crystal that generates an output at a specific highly accurate frequency. One of the most commonly used crystal frequencies is 11.0592 MHz which is the value used on the SBC. In the case of the Atmel AT89S8252, the maximum crystal speed is 24 MHz while the Dallas DS89C420 is capable of operating at up to 33 MHz. The 11.0592 MHz crystal of this SBC may be replaced with a faster crystal to achieve higher processing speeds.

If a crystal other than 11.0592 MHz is used, any software that uses the serial port and which is designed for an 11.0592 MHz crystal will have to be modified in order to generate a valid baud rate. Timing delays based on the execution time of specific instructions must also be adjusted.

The crystal should be connected between XTAL1 (pin 19) and XTAL2 (pin 18) of the microcontroller. Additionally, both XTAL1 and XTAL2 should be connected to ground via a small capacitor, normally between about 27pF and 33pF—the exact value depends on the crystal. This provides a capacitive load to the oscillator which stabilizes the output of the crystal to the desired frequency.

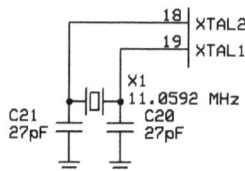

When designing and implementing a circuit that uses an oscillator, be sure that the oscillator is placed close to the microcontroller and keep the leads between XTAL1, XTAL2, the capacitors, and the oscillator as short as possible. This will avoid undesired load capacitance as well as minimize external interference on what needs to be a very precise clock signal.

14.2.2 Latch: 74LS573 (U2)

The 74LS573 is a "latch." The function of a latch is to receive and store signals on one set of pins when the latch is triggered and send those same signals out another set of pins even after the input signal has ended or changed.

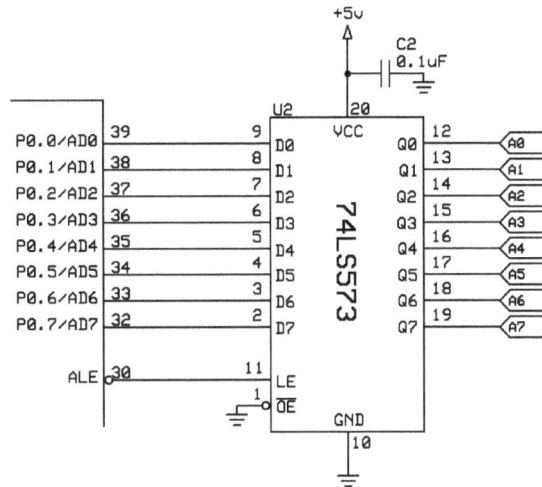

```
                                      +5v
                                       |
                                      C2
                                      0.1uF
                            U2    |20
         P0.0/AD0  39    9  D0   VCC   Q0  12   A0
         P0.1/AD1  38    8  D1         Q1  13   A1
         P0.2/AD2  37    7  D2         Q2  14   A2
         P0.3/AD3  36    6  D3         Q3  15   A3
         P0.4/AD4  35    5  D4  74LS573 Q4  16   A4
         P0.5/AD5  34    4  D5         Q5  17   A5
         P0.6/AD6  33    3  D6         Q6  18   A6
         P0.7/AD7  32    2  D7         Q7  19   A7

            ALE   30    11  LE
                          1  OE        GND
                                       |10
```

In the case of this circuit, the eight I/O lines of P0 on the AT89S8252 are attached to the input pins D0 through D7 (pin numbers 2 through 9) on the latch IC. Whenever "Latch Enable" (LE, pin 11) goes high, the input values on D0 through D7 are locked into the latch and sent to the output pins Q0 through Q7 (pins 12 through 19). Once Latch Enable goes low, it doesn't matter what values are found on the input pins, the output that was previously latched will continue to be output on Q0 through Q7.

The AT89S8252 uses this latch to store the low byte of the 16-bit address when accessing external code and data memory. Since the AT89S8252 uses 16 I/O lines (P0.0 through P0.7 and P2.0 through P2.7) to output 24 bits of data (16 bits for the address and 8 bits for data), it is necessary to multiplex eight of the sixteen lines. This means that P0 is used to transmit the low eight bits of the address *and* to output the eight bits of data. This is accomplished by first placing the low eight bits of the address on P0 and asserting the Latch Enable pin on the 74LS573 using its Address Latch Enable pin. At that point the output of the latch IC is set to the low eight bits of the address. When the Latch Enable pin is brought low, the output of the latch IC continues to be the low eight bits of the address that were previously latched into the input when Latch Enable was high. The AT89S8252 then places the eight data bits on P0.

Thus, whenever external memory is accessed, the low eight bits of the address will be found on pins Q0 through Q7 of the latch IC, the high eight bits will be found on P2 of the microcontroller, and the eight data bits will be found on P0.

The latch IC is only necessary if the circuit will be using external code or data memory or other memory-mapped devices. The Address Latch Enable (ALE, pin 30) on the AT89S8252 is connected to and drives the Latch Enable pin and is handled automatically by the microcontroller. The user program need not concern itself with latching the low byte of the address into the latch IC.

This SBC uses the 74LS573 latch instead of the more traditional 74LS373. These two latches perform the exact same function but have different pinouts. The '573 used in this design may be preferable since all the inputs are on the left side of the IC (near the microcontroller) and the outputs are on the right side (near the RAM and EPROM ICs) while the traditional '373 has inputs and outputs on both sides of the IC. Building an SBC with a '373 may be a little more confusing than with a '573 due to the additional wiring cross-overs.

14.2.3 Address Decoder: 74LS138 (U5)

The 74LS138 is a 3-to-8 multiplexer used to create the **memory map** for the SBC.

Address Range	Mapped Device
0000h – 3FFFh	32k EPROM (U3)
4000h – 4FFFh	LCD (J3) 4000h: Write command to LCD 4001h: Write text to LCD 4002h: Read command register from LCD 4003h: Read text from LCD
5000h – 5FFFh	4x4 Keypad (J5) 500Eh: Keypad row 1 500Dh: Keypad row 2 500Bh: Keypad row 3 5007h: Keypad row 4
6000h – 6FFFh	Reserved for other user-defined devices #1
7000h – 7FFFh	Reserved for other user-defined devices #2
8000h – FFFFh	32k RAM (U4)

The SBC is designed so that any access to external memory between the addresses 0000h and 3FFFh will address the EPROM IC, 8000h to FFFFh will address the RAM IC, and so on as outlined in the table above. To accomplish this the SBC must be able to take an address and assert an "enable" line for the corresponding device when an address in the specified range is referenced. For example, if any address is accessed between 8000h and FFFFh, the CS1 (pin 20) line of the 62256 RAM IC must be asserted while accessing the address 5007h should activate the fourth row of the keypad. The address decoder allows multiple devices to be memory-mapped into the SBC's external RAM address space.

The 74LS138 takes three address lines as input and, based on their state, asserts exactly one of eight output lines. For example, if the three address lines A0, A1, and A2 are all low then the Y0 output will be asserted. If all three lines are high then the Y7 output will be asserted. This is illustrated in the following table.

A2	A1	A0	Yx Line Enabled
0	0	0	Y0
0	0	1	Y1
0	1	0	Y2
0	1	1	Y3
1	0	0	Y4
1	0	1	Y5
1	1	0	Y6
1	1	1	Y7

This is accomplished as shown in the following schematic excerpt:

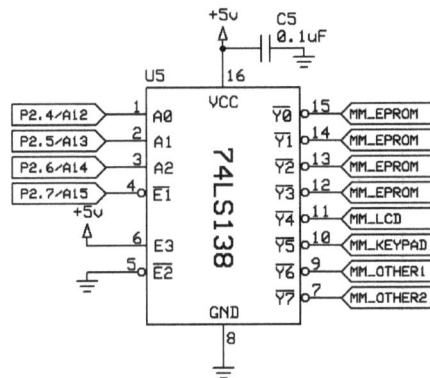

In the SBC's design, A0 of the '138 is connected to A12 of the microcontroller, A1 to A13, and A2 to A14 while -E1 is connected to A15. The -E1 line controls the '138 such that the IC only outputs a value if E1 is low. Since it is connected to A15, the '138 will only be active when A15 is not high—which means the '138 will only be active when the address is between 0000h and 7FFFh. Otherwise, the '138 will not operate and, instead, A15 will drive CS1 of the 62256 RAM (U4). In summary, this means that if the address is between 0000h and 7FFFh, the '138 will select the device that is being addressed. If the address is between 8000h and FFFFh then the 62256 RAM IC will be selected.

When any address between 0000h and 7FFFh is accessed, A15 will always be low which means the -E1 line will activate the '138. If, for example, the LCD is accessed via address 4000h, the highest bit (A15) will be clear, line A14 will be set, and A13 and A12 will be clear (since 4000h is the same as **01000000** 00000000 in binary)—thus -E1 of the '138 will be low, activating the '138, A2 will be set, and A1 and A0 will be clear so the Y4 line will be asserted which activates the LCD. This continues for each of the eight possible combinations for the address lines A12, A13, and A14. Each combination selects a different IC or device.

The three address inputs (A0, A1, and A2) to the 74LS138 are active high but the outputs (Y0 through Y7) are active low. That means that when all three input lines are low, the output Y0 will be asserted low and the rest of the outputs will be high. Similarly, when all three address inputs are high, the output Y7 will be asserted low and the rest of the outputs will be high. This is because most devices that will be attached to the decoder are activated by a low signal rather than a high signal. In the event that a device is activated by a high signal, the output of the '138 would have to be inverted with a 74LS04 or similar logic IC.

14.2.4 EPROM Code Memory: 27C256 (U3)

The 27C256 IC is a non-volatile EPROM which is generally used for program storage. A device programmer is used to download the program to the EPROM IC which is then inserted into the circuit. A socketed connection is generally used so that the EPROM can be repeatedly inserted and removed from the circuit to allow the program it contains to be updated. The contents of EPROM are not lost when the circuit loses power but, rather, are only lost when a small window on the EPROM is exposed to ultraviolet light for a period of time—normally 10 to 15 minutes. For this reason, the window of an EPROM is typically covered with a sticker to prevent stray ultraviolet light from inadvertently erasing the contents of the IC.

If the SBC is used with an AT89S8252, the EPROM IC is not necessary. The AT89S8252 includes 8k of on-chip code memory which means the program may be downloaded directly to the microcontroller with no need to store it in a separate EPROM. However, if a traditional 8052 or some other derivative that doesn't have on-chip code memory is being used in the SBC, the EPROM must be included in the circuit to supply code memory to the microcontroller.

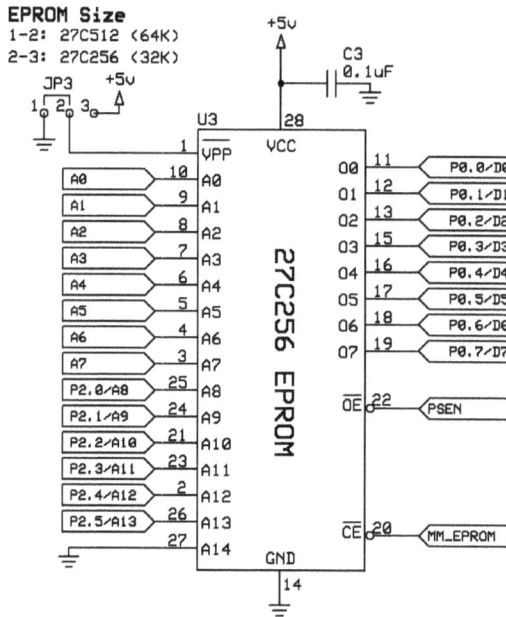

EPROM Size
1-2: 27C512 (64K)
2-3: 27C256 (32K)

JP3   +5v

U3   27C256 EPROM

Pin	Signal
1	VPP
28	VCC
10	A0
9	A1
8	A2
7	A3
6	A4
5	A5
4	A6
3	A7
25	A8
24	A9
21	A10
23	A11
2	A12
26	A13
27	A14
11	O0 — P0.0/D0
12	O1 — P0.1/D1
13	O2 — P0.2/D2
15	O3 — P0.3/D3
16	O4 — P0.4/D4
17	O5 — P0.5/D5
18	O6 — P0.6/D6
19	O7 — P0.7/D7
22	OE — PSEN
20	CE — MM_EPROM
14	GND

C3 0.1uF

When reading information from the 27C256, the address of the memory being accessed is placed on pins A0 through A15, though only lines A0 through A13 are connected to the EPROM. At that point the Program Store Enable (PSEN, pin 29) line of the AT89S8252 is asserted which brings the Output Enable

(OE, pin 22) line of the EPROM low. If the address being accessed is within the range allocated to the EPROM (0000h – 3FFFh), the '138 will assert MM_EPROM which will bring Chip Enable (CE, pin 20) low and cause the EPROM to put the contents of the requested memory address on the O0 through O7 lines of the EPROM. These lines are connected to P0.0 through P0.7 which allows the microcontroller to read the value output by the EPROM.

The Program Store Enable (PSEN, pin 29) line on the AT89S8252 is connected to and drives the Output Enable (OE, pin 22) on the EPROM. This line is handled automatically by the microcontroller, when necessary, so the user program need not concern itself with modifying these lines when accessing external code memory. In fact, there is no way to specifically instruct the microcontroller to assert the PSEN line on demand. It is only asserted by the microcontroller when it needs to fetch a program instruction byte from the EPROM or when the MOVC instruction is executed (which also fetches a byte from code memory).

14.2.5 Static RAM: 62256 (U4)

The 62256 IC is static RAM that can store up to 256 kilobits (32kBytes) of data. Data can be stored and retrieved from the 62256 by setting DPTR to the memory address and using MOVX to read from or write to memory.

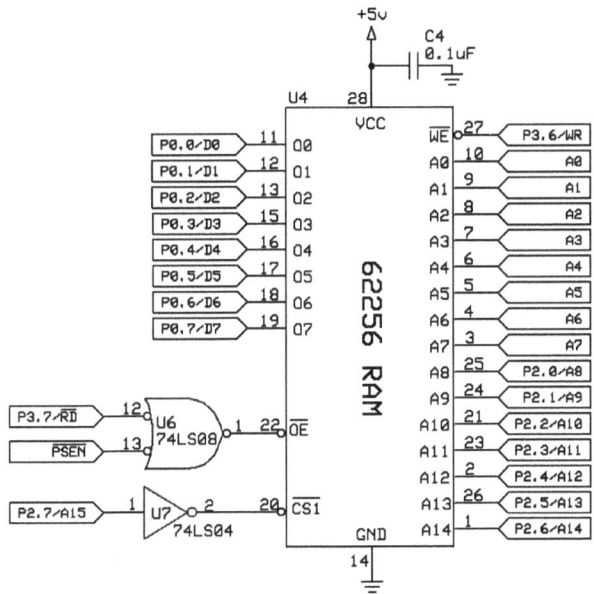

When manipulating data in the 62256, the address of the memory to access is placed on pins A0 through A15 of the microcontroller, though only lines A0 through A14 are connected to the RAM IC address bus. Address line A15 from the AT89S8252 is connected to the Chip Select (CS1, pin 20) pin of the 62256 such that the IC will only be active when A15 is high—that is, external RAM will only be active when the address being accessed is between 8000h and FFFFh.

If the microcontroller is reading from the 62256, it will bring Output Enable (OE, pin 22) low which causes the data at the requested address of the 62256 to be placed on the data lines D0 through D7 so that the microcontroller can read it on the eight I/O lines of P0. If the microcontroller is writing to the 62256, it will bring Write Enable (WE, pin 27) low which causes the data that the microcontroller has placed on the eight I/O lines of P0 to be written to the specified address of the 62256.

The Write (WR, P3.6, pin 16) line on the AT89S8252 is connected to and drives the WE pin on the 62256 while the Read (RD, P3.7, pin 17) line is normally connected to and drives the OE pin. Both of these lines are handled automatically by the microcontroller when necessary. The user program need not concern itself with modifying these lines when accessing external data memory.

Although the RD pin of the microcontroller is normally connected directly to the OE pin of the 62256, the connection is somewhat more complex in the SBC's circuit. As shown in the schematic, the RD line is OR'd with the PSEN line and if *either* line is low then the 62256's OE line is brought low. This means the microcontroller will read data from the 62256 when RD *or* PSEN is low which allows this memory to be treated as both data RAM and program memory. The RD line will be brought low when the microcontroller is accessing external RAM while the PSEN line will be brought low when the microcontroller is accessing external program memory. Since the 62256 will respond to either line, programs can be stored in RAM as data and executed as code.

The 62256 will be active only when the CS1 line is brought low by the A15 line from the AT89S8252 being inverted. This will only occur when the address being accessed is in the range of 8000h through FFFFh. This means that regardless of the RD, WR, and OE signals, the IC will only respond to memory reads and writes when the address in question is within that range.

The 62256 is only necessary if the application requires external data memory. Other RAM ICs may be chosen instead of the 62256, such as the 6264 (8k), 62128 (16k), etc. depending on the amount of external RAM necessary. The pinouts of other RAM ICs are similar but not identical.

14.2.6 LCD: Memory-Mapped (J3) and Direct Connect (J4)

There are two connectors on the SBC for an external LCD display which allows the user program to communicate with the LCD in two different ways.

When the LCD is connected to J4, the 11 I/O lines of the LCD are driven directly by 11 port pins on the AT89S8252. Specifically, port 1 drives DB0 through DB7 of J4 while P3.3 through P3.5 drive the E, RW, and RS lines. This connection will be used to initially explain how communication with the LCD works since this is direct communication with the LCD in its most basic form with absolutely no intervening logic. The signals that the program sends out on P1 and P3.3 through P3.5 are connected to the LCD directly.

When the LCD is connected to J3 it becomes a **memory-mapped device**. This is a device that is connected to the SBC in such a way that it can be read from and written to as if it were a memory device. Rather than manipulating 11 I/O lines of port 1 and port 3, the LCD may be accessed by using the MOVX instruction to read and write to memory addresses 4000h, 4001h, 4002h, and 4003h, depending on what type of communication with the LCD is required. The advantage of a memory-mapped device is that the

LCD shares the address and data bus that already exists to communicate with the EPROM and/or RAM IC so it doesn't need 11 I/O lines dedicated exclusively to LCD communication.

> The logic used to connect the LCD in memory-mapped mode will be explained in chapter 19 which discusses LCD programming since understanding the memory-mapped approach to the LCD first requires an understanding of how one communicates with the LCD in a direct fashion.

14.2.7 Keypad Connector (J5)

The keypad connector J5 is an eight-pin connector which permits the connection of a 16 key (4 columns by 4 rows) keypad. These types of keypads use four lines to select which row of the keypad to read and four lines to return the status of each of the four keys in that row.

The keypad connector on the SBC is connected in a memory-mapped fashion similar to the memory-mapped LCD connector J3. That means the status of the keypad is determined by reading the keypad as external memory with the MOVX instruction. When the program reads from address 5000Eh, the status of the first row of the keypad will be returned; reading 500Dh will return the second row; 500Bh will return the third row; and 5007h will return the fourth row.

> The logic used to connect the keypad in memory-mapped mode will be explained in chapter 18 which discusses the programming necessary to read the keypad.

14.2.8 DS1307 Real Time Clock (U13)

The SBC includes a real time clock (RTC). An RTC is an IC which keeps track of the current time—typically the hour, minute, second, day, month, year, and day of the week.

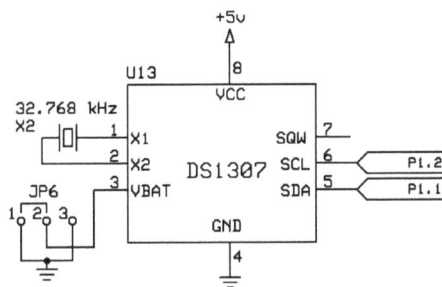

The DS1307 is attached to the microcontroller via the P1.1 and P1.2 port pins which allow the microcontroller to read and write data from and to the DS1307 using the I²C inter-chip communication protocol. Pins 1 and 2 of the DS1307 are attached to a 32,768 Hz crystal which provides the RTC with a constant time reference.

> Pin 3 of the DS1307 is available to provide a connection to the positive terminal of a battery backup to retain the block's time when the SBC loses power. It is possible to attach a battery to this pin by using JP6 such that the center pin of JP6 is the positive terminal of the battery and the two external pins are connected to the ground terminal of the battery.

14.2.9 AT25010A Serial EEPROM (U14)

The AT25010A (U14) is a serial EEPROM IC. EEPROM is memory that can be written to at any time but which will not be lost when power is removed. The AT25010A contains 128 bytes of non-volatile memory. As such, this IC is not appropriate for storing large quantities of data but may be useful for storing information such as configuration parameters, small activity logs, etc.

This part was included in the SBC because it is accessed using the Serial Peripheral Interface (SPI) which is an inter-chip protocol that is used with many other devices. The AT89S8252 microcontroller already has on-chip EEPROM so this IC could be considered redundant in terms of functionality, but its main purpose is to illustrate the use of the SPI protocol.

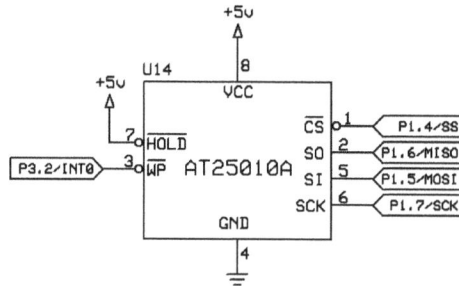

Pins 1 (Chip Select), 2 (MISO), 5 (MOSI), and 6 (SCK) are the four SPI I/O lines described in section 20.1. The HOLD line (pin 7) can be used to temporarily pause serial communication with the AT25010A without releasing CS. For the purposes of the 8052.com SBC, this functionality isn't useful so it is permanently connected to +5V so that it's never a factor in communication.

The final line is the Write Protect (WP) line on pin 3. When this line is driven low by the microcontroller the AT25010A will be in a write-protect mode; when the line is high the AT25010A will permit writes to its EEPROM. This line is connected to P3.2 on the microcontroller so the program must make sure P3.2 is high prior to attempting to write to the device. This is a safety feature to ensure that the part's EEPROM is not written to inadvertently.

EEPROMs have a finite life. After an EEPROM has been written to a certain number of times, that memory address within the EEPROM will become unreliable or completely fail. In the case of the AT25010A, the EEPROM is rated for one million write cycles. This is a large number so the EEPROM will tend to last "forever" when used for its intended purpose. However, if a program were to write to the EEPROM 50 times per second the EEPROM could start to fail after less than six hours. Thus it is important when designing software to make sure that writes to the EEPROM are made only when necessary and are not part of a routine that executes frequently.

14.2.10 Reset Circuit: RC Network, MN13811 Reset Supervisor, and Manual Reset

The reset pin (RST, pin 9) of the microcontroller is used to reset the IC when it goes high. Resetting the microcontroller causes program execution to stop, all SFRs to be reset to their default values, and program execution to restart at code memory address 0000h.

The reset pin is driven as follows:

1. Immediately after power-up, the RST line should be driven high (reset condition) briefly and then drop to low (normal execution condition) thereafter.
2. When the user presses the reset push button, the RST line should go high (reset condition) for as long as the user holds down the button and then go low (normal execution condition) when the button is released.

The power-on sequence of briefly raising the reset pin and then bringing it low has traditionally been accomplished in academic literature through the use of a a resistor-capacitor (RC) network such as that pictured in the following diagram.

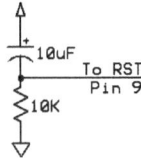

This circuit works since current flows from the power supply through the capacitor as it charges which drives the reset line high. Once the capacitor has fully charged the current flow drops and the reset line is pulled to ground by the 10K pull-down resistor. Although this is a very simple and easy-to-build power-up circuit, it is not optimum since the minimum reset period of the microcontroller is not always met. Further, the transition from a high to low level on the reset pin may not be as clean as the microcontroller requires.

A better solution is to include a **reset supervisor**. A reset supervisor is an IC which monitors the voltage and asserts the reset signal whenever it detects an out-of-specification voltage. Once the voltage has become stable, the reset supervisor will continue asserting for a specific amount of time before the reset signal is released. This doesn't just apply when the power is first applied but rather is also true if the voltage ever goes out of tolerance during normal operation. Rather than the microcontroller being subject to an unreliable power supply which may lead to undesired program execution, the microcontroller will automatically be reset until such time as the power becomes stable again. So not only does the reset supervisor assert the reset signal at power-on, it also makes sure the microcontroller is reset whenever the voltage drops below a minimum threshold.

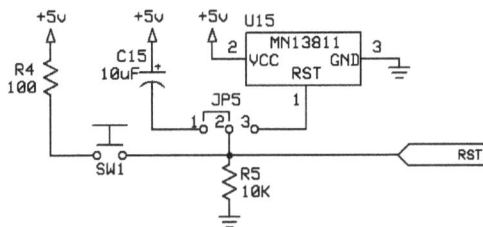

The 8052.com SBC has both a traditional RC network as well as a reset supervisor IC. Whether the RC network or the reset supervisor is used is selected by jumper JP5. When JP5 is in the 1-2 position, the RC network will be used to provide the reset signal while the MN13811 reset supervisor will be enabled when the jumper is in the 2-3 position. Jumper JP5 selects whether capacitor C15 or the output of the MN13811 is connected to the microcontroller's reset line.

Any design which requires any level of reliability should use a reset supervisor. RC networks are only appropriate for non-critical applications.

The DS89C420 has its own internal reset supervisor that performs the same function as the MN13811. For this reason if the microcontroller being used in the SBC is a DS89C420, the jumper JP5 should be completely removed—that is, neither the capacitor C15 nor the MN13811 should be connected to the reset line.

When the user presses the push button SW1, a connection is established to +5V so the reset pin is again taken to the reset condition for as long as the button remains pressed. When the button is released, the pin drops to ground which results in the microcontroller restarting execution at the newly reset state. This allows SW1 to function as a manual reset.

In a commercial or professional design, the SW1 push-button should be properly debounced. The push button does not generate an entirely clean signal when it is activated and it is possible that the low-high or high-low transition on the reset line will not meet the requirements of the microprocessor for a proper reset. This can be solved by using a reset supervisor which includes a manual reset input. Such a supervisor will properly debounce the input from the push-button and assert a reset signal that meets specifications.

14.2.11 RS-232 Transceiver: MAX232

The MAX232 IC is used to convert the voltage levels produced by the microcontroller (0V and 5V) to the voltage levels required by RS-232 interfaces (-10V and +10V). The transmit and receive pins from the microcontroller are connected to two pins of the MAX232 while another two pins from the MAX232 are connected directly to the transmit and receive pins of the RS-232 DB9 connector. A third pin is connected between the MAX232 and the DB9 to convert the DTR line to TTL voltages in order to handle ISP for the Dallas 89C420 (see section 14.2.13.2).

The Transmit 1 In (TTL-I1, pin 11) on the MAX232 is connected to Transmit (TXD, P3.1, pin 11) on the microcontroller and Receive 1 Out (TTL-O1) is connected to Receive (RXD, P3.0, pin 10). The Receive 1 In (232-I1, pin 13) is connected to pin 3 of the DB-9 connector and Transmit 1 Out (232-O1, pin 14) is connected to pin 2 of the DB-9 connector.

The schematic also illustrates that the MAX232 must be connected to four 1µF capacitors (C16, C17, C18, and C19) in addition to the normal decoupling capacitor C10. This is to conform to the required connections as specified by the MAX232 datasheet.

14.2.12 Power Connector, Rectification, and Voltage Regulator

The power connector J1 serves as the connection point for the power source coming from a typical AC or DC power supply.

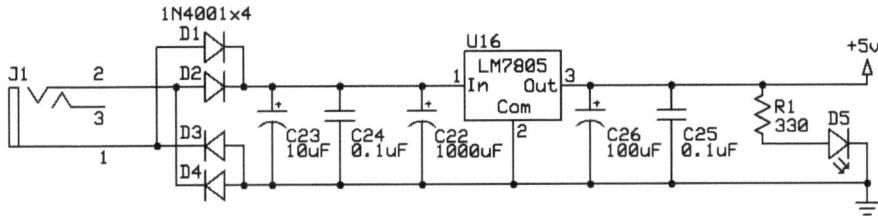

The power circuit includes decoupling capacitors (C23, C24, C25, and C26), a 5V voltage regulator (U16), and a red LED to indicate power (D5). It also includes an additional 1000µF capacitor which provides sufficient charge to handle alternating current should an AC power supply be used.

14.2.12.1 Rectification Bridge

The first section worth mentioning is the four 1N4001 diodes that connect to the J1 power jack. These four diodes are together known as a **rectification bridge**. This bridge serves two very important functions: 1) It allows the circuit to work with a DC power supply regardless of the jack's polarity. 2) When implemented with the 1000µF capacitor, it allows the SBC to also function with an AC power supply.

First, a review of why this bridge allows the circuit to work regardless of polarity is in order.

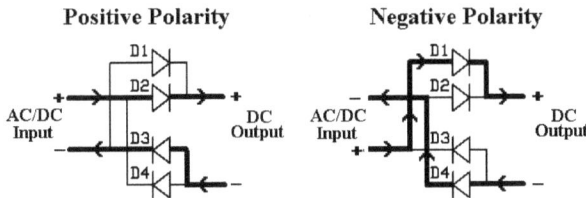

As depicted in the this diagram, regardless of which input line is positive and which is negative, the four diodes ensure that the top DC output line is always positive and the bottom line is always negative. This means that if a DC power adapter is connected to the circuit it doesn't matter which polarity is used—the bridge rectifier will "correct" the polarity as needed. Thus there is no risk of connecting a DC adapter "backwards."

The rectification bridge also allows the SBC to work with alternating current when used in conjunction with the 1000μF capacitor. Alternating current is simply a current that switches between positive and negative voltage—or polarity—in the form of a sine wave. In the following graphic, a 9V alternating current on the left can be seen in which the voltage varies from +9V to -9V. After going through the rectification bridge, the current is rectified such that the negative voltages effectively become positive voltages and the output voltage now varies between +9V and 0V.

AC Input into Rectifier Bridge

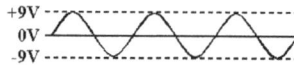

Rectified Output

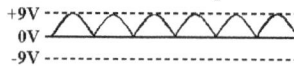

The only problem with the rectified output is that the output varies between 9V and 0V—which means the output is sometimes 0V and often below the voltage required to operate the components of the SBC. The 1000μF capacitor C22 provides sufficient filtering to maintain sufficient DC voltage throughout the cycle, which is further refined by the voltage regulator.

In reality, the input voltage to a bridge rectifier is not the same as the output. In the case of alternating current, the bridge rectifier multiplies the voltage by 1.41 and in the case of both AC and DC the voltage is reduced by 1.4V. This reduction is due to the fact that each diode reduces the voltage by 0.7V and the current travels through two diodes when being rectified: One diode going into the bridge and one diode going out. Thus, given input voltages of 12VAC and 12VDC, the output of the rectifier bridge would be:

12VAC input to bridge rectifier: 12V * 1.41 − 1.40V = 18.34VDC output
12VDC input to bridge rectifier: 12V − 1.40V = 10.6VDC output

14.2.12.2 Voltage Regulator

The output of the bridge rectifier continues through the LM7805 (U16) voltage regulator. This component takes the input voltage and converts it to a constant 5V output. The power input to the LM7805 should be at least 7.5V in order to guarantee a constant output of 5V. Using the voltage adjustments to calculate the output of the bridge rectifier mentioned in the previous section, this means that the power supply used with the SBC should be at least 6.3VAC or at least 8.9VDC.

The voltage regulator converts unused power to heat. This means that the LM7805 is often quite hot to the touch—even hot enough to burn you if you're not careful.

140

The amount of heat generated—or power dissipated—by the LM7805 depends on the excess voltage and the amount of current (amps) used by the SBC. The amount of power dissipated is calculated by multiplying the excess voltage by the number of amps consumed by the circuit. For example, if a 12VAC power supply is used then the LM7805 will receive about 18.3VDC of power. Only 5VDC is required by the LM7805 so that equates to 18.3V − 5V = 13.3V excess volts. If the SBC requires 300mA (0.3A) of current then the amount of power dissipated would be 13.3V X 0.3A = 3.99 Watts. The higher the amount of power dissipated the hotter the LM7805 will become.

As the temperature of the LM7805 increases beyond a certain level, the output of the LM7805 will start dropping below 5V. Thus to provide the highest reliability the temperature of the LM7805 should be kept as cool as possible. This is accomplished in two ways:

1. Choose the right power supply. As mentioned above, the SBC will operate on any AC power supply above 6.3VAC and on any DC power supply above 8.9VDC. Since excess voltage is what leads to a rise in the LM7805's temperature it is desirable to select the power supply with the least voltage that meets the minimum voltage required. In other words, don't use a 12VAC power supply if a 9VAC power supply is sufficient.

2. A heat sink may be attached to the voltage regulator. The heat sink, made of metal, distributes the heat that is generated by the voltage generator and distributes it over a greater area thus lowering the temperature of the LM7805 itself.

14.2.13 In-System Programming (J2, J6, U11, U12)

The SBC has the capability of performing in-system programming when the installed microcontroller (U1) is either an Atmel AT89S8252 or a Dallas 89C420. In-system programming means the program that is to be executed on the SBC may be downloaded directly to the microcontroller rather than requiring an external EPROM be installed in socket U3. This is a lot easier, faster, and requires much less wear and tear on the board since the EPROM doesn't need to be removed and re-inserted after each firmware update.

14.2.13.1 Atmel In-System Programming

The Atmel ISP is accomplished using a standard female->male DB25 cable that is connected to the PC's parallel port and to the SBC at connector J2. The VisISP-52 software (**http://www.8052.com/visisp52**) is used to download the firmware in HEX format to the Atmel microcontroller.

The circuit shown in the following schematic is used. The Atmel microcontroller accepts ISP updates by receiving the download via SPI protocol on pins P1.5, P1.6, and P1.7 while the RST pin is held high. The interface provides a reliable way for the PC to send the program to the microcontroller using this protocol.

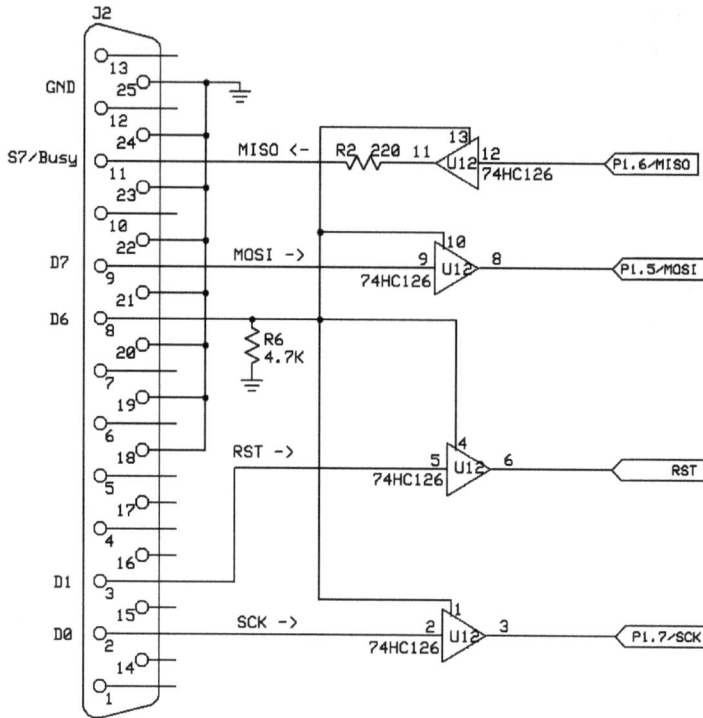

The D0 pin of the parallel port is used to transmit the SPI clock signal while D1 is used to hold the RST line high for the duration of the update. D7 is used to transmit each bit of the program to the microcontroller while the "busy" pin of the parallel port is used to receive responses from the microcontroller. Finally, D6 is used as a general "enable" for the ISP circuit. If D6 isn't high then the 74HC126 IC will not output anything (i.e. the outputs will be "high impedance") and will therefore not interfere with the microcontroller's normal operation.

When a DB25 cable isn't connected to J2, the 4.7K pull-down resistor R6 will cause the default value to be low. This means that when J2 isn't connected to a computer the ISP circuit will be disabled. Likewise, the ISP circuit will remain disabled unless the PC raises the D6 line. Also note the 220 ohm terminating resistor R2 between the output of the HC126 and the J2 pin. This keeps the signal clean and prevents reflections of the signal which could cause interference.

It is important that a 74HC126 be used rather than an 74LS126 or 74HCT126. The HC line of ICs is appropriate for driving off-board buses such as the DB25 that will connect the SBC to the PC's parallel port.

The 74HC126 (U12) may be completely removed from the circuit if the SBC is not going to be used with an Atmel AT89S8252. The U12 IC is only necessary if AT89S8252 in-system programming is planned. Likewise, the entire J2 parallel port connector can be omitted if AT89S8252 ISP isn't required.

142

14.2.13.2 Dallas In-System Programming

The Dallas ISP is accomplished using a standard female->male DB9 RS-232 cable that is connected to the PC's serial port and connected to the SBC's serial port connector J6. The following circuit is used:

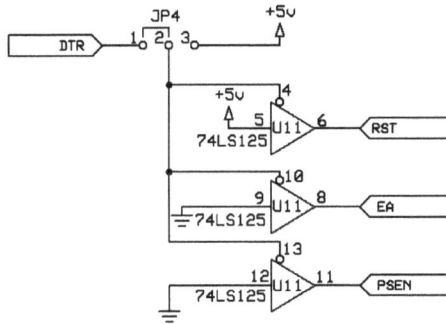

The DTR line comes from the MAX232 IC which converts the RS-232's DTR line from +10V and -10V to +5V and 0V. When the DTR line is high on the RS-232 port, the output will be 0V because RS-232 uses inverted logic. When JP4 is in the 1-2 position, a high DTR on the RS-232 will activate the 74LS125 (U11) IC. This will cause RST to be pulled high to +5V and both EA and PSEN to be pulled down to 0V. This configuration (RST high, EA and PSEN low) instructs the Dallas DS89C420 to enter ISP mode. The microcontroller will then be in ISP mode in which it can be erased or a new version of the software may be downloaded.

When the Dallas ISP mode is not going to be used—or if the microcontroller being used in the SBC is not a DS89C420—JP4 should be configured to the 2-3 position. This ties the line to +5V which permanently disables the U11 IC.

> The 74LS125 (U11) may be completely removed from the circuit if the SBC is not going to be used with a Dallas DS89C420. The U11 IC is only necessary if DS89C420 in-system programming is necessary.

14.2.14 Port Connections (J7 – J10)

The four jumper connectors J7 through J10 are direct connections to the port pin of the microcontroller. This allows the user to connect external circuits to the SBC without soldering to the board itself—the external card may be connected by using ribbon cables which can be plugged into the appropriate connector. The port connections are:

Jumper Connector	Connected to Port
J7	P1
J8	P3
J9	P0
J10	P2

Note that J7 and J8 are found on the PCB right next to each other so all the pins of P1 and P3 (typically used for user devices) can be accessed by using a single 16-pin ribbon cable which will plug into P1 and P3 simultaneously.

Likewise, J9 and J10 are found next to each other on the PCB since P0 and P2, together, represent the address bus. Again, these two ports can be accessed together by using a single 16-pin ribbon cable which will plug into P0 and P2 simultaneously.

14.2.15 Power, Ground, and other Signal Connections (J11)

Jumper connector J11 is an 8-pin connector which provides access to +5V, ground, and four other signal lines that are often needed when connecting external memory-mapped devices. The pin-out of J11 is:

J11 Pin Number	Signal
1	Ground
2	+5V
3	PSEN
4	P3.6/WR
5	P3.7/RD
6	MM_OTHER1
7	MM_OTHER2
8	Ground

The P3.6/WR and P3.7/RD lines which are available on this connector are also available on J8. They are duplicated on this connector since if a memory-mapped device is being developed on an external circuit board that is to be connected to the SBC, such a circuit would require only P0 and P2 along with the RD and WR signals from P3. Such a circuit would therefore only require J9 and J10 (for the address and data bus), J11 (for power and the RD/WR signals), and J12 (for the low byte of the address bus). If RD and WR were not available on J11, it would also be necessary to connect the entire J8 connector to access the lines on P3.

The MM_OTHER1 line on pin 6 is connected to pin 9 of the '138 (U5). This line will be asserted when the microcontroller is accessing an address between 6000h and 6FFFh. An external memory-mapped circuit can use this line as an activation signal. Likewise, the MM_OTHER2 line on pin 7 is connected to pin 7 of the '138. This line will be asserted when the microcontroller is accessing an address between 7000h and 7FFFh. This line, too, may be used to activate a second external memory-mapped circuit attached by the user.

14.2.16 Latched Low Byte of Address Bus (J12)

Jumper J12 holds the eight low bits of the address being accessed by the microcontroller as latched by the '573 (U2). This can be used by external memory-mapped devices as the low byte of the memory address.

14.2.17 Power Jumper (JP7)

Jumper JP7 is a 3-pin connector which provides access to the circuit's ground and +5V power lines. This can be used to power external circuits instead of connecting to J11 if the other signal lines on J11 aren't needed by the external circuit.

14.2.18 Other Connections

In addition to the connections described in the schematic, the following connections are necessary:

IC	Ground	+5V with Bypass
U6	7	14
U7	3,7	14
U8	1,2,7,12,13	14
U11	1,2,7,8,12	14
U12	7	14

14.2.18.1 Bypass Capacitors

A 0.1uF (100nF) bypass capacitor is placed next to the +5V pin of each integrated circuit on the board. As the internal gates of the ICs switch on and off, momentarily higher current may be needed. In these cases, the bypass capacitor will have a charge stored up and will be able to provide it to the IC quickly when required.

DEVELOPMENT TOOLS

CHAPTER 15: SOFTWARE DEVELOPMENT TOOLS

Development of an 8052-based device involves two major components. The first is the hardware design itself which was covered in the previous two chapters in the form of a functioning SBC. The second is the use of development tools to create the software (also called 'firmware' when it is downloaded to an embedded system) that will run on the microcontroller. These software tools normally are applications that run on a Windows-based PC and allow the developer to create a program in an editor, compile or assemble it into an output file suitable for loading into the microcontroller or EPROM, and often a simulator which allows the developer to test the code on the PC itself before downloading the code to the actual hardware.

Each of these components can be acquired individually or an as integrated solution.

Some developers choose to acquire a specialized editor to write the program code, use an assembler or compiler that is executed from the command prompt to compile the program, and use a separate stand-alone simulator to simulate the code before beginning hardware testing. Since each of the components are completely separate and may come from different software companies, the integration is often somewhat reduced. The development process may not be seamless since the editor may not be designed to correctly invoke the assembler and the simulator may not be capable of loading symbol information from the compiler. On the other hand, the developer may pick and choose the exact applications that fit the needs of the project meaning the best editor can be chosen, the best compiler can be used, and the best simulator can test the code. The main drawback to this approach is the potential lack of integration in the components and the fact that each part must be obtained separately. Additionally, since each application may be from a different software company it is more probable that incompatibilities may exist between the programs which may further complicate the development process.

The alternative is to use an integrated system that combines all three of these components into a single application. Such tools are called **Integrated Development Environments** (**IDE**s). The advantage of IDEs is that all the functions necessary to develop software for the microcontroller are contained within a single package. Program code can be written in an editor that is part of the software, a single keystroke can often cause the package to compile the code, and yet another single keystroke can begin the execution of the code that was just compiled in the integrated simulator. There's no switching between at least three applications because all of the functions are contained within the IDE. The disadvantage to IDEs is that the developer is normally stuck with whatever is part of the IDE—that is, the developer may prefer a different editor than the one included in the IDE. Likewise, a stand-alone simulator may have more features than the simulator included in the IDE.

Nevertheless, most developers use IDEs and using one is certainly much more straight-forward for the typical student, hobbyist, or professional that is just beginning to work with the 8052 and may not have the time or the technical knowledge to choose individual components for the development process.

The rest of this chapter will explain how to use the **Pinnacle 52 Integrated Development Environment** from Vault Information Services to develop software for the 8052. The code written in Pinnacle 52 may be used with any 8052-compatible microcontroller, including the AT89S8252 that was used in the SBC in the previous chapter. While this chapter will illustrate the process step-by-step in Pinnacle 52, the process is generally similar regardless of which IDE is used. Other IDEs may name the features differently and use different menu and windows structures but the overall concepts should apply to any 8052 IDE. If attempting to use other development tools it is highly recommend that the accompanying manual or online help be read so that the development tools may be completely understood.

In the interest of full disclosure, it should be mentioned that the author of this book is also the author of the Pinnacle 52 IDE. Pinnacle 52 will be covered in this chapter because it was written by this author, the author is obviously very familiar with it, and because Pinnacle 52 is a very low-priced product that is within the budget of even most students. This should not be interpreted as meaning that Pinnacle 52 is the only—or necessarily the best—development environment for the 8052. Many development tool sets exist for the 8052 with varying feature sets and prices. But Pinnacle 52 is more than adequate to explain the concepts of software development. Additionally, a demo version of Pinnacle 52 is available free of charge at **http://www.vaultbbs.com/pinnacle** and may be used to create programs up to 2k in length. This is sufficient to generate the programs that will be covered in the rest of this book without requiring the reader purchase other development systems. If the user subsequently writes longer programs that require more than 2k of code memory it may be reasonable to consider purchasing the full professional version of Pinnacle 52 which expands the maximum code memory size to 64k which is the maximum size permitted by an 8052 microcontroller.

15.1 Introduction to Pinnacle 52 Integrated Development Environment

Pinnacle 52 IDE is a development environment that will run on any Windows operating system. It includes an integrated editor in which 8052 assembly code may be written by the developer. The code which is written can then be assembled into a format that can be downloaded to a microcontroller or EPROM. Further, the same code can be executed in Pinnacle's integrated simulator which allows the developer to execute the program code one instruction at a time and see what impact each instruction has on all 8052 registers, internal RAM, external RAM, etc.

Once Pinnacle 52 has been installed on the computer, the IDE may be started by launching the program icon normally in Windows.

15.2 Pinnacle 52 Environment Overview

When Pinnacle 52 is launched it will open the main environment window. The environment includes a menu bar with the menus File, Edit, View, Execute, Simulator, Project, Tools, Options, and Help. Below the menu bar is a toolbar that includes typical Windows-style buttons to quickly open and save projects as well as some development-specific options that will be discussed later. Below that is an information bar that presents data that is very useful when simulating an 8052 program on the PC.

All functions of Pinnacle, including editing, assembling programs, and simulation, are performed within this environment.

As can be seen in the following screen shot, the development environment includes a combination of editing, assembling, and simulation features. For example, the buttons in the toolbar include functions to create, load, save, and print editor files. The next three buttons reset a simulation, start a simulation, and stop a simulation. The rest of the buttons include quick access to breakpoints, single-step, procedure-step, and skip step instructions, plus access to an analysis screen that will provide useful information on the program currently being simulated. As the mouse is moved over the buttons of the toolbar, the function of that button will be displayed in the helpbar at the bottom of the screen.

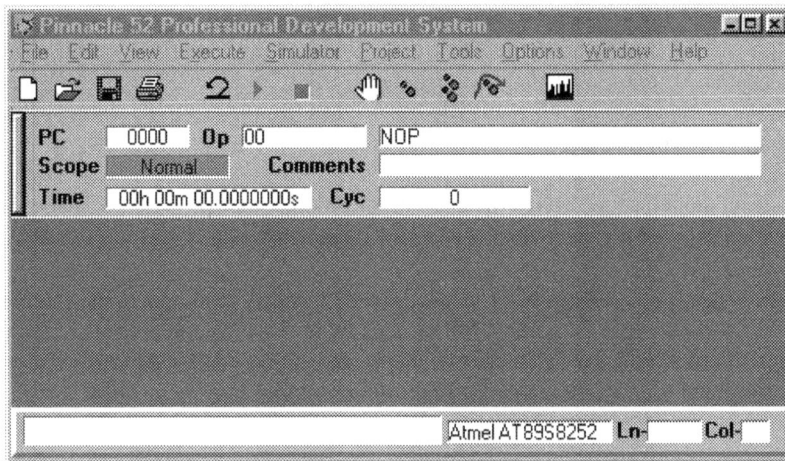

The entire section below (with "PC" and "Scope" labels) all provide additional information related to the simulation in progress.

Each feature will be covered in detail in the remainder of this chapter.

15.3 Creating, Loading, and Saving Editor Files

The integrated editor is used by the developer to write and edit assembly language programs. A new editor file can be created by either pressing the "New File" button (the left-most button on the toolbar) or by clicking on the "File" menu and selecting "New". This will open a blank editor file at which point the developer may enter the program into the editor window.

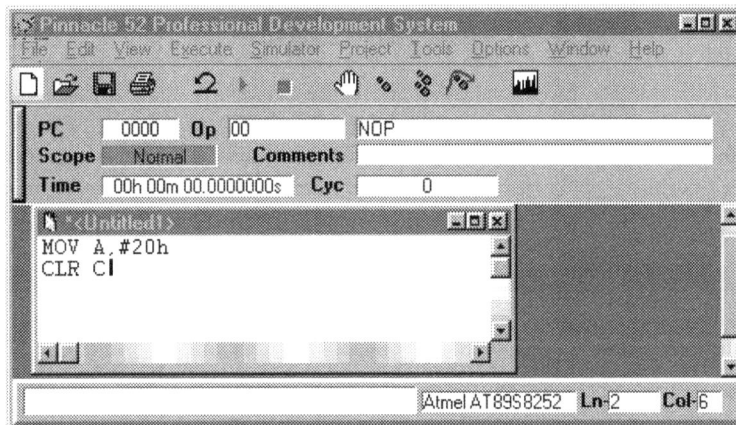

As text is entered, 8052 assembly language instructions and SFRs will be highlighted in a different color. Additionally, the line and column on which the cursor is positioned will always be visible in the help bar at the bottom of the environment. In the previous screen shot the cursor is on line 2, column 6.

An editor file may be saved at any time during the editing of a file. To do so, either click on the "Save File" toolbar button (third button from the left that looks like a floppy disk) or click the "File" menu and select the "Save" option. If the file hasn't previously been saved, Pinnacle 52 will ask the user to select a filename under which to save the file. Once the file has been saved, the "Untitled" name in the editor's title bar will change to the name of the file.

An asterisk to the left of the editor filename in the editor's title bar indicates that changes have been made to the file since the last time the file was saved.

Editor files may be loaded into the editor by pressing the "Open File" button on the toolbar (the second button from the left that looks like a file folder) or by clicking on the "File" menu and selecting the "Open" option. Doing so will prompt the user to select the file to open in the editor. Multiple editor files may be opened simultaneously and each file will have its own editor window.

15.4 Assembling a Program

Once an assembly language program has been created and saved (normally in the editor), the next step is to assemble the program. The following screen shot shows a very simple assembly language program which will execute three distinct loops infinitely. Note that there is an intentional syntax error on the "LoopR6" line.

```
E:\MISC\LOOPTEST.ASM
Start:  MOV R5,#10
LoopR5: MOV R6,#10
LoopR6 :|    MOV R7,#00h
LoopR7: DJNZ R7,LoopR7
        DJNZ R6,LoopR6
        DJNZ R5,LoopR5
        SJMP Start
```

The next step is to assemble the program. With the editor window containing the code as the active window, click on the "Project" menu and select the option "Compile & Link LINKTEST.ASM" (or simply press ALT+F2 from the editor window). This will open a new window within which the assembler will execute and report any errors. Using the above example, the following assembly output window will be displayed:

```
Output
Initializing Compiler...
Compiling E:\MISC\LOOPTEST.ASM
E:\MISC\LOOPTEST.ASM(4): Error [E2012]: Syntax error: LoopR6
E:\MISC\LOOPTEST.ASM(6): Error [E2000]: Invalid value or unknown symbol: LoopR6
Build terminated. 2 error(s), 0 warning(s)
```

The first line of the output simply indicates the Pinnacle assembler/compiler is being initialized while the second line indicates that Pinnacle is compiling the LOOPTEST.ASM program. The next two lines indicate errors in the program.

The first error (E2012) is a "syntax error: LoopR6". The number in parenthesis (4) indicates the line number on which the error was detected. In this case the problem is that the colon following the LoopR6 label does not immediately follow the label is separated by a space. Since labels must always be immediately followed by a colon, the label isn't recognized as a label but rather as an instruction. Since no assembly language instruction exists called "LoopR6", this is a syntax error.

The second error (E2000) indicates an "Invalid value or unknown symbol: LoopR6" on line 6 of the program. This is the line that reads `DJNZ R6,LoopR6`. This line is referencing the LoopR6 label but since the line of the program that defines the label wasn't compiled due to a syntax error, the label doesn't actually exist. This causes line 6 to fail.

This shows that a single error can cause multiple errors to be reported by the compiler.

Always start correcting problems with the first error reported by the compiler. Since many compiler errors can be generated by a single mistake in the code, it is always necessary to correct the first error first. It is possible that by fixing the first error that all the other errors will be fixed automatically.

Once line 4 has been fixed by removing the space between the label and the colon and the program is recompiled as described above, the following output window results:

```
Output                                              _ □ ×
Initializing Compiler...
Compiling E:\MISC\LOOPTEST.ASM
Linking...
Outfile created: E:\MISC\LOOPTEST.HEX (14 bytes)
Build complete. 0 error(s), 0 warning(s)
```

Again, the first two lines indicate that the compiler is initializing and compiling the LOOPTEST.ASM program. But this time instead of reporting errors Pinnacle 52 reports `Linking...` followed by `Output file created: E:\MISC\LOOPTEST.HEX (14 bytes)`. This tells the user that the program was successfully compiled and linked, and that the output file LOOPTEST.HEX was created. The output file in question contains 14 bytes of 8052 machine code that would be downloaded to the microcontroller.

Finished programs that can be downloaded to an 8052 microcontroller or EPROM normally have a file extension of "HEX" which is the 3-letter abbreviation for the format that is used: "Intel Standard HEX File." This format tells the microcontroller what data should be downloaded to the microcontroller or EPROM and at what address each byte should be placed. It is considered an industry standard and virtually all 8052-related hardware and software support this format. These are essentially the "EXE" files of the 8052 industry.

Another format that is sometimes used is the BIN format, which means "binary." In this format the file contains only the data to be downloaded but omits the address information. This is desirable in some circumstances but normally the HEX format is preferred.

Once Pinnacle successfully compiles and links a program into a HEX file, the resulting file can either be downloaded directly into the microcontroller (if it has on-chip program memory) or to the external EPROM. The same file can also be loaded into a simulator or emulator for further testing.

Since Pinnacle has an integrated simulator, the HEX file is automatically loaded into Pinnacle's simulator every time a program is compiled.

15.5 Loading a Program for Simulation

Programs may be simulated within Pinnacle either by loading a HEX file into memory or by compiling a program in the editor.

To manually load a HEX file that has been created in the past by Pinnacle or that has been created by some other compiler, click on the "File" menu, select "Open," and then use the file selector dialog window to select the HEX file in question. This will load the HEX file into the simulator's memory.

When using the File->Open dialog, Pinnacle 52 defaults to displaying files with the ASM extension which are normally assembly language source code. To make it easier to search for HEX files, select "Intel HEX" in the "List Files of Type" listbox at the bottom of the dialog window to display HEX files.

When using Pinnacle 52 to compile programs (as explained in the previous section) the resulting HEX file will automatically be loaded into Pinnacle's simulator every time a program is compiled and linked successfully. This saves the user the effort of manually loading the HEX file that was just created and speeds the development process.

15.6 Simulating a Program in Pinnacle 52

Programs may be simulated within Pinnacle 52 either at high-speed for quick simulation of a large amount of code or by executing one instruction at a time. The simulator provides handy information in the simulation information bar just below Pinnacle's toolbar.

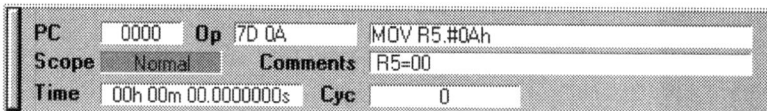

PC	0000	Op	7D 0A	MOV R5,#0Ah
Scope	Normal	Comments		R5=00
Time	00h 00m 00.0000000s	Cyc		0

The **PC** field ("0000" in this example) indicates the address of the instruction that is about to be executed. Since this is the first instruction of the program and nothing has been executed yet, 0000 represents address 0000 of code memory.

The simulator may be set to execute any address in code memory by clicking on the current address shown in the PC field. Pinnacle will then prompt the user to enter a new code address.

The **Op** field ("7D 0A" in this example) are the bytes of code memory that represent the instruction that is about to be executed. Since this instruction begins at 0000 it is understood that code memory address 0000h contains the byte 7D and address 0001h contains the byte 0A. These two bytes together form the assembly language instruction that is displayed just to the right. In this example, MOV R5, #0Ah. This is an assembly language representation of the bytes at the address 0000 which are about to be executed.

The **Scope** field ("Normal" in this example) indicates whether the program is in normal mode or is currently executing an interrupt. This field will always be "Normal" with a green background when the program is in non-interrupt mode and will have a red background and indicate the name of an interrupt when an interrupt is being executed (such as "Timer 0", "Serial", etc.).

The **Comments** field provides additional information that may be useful in the context of this instruction. Since this example is going to modify the value of the R5 register it may be useful to know what the current value of R5 is—the comments field indicates that the current value is 00. The contents of the comments field will change depending on the instruction that is to be executed.

The **Time** field indicates how much time has elapsed in the simulation. Since the simulation does not execute at the same speed as it would in a real microcontroller and because the simulation can be paused by the user at any time to inspect the values of registers, the time field will indicate how much time would have elapsed had the code been executed on a real microcontroller. This is useful for determining how long code would take to execute in a real environment or to benchmark certain sections of code.

The time field may be reset to zero at any point by right-clicking the time field and selecting the "Reset" option.

The **Cycles** field indicates how many instruction cycles have been executed by the current simulation. This is similar in purpose to the time field just described but expresses time in instruction cycles.

Note that **Cycles** is expressed in instruction cycles, not oscillator cycles. With a standard 8052 that requires 12 oscillator cycles for each instruction cycle, the number of cycles must be multiplied by 12 to calculate the number of oscillator cycles.

To execute a simulation one instruction at a time, the user may either press the button on the toolbar that looks like a single footprint or press the F8 key. The simulator will execute the instruction that is currently displayed in the simulator information bar and then stop to allow the user to inspect the new information.

A simulation may be executed at high-speed by pressing the button on the toolbar that looks like a "play" button (a triangle pointing to the right). The simulator will quickly execute instructions until the user interrupts simulation by pressing the stop button (the button with the square just to the right of the "play" button on the toolbar).

The simulation may be reset at any time by pressing the button on the toolbar that looks like an arrow wrapping back to the left, just left of the "Run" button. This resets all 8052 registers and memory to their preset values, resets the program counter to 0000, and clears the time and cycles fields back to zero.

Additional simulator functions are found in the "View", "Execute", and "Simulator" menu options.

15.6.1 Pinnacle's View of Registers and SFRs

The "View" menu consists of a list of options, each of which will display a corresponding window within the Pinnacle environment that will display some aspect of the simulation in progress.

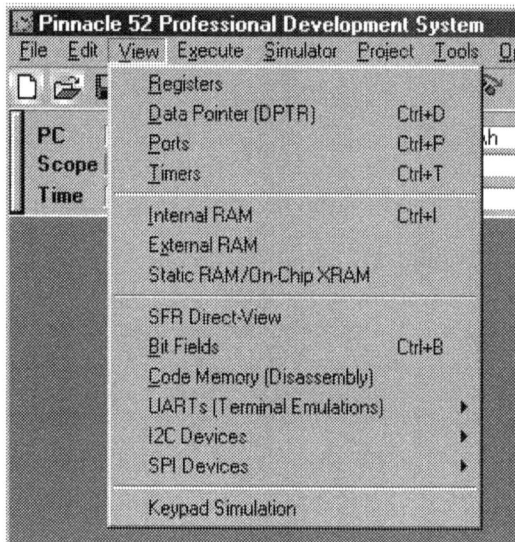

This menu looks intimidating at first but quickly becomes second nature. Since each of these options opens up a window, the follow section will review each option and, thus, each corresponding window.

15.6.1.2 Editing Registers and SFRs

In each of Pinnacle's simulator windows, the registers or data being displayed can be modified by double-clicking on the field in question. For example, double-clicking on the accumulator register in the Registers window (see next section) will open the byte-editing window show on the following page.

The value of the byte in question may be edited in this window. A hexadecimal value may be entered into the text field at the top or the individual bytes may be turned on or off in the checkboxes below.

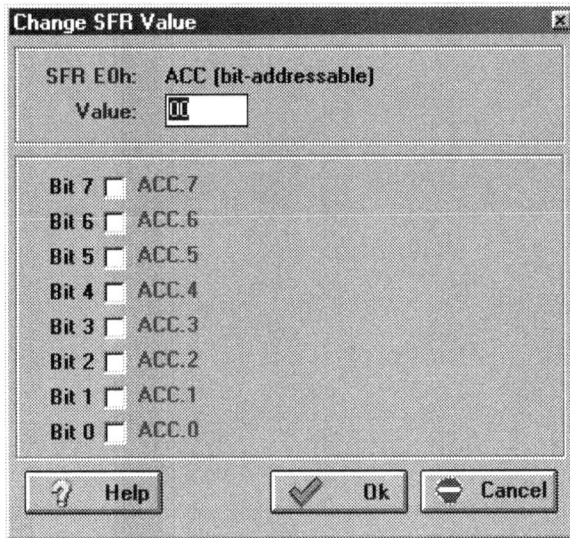

The values of registers, memory, and SFRs, can also be modified quickly by right-clicking the value in question which will bring up the following pop-up menu:

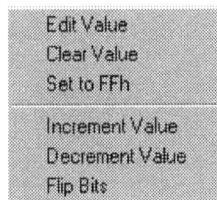

The **Edit Value** option will open up the previous window that allows the user to edit the byte by entering a new value or setting or unsetting the individual bits. The **Clear Value** will set the value to zero while **Set to FFh** will set the value to FF. **Increment Value** and **Decrement Value** will increase or decrease the value, respectively, while **Flip Bits** will update the value such that all bits that were 1 will become 0 while all bits that were 0 will become 1.

15.6.1.3 Registers Window

The **Registers window** displays the current status of the microcontroller's primary and most commonly-used registers. The Registers window is shown on the next page.

The Registers window includes the accumulator (ACC), B, registers R0 through R7, stack pointer (SP), interrupt enable (IE), interrupt priority (IP), power control (PCON), program status word (PSW) and serial control (SCON). In addition, the six flags of the PSW register are displayed on a flag-by-flag basis. Double-clicking on any of the flags will cause that flag to toggle (a zero will become a one and a one will become a zero).

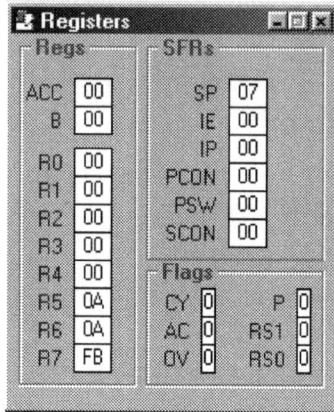

15.6.1.4 DPTR Window

The **DPTR window** displays the data pointer—or pointers—of the microcontroller.

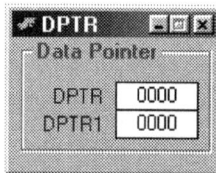

In this example, two data pointers are shown because the selected microcontroller is the AT89S8252 which provides two data pointers. Only one data pointer will be shown if the selected microcontroller only has one data pointer. Likewise, additional data pointers will be displayed if the microcontroller offers more than two data pointers.

15.6.1.5 Ports Window

The **Ports window** is a little more complicated than most windows since the ports of an 8052 microcontroller have an output latch and an input value.

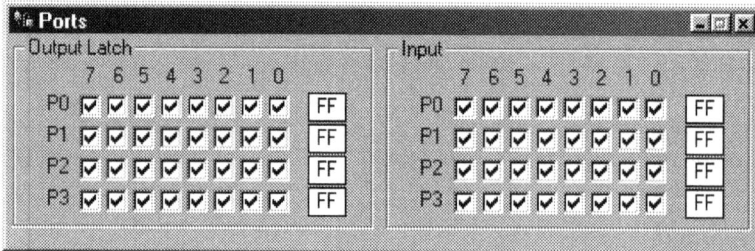

Writing a 1 to a port actually writes the 1 to the output latch. If the pin is immediately read, it will return the input value on that line which may be either 1 or 0. If the output latch is 0, however, the value read will always be zero.

For this reason the Ports window displays two sets of bits and values for each port. The values on the left side of the window reflect the current value of the output latch while the values on the right side of the window reflect the current value being detected by the microcontroller on that line.

When modifying the value of bits in the ports window to simulate the changes on the I/O lines, always modify the value of the input and not the value of the output latch. Only modify the values in the output latch side of the window if you really mean to modify the values that have previously been written to the port by the user program. The output latch reflects values written to the port by the user program while the input values reflect the status of the lines as affected by components external to the microcontroller.

When a program reads a port (P0, P1, etc.) or a port pin (P0.0, P0.1, etc.), the value that is returned is actually the value of each output latch bit logically ANDed with the value found on the input. This is why a port must be set to FFh to be able to read the pins on that port—if the output latch bits are zero then nothing can be read from those pins since any read will be ANDed with zero on the output latch and return a zero.

As a general rule of thumb, the values that an 8052 program writes to the ports are written to the output latch and affect the I/O line as "seen" by other external devices. The changes on any I/O line can also be affected by external devices and those are simulated as the "Input" value on the right side of the window. These values will be readable by the program only if the output latch for that pin is 1.

15.6.1.6 Timers Window

The **Timers window** displays the SFRs related to the microcontroller's timers.

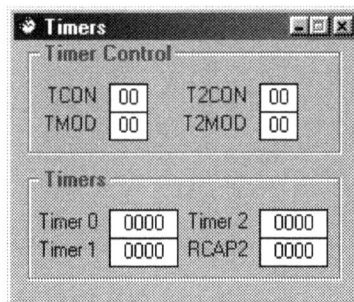

Note that the SFRs related to timer 2 will only be visible if the selected microcontroller has a timer 2.

15.6.1.7 Internal RAM (IRAM) Window

The **Internal RAM window** displays the contents of the 256 bytes of internal RAM of the microcontroller. The left-hand column (00, 10, 20, etc.) indicates the high-order nibble of the internal RAM address while each row (0, 1, 2, etc.) indicates the low-order nibble of internal RAM. Thus column 6 in row 70 corresponds to internal RAM address 76h.

The internal RAM addresses that have a green background reflect the addresses that are currently selected as the active register bank. The address that has a red background reflects where the next value to be pushed onto the stack will be placed at. The addresses with a yellow background (20h through 2Fh) reflect the 16 bytes of bit memory.

The contents of internal RAM may be displayed as hex (as shown above) or as ASCII. If displayed as ASCII then the character corresponding to the ASCII value of that internal RAM address will be displayed instead of the numeric hex value. This can be useful when reviewing internal RAM for text strings, especially those that have been input by a user over the course of the simulation.

> The Internal RAM window will display fewer registers if the microcontroller that is being used offers fewer than 256 bytes of memory. The original 8051, as well as some modern low-end derivatives, offer only 128 bytes of internal RAM in which case this window will only display rows from 00 through 70 (corresponding to addresses 00h through 7Fh).

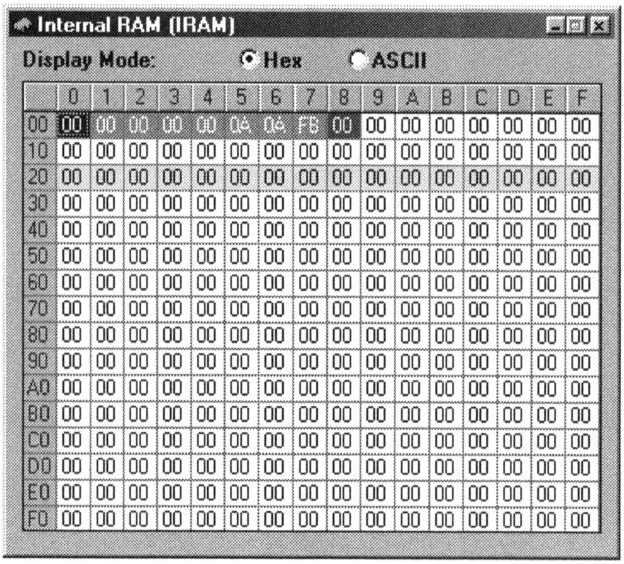

15.6.1.8 External RAM (XRAM), Static RAM (SRAM) Window

The **External RAM window** and **Static RAM window** look and function identically to the Internal RAM window just described but they correspond to off-chip data RAM and on-chip RAM, respectively. The Static RAM window will only be available as an option for those microcontrollers that feature.

15.6.1.9 SFR-Direct View Window

The SFR-Direct View window provides a single window that displays all 128 possible SFR locations. In some cases it may be easier to open this window to have a total view of all SFR values than to open the registers, timers, DPTRs, and ports window.

In this window, the registers with a green background are valid while those with a red background are not valid for the currently selected microcontroller.

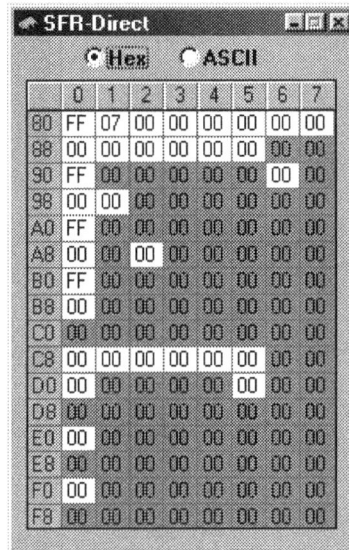

15.6.1.10 Bit Fields Window

The Bit Fields window provides the developer with a way to view and manipulate the 128 user bits that occupy internal RAM addresses 20 through 2Fh on a bit-by-bit basis.

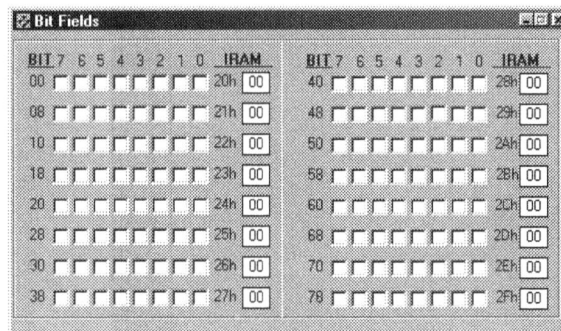

Changing the value of a bit in the window will change the corresponding bit at that bit's internal RAM address. For example, clicking the checkbox in the upper left-hand corner will set internal RAM address 20h to 80h since bit 7 of address 20h will then be set. Likewise, clicking on the checkbox in the lower right-hand corner will set internal RAM address 2Fh to 01h since bit 0 of address 2Fh will be set.

Any changes made to the bits in this window will affect internal RAM and, in turn, change the respective value in the Internal RAM window. Likewise, any change made in the Internal RAM window to addresses 20h through 2Fh will change the bits displayed in this window.

15.6.1.11 Code Memory (Disassembly) Window

The **Code Memory Window**, also known as the **Disassembly Window**, displays the current contents of code memory in disassembled form. A "disassembly" refers to the process of taking data from code memory and displaying it as the assembly language instructions it represents.

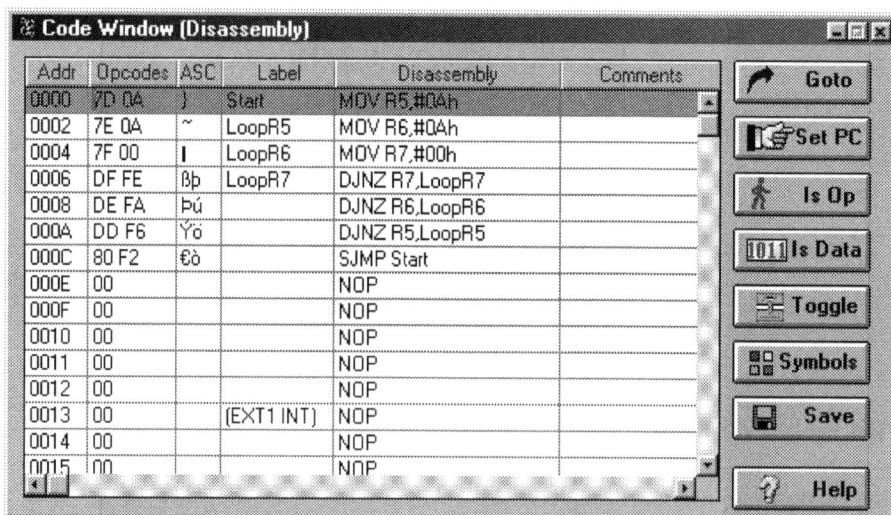

Since code memory contains the compiled program, which is nothing more than a sequence of bytes in program memory, it would not be very useful to simply display those bytes to the user. Instead, Pinnacle takes those bytes and displays them as assembly language instructions. In this manner the user has a better idea of what the bytes in code memory represent.

The screen shot above shows the example program from earlier in this chapter in disassembled form.

The **Addr** column indicates the address at which the corresponding instruction is found. For example, the third line indicates an address of 0004 so that instruction is found at code memory address 0004h.

The **Opcodes** column shows the actual bytes that make up the given instruction. This will consist of either one, two, or three hexadecimal values between 00 and FF. Since all the instructions used in the example program are 2-byte instructions, each instruction consists of two bytes and each address in the Addr column increases by two for each instruction. Note, however, that the NOP instructions that are found after the program only consist of a single byte since each NOP only requires a single 00 byte.

The **ASC** column displays the opcode bytes in ASCII form. When displaying the ASCII version of 8052 instructions this will usually appear to be unintelligible garbage as is the case above. However, when there is ASCII data, tables, or strings in code memory, these will be easily visible in the ASC column.

The **Label** column displays the label of the specific address in memory on that line. The label will only be available if the program was compiled from within Pinnacle. If the program being simulated was loaded from a HEX file created by another program, the label will always be blank since label information is not contained in a HEX file.

The **Disassembly** column displays the instruction that the bytes at this code memory location represent. This essentially shows the same information as is displayed in the simulator information toolbar that was described earlier, but shows the information for many addresses at once rather than just showing the current instruction.

The **Comments** column displays any comments that were on the same line as that instruction in the original source code. Like the label column, this column will only display comments if the program was compiled from within Pinnacle.

The line in the code memory window with a green background indicates the next instruction that will be executed by the simulator. Lines in code memory with a red background indicate instructions that have a breakpoint. Breakpoints will be covered in section 15.6.3.2.

The code memory window also includes a number of buttons on the right side of the screen.

Goto: This button will open a window that allows the user to jump to a specified address in code memory. This may be a faster way to view a specific address of code memory than scrolling through it with the scrollbar or page-up/page-down. This will only change the display within the disassembly window, not the simulator's program counter.

Set PC: Pressing this button will set the program counter to the address at which the cursor is currently located within the code memory window. When simulation is resumed, execution will continue at the address that was set.

Is Op / Is Data: Bytes in code memory can either represent 8052 instructions or user data or text strings. When a program is loaded, Pinnacle will make its best effort to determine whether each byte of code memory represents an instruction or user data. This determination may be further corrected when the program is executed. Nevertheless, it is not possible for Pinnacle to always make this determination correctly. When reviewing code memory, if the user discovers a line that has been incorrectly determined to be data, the **Is Op** button may be used to force the disassembler to treat it as an operation (instruction). Likewise, if a line has been incorrectly determined to be an instruction, the **Is Data** button may be used to force the disassembler to treat it as data.

Toggle: The toggle button will either set or unset a breakpoint at the address of code memory on which the cursor is currently located. If the cursor is on an address that doesn't have a breakpoint, one will be set. If the address has a breakpoint, it will be removed.

> Breakpoints are "stop points" that cause program simulation to stop as soon as they are reached. They are normally set just before points in code that need to be tested. The program is quickly simulated by running it (with the F5 key) and program simulation stops when the breakpoint is reached. The user then executes one instruction at a time using the F8 key. This allows the developer to quickly simulate code that is known to work and concentrate on reviewing untested code on an instruction-by-instruction basis.

Symbols: This button opens the Symbols window. The symbol window provides a single point at which all labels and symbols may be viewed and edited.

Save: The Save button allows the current disassembly to be saved to a file. This is normally only used when reverse engineering a program in a commercial environment and will generally be of no interest to others. This feature is only available in the professional, purchased version of Pinnacle 52.

15.6.1.12 UART (Terminal Emulation) Window

The **UART/Terminal Emulation window** provides a way for the developer to simulate the activity that will be found on the serial port without actually requiring a free serial port nor requiring any of the external devices that will normally be connected to the serial port.

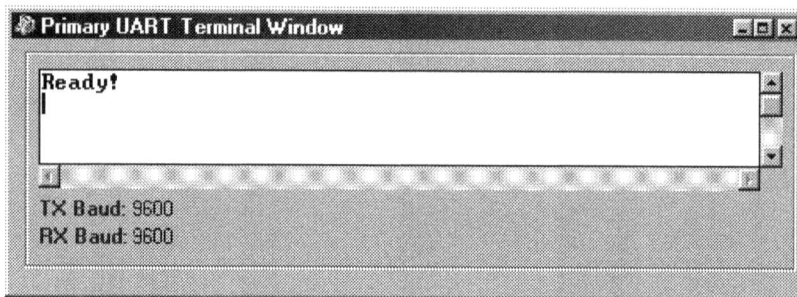

The UART window consists of a free-from field at the top where information can be either displayed or typed. Below the input/output field, the transmit baud rate and receive baud rate will be displayed. These will normally be the same value but can be different if timer 2 is being used as a baud rate source.

Any information that the 8052 program being simulated sends to the serial port by way of the SBUF SFR will be displayed in the I/O field. Likewise, anything the user types into the I/O field will be sent to the 8052 program as if it had been received by the UART—thus pressing the 'A' key in that window will set the Receive Interrupt flag and set SBUF to 0x41 which is the ASCII value for 'A'. Of course, this will only occur if the Receive Enable (REN) bit of SBUF is set.

15.6.2 Execute Menu

The "Execute" menu option provides a number of specific features useful for executing certain sections of code.

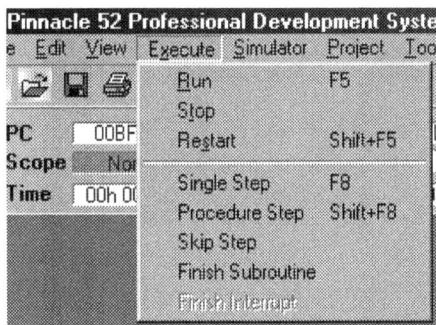

The **Run**, **Stop**, **Restart**, and **Single Step** options correspond to the corresponding functions on the toolbar that were already described in section 15.6.

The **Procedure Step** function, also accessible by pressing Shift+F8, will execute a single instruction just like Single Step. However, if that instruction is an LCALL or ACALL it will also execute all the instructions in that subroutine.

The **Skip Step** function, as the name implies, will skip the current instruction and set the program counter to the address of the next instruction. This is useful for skipping an instruction or a call to a subroutine that is known to cause problems or which may be time-consuming or unnecessary to simulate.

Finish Subroutine will execute the program as if the Run option had been selected but will automatically stop when it encounters the next RET instruction. This is useful to finish the currently running subroutine quickly.

The **Finish Interrupt** function functions identically to the Finish Subroutine function just described but will execute the program quickly until it encounters the next RETI instruction. This is useful to finish the currently running interrupt service routine quickly. This option will only be available if the simulator is currently simulating an interrupt.

15.6.3 Simulator Menu

The "Simulator" menu option provides additional control over the simulation environment.

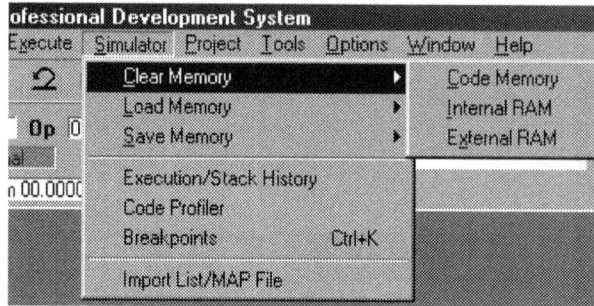

The **Clear Memory** option is useful for immediately zeroing out the entire contents of code memory, internal RAM, or external RAM. The option is invoked by moving the mouse pointer over the "Clear Memory" menu option and then selecting which memory area is to be cleared.

The **Load Memory** and **Save Memory** options are also activated by moving the cursor over the menu option and also result in an identical submenu being displayed. These options provide the developer with the capability of loading or saving the specified type of memory. This can be useful if memory is pre-loaded with specific data before each execution. It can also be useful for saving the state of memory, when a given situation presents itself, for subsequent analysis.

The **Import List/MAP File** is used when a compiler or assembler other than Pinnacle is being used but the Pinnacle simulator is being utilized. Loading the list/map file allows Pinnacle to show symbols in the code disassembly that would otherwise not be available. This option may be ignored if Pinnacle is being used to assemble programs.

15.6.3.1 Execution/Stack History

The **Execution/Stack History** window is a useful tool in determining the current state of the program in the simulation and what has recently been executed that resulted in the current simulator situation.

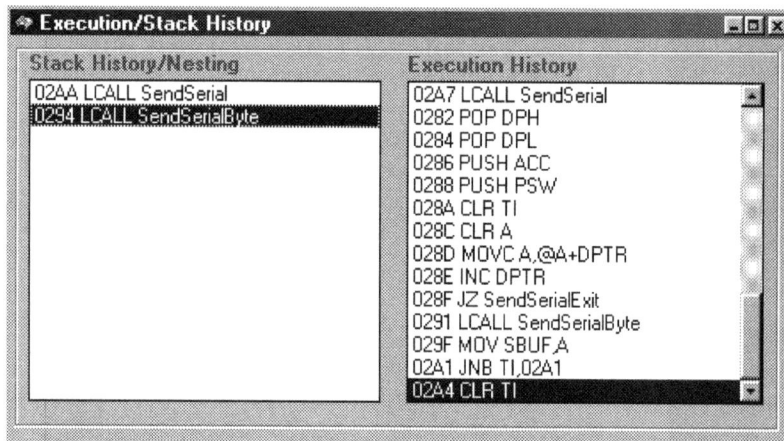

The **Stack History/Nesting** list shows the nested subroutine calls that are pending in the simulation and the address that the program will return to when the subroutine's RET is executed. In the above example the main program first called the `SendSerial` subroutine which, in turn, called the `SendSerialByte` routine. When `SendSerialByte` is done and returns with the RET instruction it will return to 0294. When `SendSerial` is complete, it will return to the main program at 02AA which is the instruction after the `LCALL SendSerial` that originally made the subroutine call. This feature is very useful when diagnosing stack problems. For example, a subroutine that never executed a RET will never be removed from the list.

Due to the way in which the Stack History/Nesting feature works, the last entry in the list is always the currently-executing subroutine. In other words, since the last subroutine in the list above is SendSerialByte, it is known that the simulator is currently executing the SendSerialByte subroutine. As soon as SendSerialByte finishes and returns with RET, this entry will be removed from the list.

The **Execution History** is a simple list of the most recently executed instructions. The top of the list is the oldest instruction while the last instruction in the list is the most recently executed instruction. In this example, the main program called the SendSerial subroutine from 02A7. The next instruction was executed at 0282 which indicates that the SendSerial is at address 0282 in code memory. Execution continued until the LCALL SendSerialByte at 0291 which caused execution to jump to 029F.

Often, when debugging a program in the simulator, one is left asking how the simulator got to the point it did. The execution history list provides an excellent way to review the sequence of execution that led to the current situation.

The execution history list only displays the 500 most-recently executed instructions.

15.6.3.2 Breakpoints

The **Breakpoints** option displays the breakpoints window. Breakpoints are addresses at which the simulator will automatically stop execution.

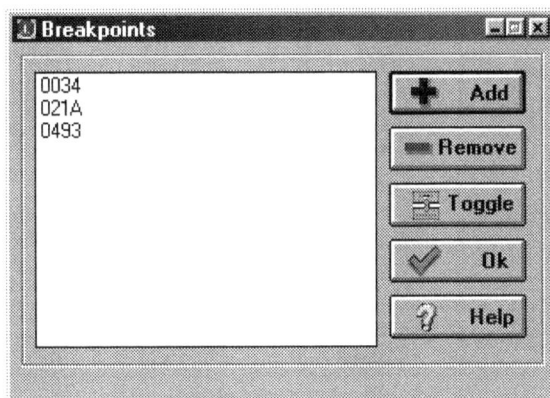

In this example there are breakpoints set at code memory addresses 0034, 021A, and 0493. That means when the simulator is executing the program it will automatically pause when it reaches any of these addresses. SFRs, memory, or any other aspect of the simulation may then be reviewed to verify that the program is operating as expected.

The **Add** button allows the user to add a new breakpoint. The address entered must be a hexadecimal code memory address. The **Remove** button removes the currently selected breakpoint. The **Toggle** button will temporarily disable a currently active breakpoint or re-enable a breakpoint that has previously been disabled. A disabled breakpoint will be appended with the notation "[DISABLED]". Using the toggle button is more convenient than repeatedly adding and removing a breakpoint entirely.

15.6.3.2 Code Profiler

The **Code Profiler** window is a powerful utility that can help the developer find bottlenecks and inefficient sections of code in the program being simulated. To use the code profiler, the program should first be simulated completely and let run as long as possible. The longer the simulation runs, the more accurate the analysis of the code profiler will be.

Once the program has been allowed to run as completely as possible in the simulator, the code profiler may be opened. The code profiler consists of five separate screens: **General Info**, **Instruction Usage**, **Interrupt Usage**, **CPU Utilization**, and **Subroutine Usage**. Each screen is selected in the code profiler window by pressing the radio button corresponding to the screen.

The **General Info** screen of the code profiler is shown below.

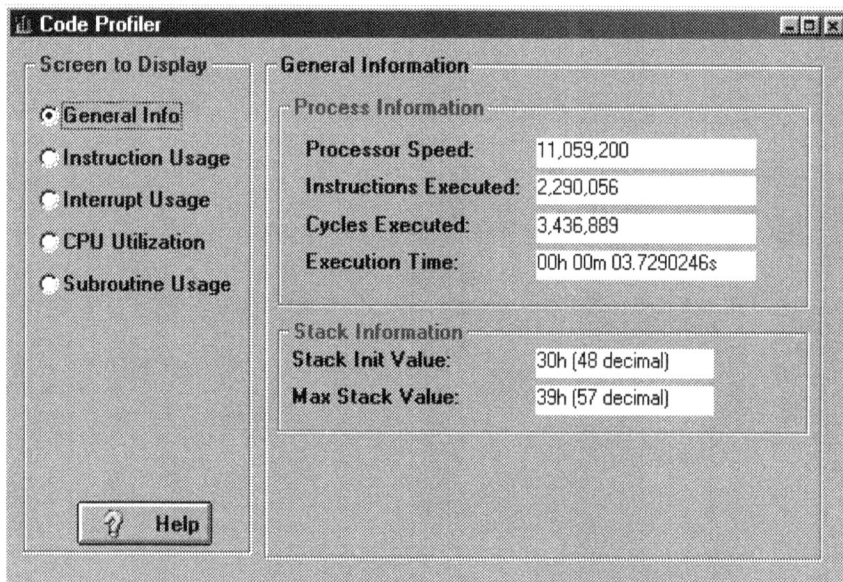

This screen provides general information about the simulation. The **Processor Speed** is indicated as 11,059,200 which corresponds to an 11.0592 MHz oscillator. The **Instructions Executed** indicates precisely how many instructions have been executed in the current simulation. The **Cycles Executed** is a better measure since each instruction may take a different amount of time to execute. **Execution Time** indicates how long the same code would have taken to execute on a real processor.

The **Stack Init Value** indicates the value that the stack began at. This is usually one greater than the value that SP is first set to by the program or will be 08h if the stack pointer was not initialized. Finally, the **Max Stack Value** indicates that the highest the stack grew to during the simulation. This information can be used to estimate whether there is enough room in memory for the stack operations in which the program is engaging.

The **Instruction Usage** screen provides information about the instructions that the simulator has executed.

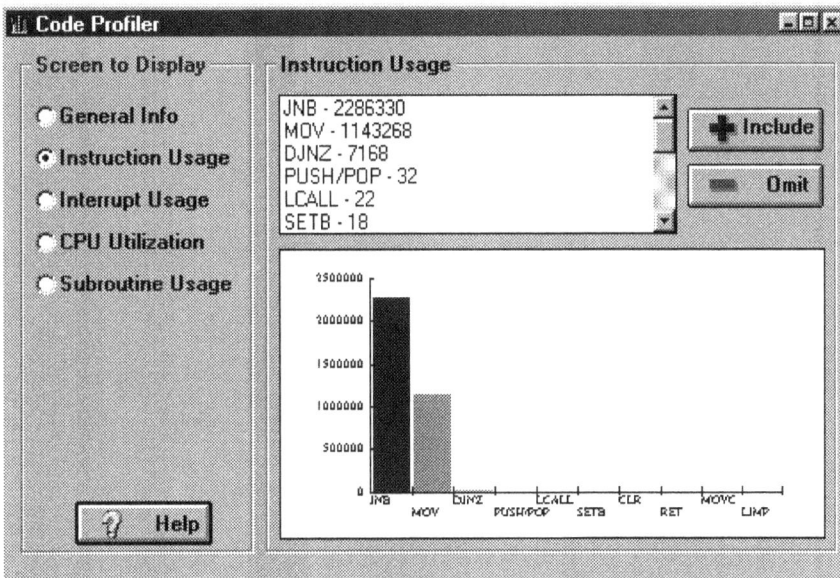

The listbox at the top of the screen lists all the instructions that have been executed in the simulator and how many times each one was executed. In this example, the JNB instruction has been executed 2,286,330 times while the MOV instruction has been executed 1,143,268 times. The graph below provides a visual representation of the number of times each of the most-used instructions has been executed. The **Omit** button will omit the selected instruction from the graph while **Include** will include an instruction that has previously been omitted.

This feature can be useful in determining what instructions are consuming the most time in a program. For example, if it were found that the slow MUL instruction were being executed a significant number of times, it might be useful to review whether or not a more efficient approach to the problem exists.

The **Interrupt Usage** screen provides information about the interrupts that have been executed.

In the example on the next page, the microcontroller was in "normal" (non-interrupt) mode for 506,098 instruction cycles, it spent 200 instruction cycles executing the serial port interrupt, and spent 70 cycles executing the timer 0 interrupt. The graph only shows the serial port and timer 0 statistics because the "Normal" mode was omitted by using the **Omit** button.

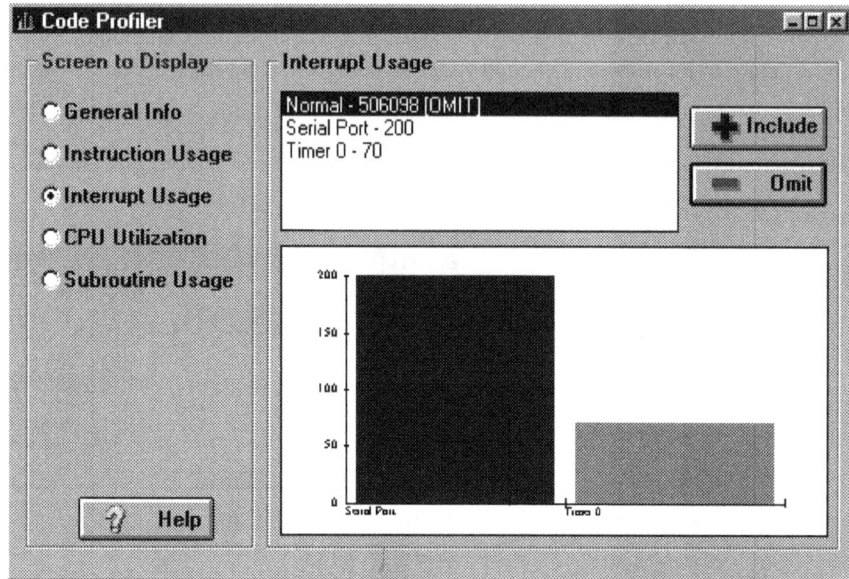

This feature provides the developer with an excellent method of determining just how much time the interrupt system is taking from the system. Since interrupts should be as short and as fast as possible this type of information can quickly draw attention to bloated or inefficient interrupt routines.

The **CPU Utilization** screen provides even more information about the nature of the program that is being simulated.

In this example, most of the microcontroller's time is spent doing nothing. 5.65% of the time is spent executing NOP instructions, 58.547% is spent in idle cycles, and only 35.803% of the processor's time is spent doing any significant work or calculation. An idle cycle refers to any instruction which loops back to itself. For example a `JNB TI, $` loop that waits in a loop until a character is sent successfully on the serial port would count towards idle cycles since it is just repeating the same instruction over and over. While there are always exceptions, it is generally a good idea to keep CPU utilization as high as possible and spend as little time as possible on idle cycles. This is especially true if the developer is attempting to optimize a program to run as quickly and efficiently as possible.

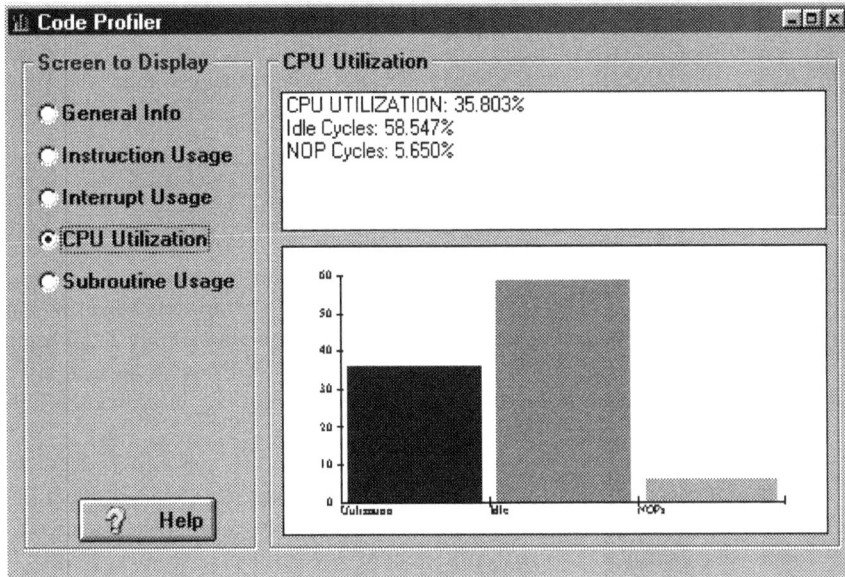

The last screen of the code profiler is **Subroutine Usage** which provides a large amount of information about each subroutine that was called in the program and which is shown on the next page.

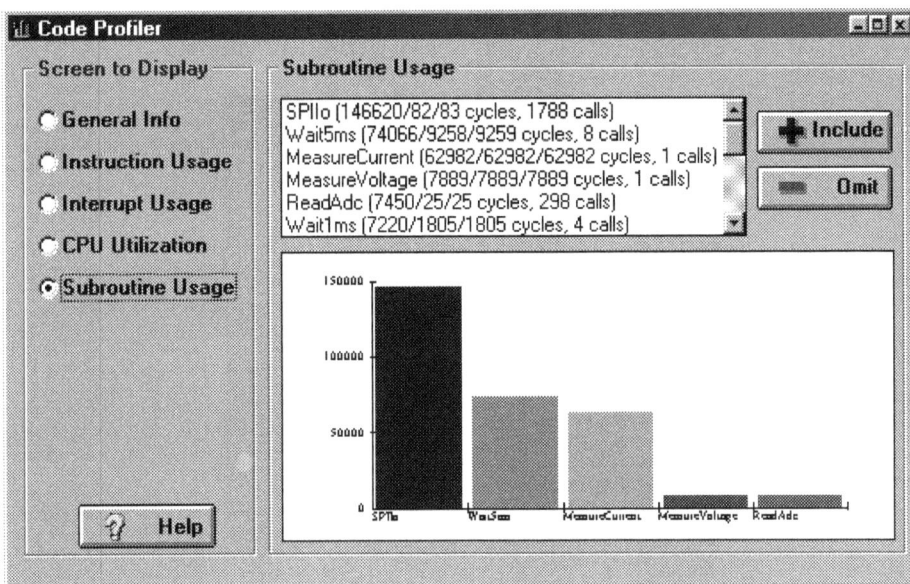

In this example the SPIIo subroutine was called a total of 1788 times. The first number in the parenthesis (146,620) indicates the total number of instruction cycles spent in the subroutine. The second number indicates the minimum number of instruction cycles required to execute one instance of the subroutine while the third number indicates the maximum number of instruction cycles to execute the subroutine.

💡 The instruction cycles displayed for each subroutine reflect only the time spent in that specific subroutine. For example, if the program calls the SPIIo subroutine which in turn calls the Delay subroutine, the time spent in Delay will be displayed under the Delay entry. It will not be displayed as part of the time spent in the SPIIo subroutine even though the Delay routine was called by SPIIo.

15.7 Concept of Projects in Pinnacle 52

Up until now, this chapter has described the concept of writing a single assembly language file in the editor, saving it, assembling it, and simulating it. While this is often a completely adequate approach it is often desirable to have multiple source files combined into a single output file by the assembler. This allows for the separation of conceptual parts of a program into various component files. It also allows the creation of code libraries which can be included in future projects simply by including the library source file.

The ability to take multiple source files, assemble them, and automatically combine them into a single output file is known in Pinnacle as "projects." A project consists of a list of component source files as well as a grouping of additional parameters that may be different for each project.

Creating a project is slightly more complicated than working with a single source file. If the program is just a quick test, or if the program is relatively small, then usually it isn't worthwhile to go through the extra steps of creating a project. If, however, the program is larger or the extra features of a project are desired, then it may be necessary to create a project.

All aspects of a project are managed from the Projects menu which is shown below.

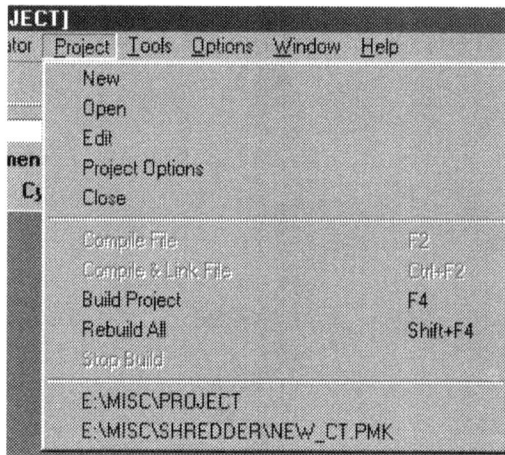

15.7.1 Creating Projects

Projects are created by selecting the **New** option of the project menu. This will open the following window.

Every project has a "project file"—normally with a PMK file extension—that is used to store the information and configuration related to the project. When creating a new project, a new filename must be selected. The filename (and directory) may either be typed directly into the **Project Filename** field or may be selected by clicking on the filename selector button immediately to the right of the field.

The **Name/Description** is an optional field that may be used to attach a name, description, or comment to the project for future reference.

Once the project filename and, optionally, the name/description have been provided, the user clicks on **Ok** to create the project. At this point the project is empty but is ready to be configured as will be described in the next section.

15.7.2 Editing Projects

Projects are, primarily, lists of source files that together constitute a single project. The source files that are to be part of the project must be specifically added using the **Edit** option in the Projects menu. Selecting the Edit option will open the **Edit Project** window.

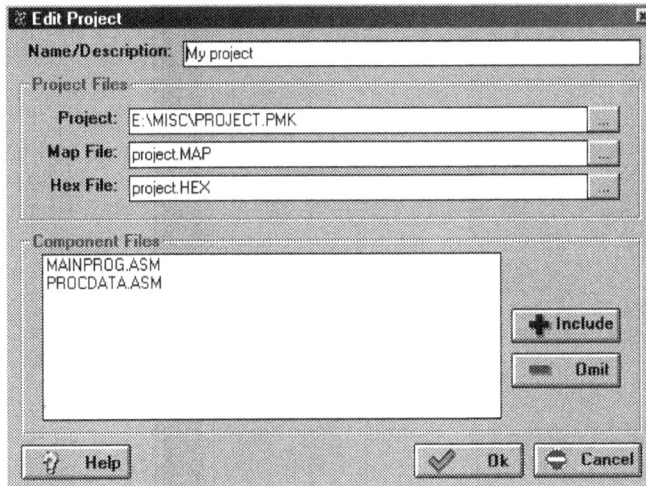

The **Name/Description** field at the top of the window is the same as was found in the **New Project** window. This field is optional but may be used to enter a brief description of the project.

The **Project** field indicates the filename of the currently loaded project.

The **Map File** is created whenever a project is built. That is to say, when the project is assembled a map file is created for that particular assembly. The map file contains general address information regarding symbols and code and can be used to help determine how and where certain symbols are located in memory. By default, the map file will have the same name as the project but with a MAP extension. If no directory is specified for the map file then it will be created in the same directory as the project file.

The **HEX File** is the output of any build. This is the file that is created and which can either be simulated in Pinnacle or downloaded directly to a microcontroller or EPROM. By default, the hex file will have the same name as the project but with a HEX extension. If no directory is specified for the hex file then it will be created in the same directory as the project file.

The bottom section of this window is where **Component Files** may be added to the project. When a project is first created there will be no component files. The **Include** button should be used to add assembly language source files to the project. Once they are selected they will appear in the list of component files and will automatically be assembled and integrated into the HEX file every time the project is built. The **Omit** button may be used to remove unneeded files from the project's component list.

It is important to note that not just any assembly language source program may be added to a project. Since various source files are pooled together to create a single output file there are some restrictions and additional concepts that must be taken into account when adding files to a project so that each source file doesn't try to locate itself at the same code memory address. These concepts will be discussed in the **Relocatable Code** section later in this chapter (see section 15.8).

15.7.3 Building Projects

Building a project refers to the process of having Pinnacle assemble all component files and link them together into a HEX file. This process is extremely easy but doesn't use the same menu options as were used to assemble and link a single assembly language program earlier in this chapter.

Building a project is accomplished by opening the Projects menu and selecting the **Build Project** option or by pressing the F4 key anywhere within Pinnacle. This will assemble all the files that have changed since the last build, link them together, and produce a new HEX file. Only the source files that have changed will be re-assembled. This is useful in large projects since there is no reason for Pinnacle to assemble all the source files if only one has changed. If, for some reason, it is desirable to have Pinnacle re-assemble even the source files which haven't changed, the **Rebuild All** option may be used. This option re-assembles all files regardless of whether or not they've changed.

When working with projects be very careful to build them with the **Build** command and not the **Compile & Link File** command. The Build command will build the entire project while the Compile & Link File command will only compile and link the currently selected editor file.

15.7.4 Project Options

Pinnacle 52 provides many options to control and configure the manner in which projects are built from within the **Project Options** window. Additionally, many of the general configuration parameters for Pinnacle 52 are found within the Project Options window even if the developer is only working with a single assembly language file rather than a complete project.

The Projects Options window is accessed by selecting the Projects menu and then selecting **Project Options** which will open the Projects Options window. Within the window, individual configuration categories are selected by clicking on the category name in the **Configuration Category** listbox on the left side of the window. When the window first opens the **Microcontroller Options** window will be selected.

15.7.4.1 Microcontroller Options

The **Microcontroller Options** category includes parameters that select and configure which microcontroller is being used by the developer.

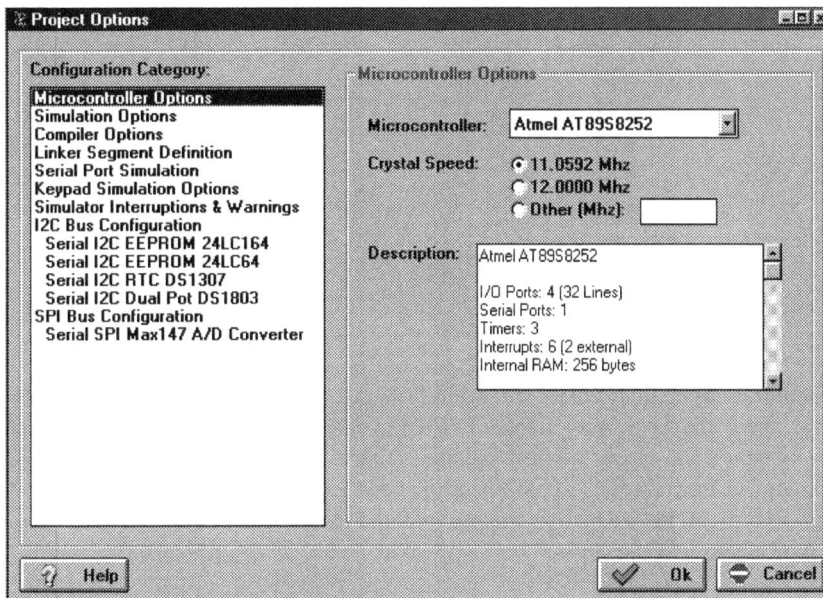

The **Microcontroller** listbox allows the user to select which microcontroller will be used by Pinnacle. Selecting the right microcontroller allows Pinnacle to better simulate the controller in question and also enables the SFRs that may be specific to that particular derivative.

The **Crystal Speed** option configures the speed at which the program will be simulated. This effects the simulator's baud rate calculation and also the calculation of total simulation time (hours, minutes, seconds) that is displayed in the information bar. Two common crystal frequencies are provided and may be selected by choosing one of these options, or by entering a custom crystal speed into the **Other** field. The number entered should be entered in megahertz.

The **Description** field provides general information about the microcontroller which has been selected, such as its most notable features.

15.7.4.2 Simulation Options

The **Simulation Options** category includes parameters relating to the simulator's performance and operational preferences.

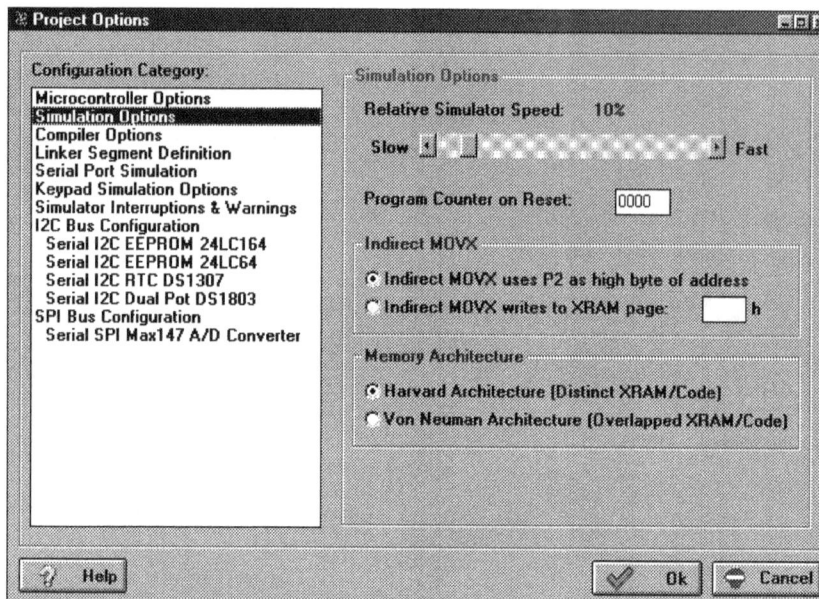

The Relative **Simulator Speed** is used to configure how fast the simulator should run. The higher the relative simulator speed the faster the simulator will run but the slower it will be to respond to user interaction such as typing and mouse clicks. A lower relative simulator speed will cause simulations to execute more slowly but Pinnacle will be more responsive to user input. Unless the program is very large and large number of instructions are being simulated without user interaction, a setting of between 5% and 10% is usually appropriate.

The **Program Counter on Reset** indicates where the simulator should start executing program code when the simulation is reset. This will almost always be 0000h. However, some evaluation tools from third parties create program code at other addresses so this option can be used to start executing code wherever it may be found in memory.

The **Indirect MOVX** setting is used to configure what external memory address is accessed when the MOVX @Rx instruction is simulated. Since the MOVX @Rx instruction only outputs the low byte of an address the simulator must know what it should use for the high byte of the address. This varies depending on how the real circuit is designed. If **Indirect MOVX uses P2 as High Byte of Address** is selected, then whenever the MOVX @Rx instruction is executed the current value of P2 will be the high byte of the data memory address. If the **Indirect MOVX writes to XRAM page** number is selected then the hexadecimal value entered in that field will be used as the high byte of the data memory address.

The **Memory Architecture** setting allows the user to select whether the current simulation is executing in a Harvard or Von Neumann memory configuration.

15.7.4.3 Compiler Options

The **Compiler Options** category, show on the previous page, allows the user to configure parameters which impact Pinnacle's compiler/assembler.

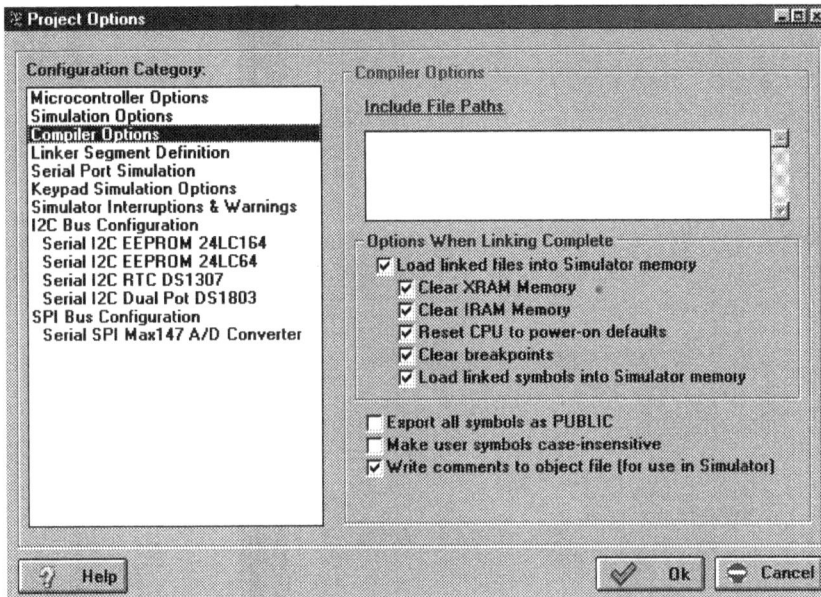

The **Include File Paths** lists the directories that Pinnacle will look in when trying to find a file that has been included with the $INCLUDE directive. By default, this field is blank and Pinnacle will only look for include files in the same directory of the file that included it.

Options When Linking Complete controls what Pinnacle will and won't do automatically after a successful compile and link. If the **Load Linked Files into Simulator Memory** option is selected then whenever a link is completed successfully the resulting file will be immediately loaded into the simulator's memory. The remaining options refer to additional actions that Pinnacle can take automatically after the linked file is loaded into the simulator's memory. Normally all of these options should be enabled.

The **Export All Symbols as PUBLIC** is a feature that will export all symbols in all files as public symbols as if each symbol had been defined with the PUBLIC directive. This option should be selected if it is desirable to have all symbols appear in the code disassembly window of the simulator. Only public symbols are visible in the code disassembly window. See section 15.9 for more information on public symbols.

Make User Symbols Case-Insensitive adjusts the assembler's configuration so that a symbol written in lower-case is identical to a symbol written in upper-case. If this option is not enabled then the symbol MYSYMBOL will be considered different than MySymbol.

The **Write Comments to Object File** is a feature that allows the comments from the source program to appear in the comments column of the disassembly window. Generally this option should be enabled. The only reason to disable this option is if the resulting object file will be distributed to third parties and confidentiality is such that disassembly is specifically to be discouraged.

15.7.4.4 Linker Segment Definitions

The **Linker Segment Definitions** category allows the user to configure each of the segments that will be used in the program. This is related to the concept of relocatable code which will be discussed in section 15.8. For now this section may not make too much sense; it may be a good idea to skip this section until the topic of relocatable segments is covered.

The **Code Segments**, **XRAM Segments**, **IRAM Segments**, and **BIT** Segments define the various types of memory in which segments may be created. The format of each of these fields is identical with the only difference being what type of memory they relate to.

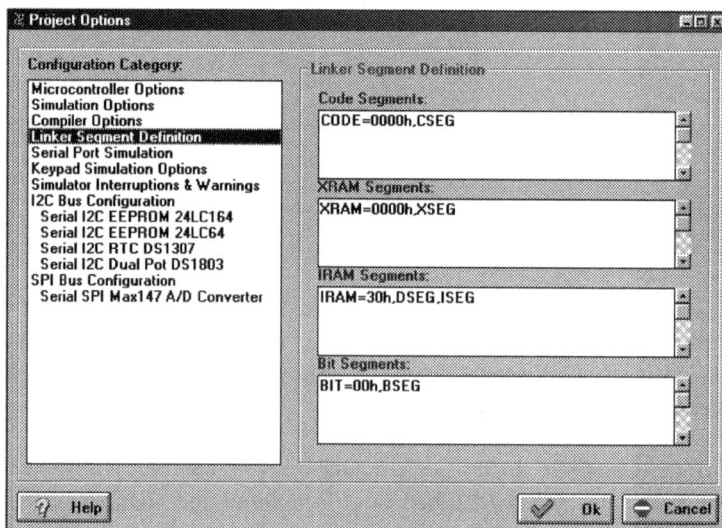

The first entry in any of these fields is always a segment name followed by an equals sign followed by an address. This indicates where that named segment should begin in memory. In the example above the CODE segment is declared to begin at 0000h which is consistent with the beginning of code memory. The next entry is simply CSEG. Since no specific address is provided any code placed in the CSEG segment will be located in code memory immediately following the last byte of the CODE segment.

Entries in each field may be separated by commas or may be placed in the field, one per line. If a segment is included in the field and is not assigned a specific address, it will be located in memory immediately following the previous segment.

178

15.7.4.5 Serial Port Simulation

The **Serial Port Simulation** category, show on the next page, allows the user to configure how the serial port (UART) will be simulated within Pinnacle.

The **Simulate Serial Transmit Delay** option configures whether or not the TI bit will be set immediately when a character is written to SBUF or delayed to simulate the amount of time it would normally take to send the character. As mentioned in the Serial Communications chapter (chapter 9), the sending of serial data is not instantaneous. It takes a certain amount of time to fully transmit the individual bits of serial data that are sent for each byte. If the Simulate Serial Transmit Delay option is selected then the simulator will not set the TI byte until an amount of time has passed that reflects how long it would actually take to transmit the character. This allows a more accurate simulation of serial communications. If this option is not selected then the TI byte will be set by the simulator immediately upon a value being written to SBUF. This reflects a less accurate simulation but the simulation will run faster.

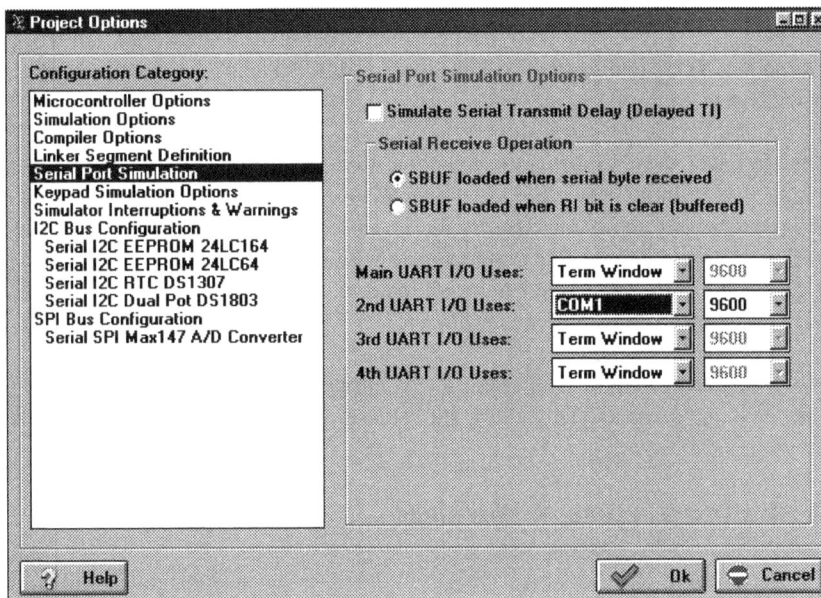

Normally it is acceptable to leave the Simulate Serial Transmit Delay option disabled in the interest of a faster, more agile simulation. This option only needs to be enabled if the program being simulated is very time-critical. For example, if the program was counting on the serial transmit delay to do other calculations while it waited then it may be necessary to simulate this delay by enabling the option.

The **Serial Receive Operation** is similar to the previous option in that it determines how accurately the reception of serial data will be simulated by Pinnacle. If **SBUF Loaded when Serial Byte Received** is selected then SBUF will be updated immediately when a byte is received. If a previously received byte has not been read by the simulated program by the time the next byte is received, the original byte will be lost.

If **SBUF Loaded when RI is clear** then the serial data received by Pinnacle will only be placed in SBUF once the RI bit has been cleared. This means that if the serial data is being received faster than Pinnacle can simulate the program, no data will be lost since incoming serial data is only passed to the simulated program when the simulated program clears the RI bit.

>   Normally it is acceptable to select **SBUF Loaded when RI is clear**. It is very possible
>   that a program that would normally work on a real microcontroller will not be simulated
>   fast enough by Pinnacle to keep up with serial input. In this case the only way to
>   simulate the program is to select this option. The **SBUF Loaded when Serial Byte
>   Received** may be worthwhile if the computer on which Pinnacle is running is so fast that
>   the program can be simulated at or near real-time speed.

The **Main UART I/O Uses** (and the similar settings for the 2nd, 3rd, and 4th UARTs) configures from where Pinnacle will receive and transmit simulated serial data. If **Term Window** is selected, as in the Main UART I/O above, then input and output will be provided by the terminal window (see section 15.6.1.12). Anything the user types into the terminal window will be sent to the program being simulated as if it had been received via the serial port. Likewise, anything the program sends to the serial port will be displayed in the terminal window. This provides a very convenient way to simulate serial communications without actually using a serial port.

The other option is to select a serial port on the computer as is demonstrated with the 2nd **UART I/O Uses** field in the example above. In this case, any serial data received on the selected serial port will be sent to the simulated program and any serial data sent by the simulated program will be sent via that serial port. This is useful for simulating programs that communicate with external devices via the serial port. These devices can be connected to the PC's serial port and the simulated program may communicate with the device as if it were executing on a real microcontroller.

>   When a communications port is selected, the baud rate must also be selected. The PC's
>   baud rate will be determined by the baud rate selected, not by the simulated 8052
>   program.

15.7.4.6 Simulator Interruptions & Warnings

The **Simulator Interruptions & Warnings** category, show on the next page, provides configuration options which control when the simulator will generate warning messages or suspend simulation.

The **NOP Behavior** provides a mechanism by which the user may request that the simulator automatically stop executing the program if more than a certain number of consecutive NOP instructions are executed. While it is common for a program to execute a few consecutive NOPs for timing purposes, the execution of a large number of NOPs is often an indicator that the program that has jumped outside of the program code and is executing "nothing" in the rest of code memory. The **NOP Behavior** can be configured to automatically cause the simulator to stop simulating the program if an excessive number of NOPs have been executed.

The **Undefined Instruction Behavior** allows the user to configure the behavior of the simulator when it attempts to simulate the execution of the undefined instruction. Since the A5h opcode is not used by the

8052 instruction set, its behavior is undefined—so Pinnacle allows the user to define what action to take if the undefined instruction is executed. If **Simulate Undefined Instruction** is selected then Pinnacle will attempt to simulate the instruction and proceed normally. If **Break and Report Error** is selected then the simulator will immediately stop and bring up an error window letting the developer know that the undefined instruction was executed. If **Dump Current Simulation Information to UART0** is set then whenever the undefined instruction is executed, a number of parameters will be dumped to UART0. This can be useful in performing diagnostic timing in code by placing an A5h instruction at the beginning of the code and another A5h instruction at the end of the code and comparing the diagnostic information of each.

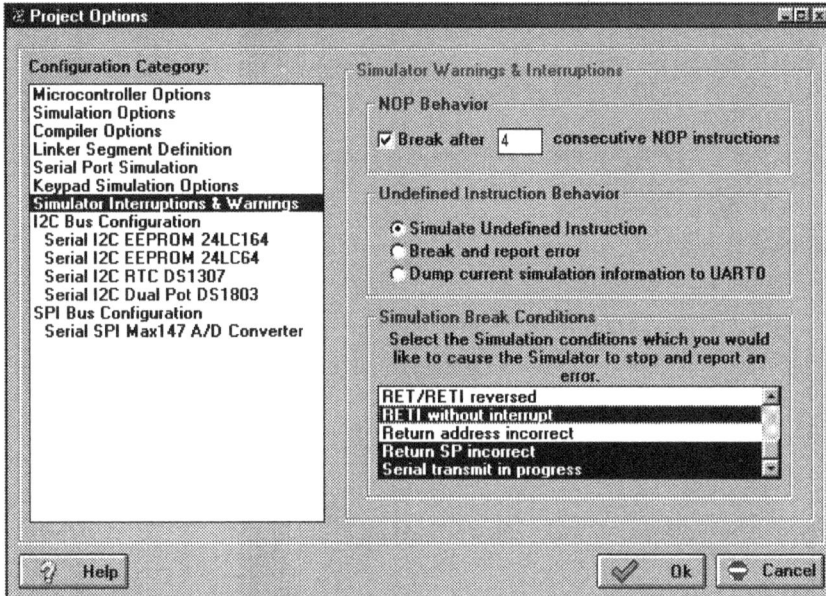

The **Simulation Break Condition** allows the user to select which events will cause the simulator to stop executing and bring up an error message alerting the developer to the problem. In the example above the **RETI Without Interrupt** option is selected which means an error window will be provoked if the simulated program attempts to execute a RETI without being in an interrupt. Many other break options are available and the developer may select which of these situations should result in an error message and which should be ignored with no warning to the developer.

15.7.4.7 Other Project Configuration Options

In addition the configuration categories just discussed, the Project Options window includes a number of other configuration categories such as **Keypad Simulation Options**, **I2C Bus Configuration**, and **SPI Bus Configuration**. These categories configure specific aspects of the simulation which allow the simulation of communication with external devices. These options are beyond the scope of this introduction to Pinnacle 52.

15.8 Relocatable Code in Pinnacle 52

Relocatable code refers to code whose address in code memory is not specified in the assembly language program but rather is determined by the linker when the project is built. Relocatable code is a very important concept if projects consisting of multiple assembly language files are to be used in Pinnacle 52.

15.8.1 The Problem of Absolute Code in Multiple Source Files

First recall how the address of assembly language instructions in code memory has been established previously. As mentioned earlier in section 11.6.1, the ORG directive is used to specify the address at which program assembly is to continue. For example the directive ORG 0000h causes the instructions that follow to be located beginning at code memory 0000h.

A problem arises when combining multiple assembly language source files. Consider the following two files:

FILE1.ASM
```
        ORG 0000h
        LJMP MAIN
```

FILE2.ASM
```
        ORG 0000h
        MAIN: MOV A,#10h
        MOV R7,#10h
```

A problem exists since both FILE1 and FILE2 include an ORG 0000h directive which means both programs will try to locate themselves at the same code memory address.

One option would be to change the ORG in FILE2 to 0003h. Since FILE1 consists only of an LJMP instruction of 3 bytes it generates code for addresses 0000h, 0001h, and 0002h. By adjusting FILE2's ORG to 0003h there would no longer be a conflict. But what would happen if additional instructions were added to FILE1 after the LJMP? FILE1's code would then grow longer and once again FILE2's code memory address would conflict with the code in FILE1. The only solution would be to adjust FILE2's ORG address every time FILE1 grows. But if there were dozens of source files each of which began where the previous file ended each of these files would have to be modified every time the length of code in another file grew. Clearly this is an inefficient way to handle multiple files.

Another option would be to modify FILE2's address to 0050h which would allow FILE1 to grow without having to modify FILE2. The problem with this approach is that it would waste the 77 bytes of code memory from 00003 through 004Fh. Further, if more code than expected is added later to FILE1 it could end up consuming all 77 bytes of memory and, once again, the issue of changing FILE2's address would come up.

The solution to this problem is to let Pinnacle automatically determine the best address for the code in each file. That is what relocatable code is: code that can be located anywhere in memory that the linker sees fit.

15.8.2 Relocatable Segments

The entire basis of relocatable code revolves around the concept of **segments**. A segment is a contiguous section of memory in which code can be placed.

Code that is written to be relocatable does not use the ORG directive to tell the assembler at what code memory address to place the code. Instead, it uses the **RSEG** directive to tell the linker to which segment the code belongs. When the project is built, the linker will then automatically place the code at the next available memory address within the specified segment.

Modifying the two source files above to be relocatable would result in the following:

FILE1.ASM
```
     RSEG CODE
     EXTERN MAIN
     LJMP MAIN
```

FILE2.ASM
```
     RSEG CSEG
     PUBLIC MAIN
MAIN: MOV A,#10h
     MOV R7,#10h
```

The ORG directives have been replaced with RSEG directives. The code in FILE1 will be placed in the CODE segment while the code in FILE2 will be placed in the CSEG segment.

When a new project is first created Pinnacle automatically creates the entry CODE=0000h, CSEG in the Linker Segment Definition (see section 15.7.4.4) window as code segments. This means that the CODE segment begins at 0000h and the CSEG segment begins immediately following the end of the CODE segment.

When a project is built with the two relocatable source files above, the contents of FILE1 will be located at code memory 0000h through 0002h since the CODE segment is defined in the linker segment definition as starting at 0000h. Since the contents of FILE2 belong to the CSEG segment and CSEG is defined in the linker segment definition to immediately follow the CODE segment, the contents of FILE2 will automatically be located immediately following the end of the CODE segment. Since the CODE segment ended at 0002h, the contents of FILE2 will end up being located at 0003h which is the next available code memory address.

The benefit of relocatable segments is that if additional code is added to FILE1 then rebuilding the project will automatically relocate the contents of FILE2 accordingly.

Multiple source files may place code in the same segment. For example, the following file could be added to the above project with absolutely no problem:

FILE3.ASM

```
          RSEG CSEG
          PUBLIC MYADD
          MYADD: ADDC A,#10
          RETI
```

FILE3 includes a subroutine called `MYADD` which is located in the `CSEG` segment which is the same segment to which the contents of FILE2 also belong. Even though both FILE2 and FILE3 use the same segment, no conflict exists. The code in FILE3 will simply be located in code memory in the `CSEG` segment after the contents of FILE2.

When multiple source files place code in the same segment there is no way to predict which will actually be located in memory first. In the example above, the MAIN routine of FILE2 could come before the MYADD routine of FILE3 or it could come after. There is no way to control the order of code within a segment. If the developer must be certain of the order or location at which code appears in memory then a different segment name must be assigned to each section of code. This is generally only necessary for the initial jump vector (0000h) and the interrupt service routine vectors as described in the next section. The rest of the program normally can be located anywhere in memory and is a perfect opportunity to use relocatable segments.

15.8.3 Initialization and Interrupt Vectors with Relocatable Code

Since relocatable code may be located anywhere in memory that the linker sees fit, the issue of the initialization vector (0000h) and the interrupt service routine vectors (0003h, 000Bh, etc.) becomes important. If relocatable code can be placed anywhere by Pinnacle, how can the developer make sure that the code for timer 0 interrupt is always located at address 0000Bh?

The answer is surprisingly simple. All that is necessary is to create a segment for each location where code must be specifically located. For example, consider the following non-relocatable code:

```
          ORG 0000h
          LJMP InitProgram
          ORG 000Bh
          LJMP Timer0Interrupt
          ORG 0050h
InitProgram:
          ... main program goes here ....
Timer0Interrupt:
          ... timer 0 interrupt code goes here ...
```

In this example the `LJMP InitProgram` must be located at 0000h since that's the address the microcontroller will start executing when it resets. Likewise `LJMP Timer0Interrupt` must be located at 000Bh since the microcontroller will always execute the interrupt service routine at 000Bh whenever the timer 0 interrupt is triggered. The actual locations of `InitProgram` and `Timer0Interrupt` in memory are not important; they could be anywhere and would work just fine. So with relocatable segments, Pinnacle must be configured so that the `LJMP InitProgram` instruction is always at 0000h, the `LJMP Timer0Interrupt` instruction is always at 000Bh, and the rest of the code can be put anywhere the linker chooses to locate it.

This is accomplished by defining a relocatable segment for the first LJMP that is specifically at 0000h, defining a relocatable segment for the second LJMP that is specifically at 000Bh, and having Pinnacle locate the rest of the program wherever it can after the second LJMP. This requires altering the configuration of Linker Segment Definitions (see section 15.7.4.4) in Project Options such that the code segments field is defined as INITCODE=0000h,T0ISR=000Bh,CSEG. This means that code in the INITCODE segment will be placed starting at 0000h, code in the T0ISR segment will be placed starting at 000Bh, and code in the CSEG segment will be placed in memory immediately following the T0ISR segment. The program above would then be modified to remove the ORGs and replace them with respective RSEGs:

```
                RSEG INITCODE
                LJMP InitProgram

                RSEG T0ISR
                LJMP Timer0Interrupt

                RSEG CSEG
    InitProgram:
                ... main program goes here ....
    Timer0Interrupt:
                ... timer 0 interrupt code goes here ...
```

The above example will result in LJMP InitProgram being located at 0000h, LJMP Timer0Interrupt being located at 000Bh, and InitProgram beginning at 000Eh which is the first available address of code memory following the T0ISR segment.

As the code above suggests, a single source file may contain multiple relocatable segments. In the example above, one source file contains code in the INITCODE, T0ISR, and CSEG segments.

15.8.4 When to Use Relocatable Segments

When to use relocatable segments and when to use absolute segments is worth mentioning. There is no right or wrong answer. Relocatable segments can be used always or they can be used never. It is mostly a matter of programmer preference as well as design goals.

If a very simple program is being written for which there is no possibility that it will later grow then it often makes sense to just create a single assembly language source file in the editor, specify the absolute address with ORG directives, and use the **Compile & Link** file option to quickly generate an output file. This avoids the hassle of creating a project, defining segments, and creating multiple source files. If the purpose of the program is just to test a small segment of code quickly then it may not make sense to spend any time creating a project and using the right segments.

On the other hand, once the concepts of relocatable segments and projects are mastered it becomes extremely easy and fast to create projects and the use of relocatable segments becomes second nature. Often even a small program that wasn't intended to be anything more than a quick test can evolve to become a much larger application. By using relocatable segments from the beginning, it won't be necessary to go back and convert absolute code into relocatable segments when the program grows too large.

An additional benefit of relocatable segments is the possibility of code modularity. It is possible to create a library of source code in which each file contains one subroutine. When that subroutine is needed in the future, one must only add the file to the new project and rebuild. This encourages the use of "objects" of self-contained code. Once the code has been tested it can be used and reused in subsequent projects.

15.9 Multiple Source Files in Pinnacle 52

The use of multiple source files requires the understanding of **public** and **external** symbols. A public symbol is a symbol contained in a source file which can be shared and accessed by the code in other source files. An external symbol is a symbol from another source file that has been publicly shared and which is to be utilized in the current source file.

When Pinnacle 52 assembles a source file it must resolve every symbol that that file contains. But what happens if one file calls a subroutine that is contained in another file? When the first file is assembled Pinnacle will report an error that it was unable to resolve the name of that subroutine since, within that file, the subroutine doesn't exist.

This problem is resolved by using the **EXTERN** directive to tell Pinnacle that the specified symbol is defined in another source file. Likewise, the other source file that contains the symbol must specifically tell Pinnacle that that symbol is to be made available to other modules by using the **PUBLIC** directive.

In the example program in 15.8, FILE1 included the instruction EXTERN MAIN which told Pinnacle that the MAIN subroutine would be found in another source file. FILE2 contained the instruction PUBLIC MAIN which specifically told Pinnacle to share the MAIN symbol with other modules. If either of these instructions had been omitted, Pinnacle would have reported "unable to resolve external" when it attempted to build the project.

The use of EXTERN and PUBLIC is necessary whenever multiple source files are used in a project regardless of whether the segments involved are absolute or relocatable.

186

CHAPTER 16: HARDWARE DEVELOPMENT TOOLS

Once an 8052 application has been built with a tool such as Pinnacle 52 (see chapter 15), the next step is to move the compiled program to some external hardware. This normally involves the use of one of three different hardware solutions.

1. **Device Programmer.** If the microcontroller that is being used includes flash program memory then the compiled program can be downloaded directly to the microcontroller. Otherwise, the program is normally downloaded into an EPROM which is read by the microcontroller in the circuit at run-time. In either case, a device known as a **device programmer** is used to download the program from the computer to the integrated circuit. The IC is then placed within the circuit itself where the program is executed by the microcontroller. This is the manner in which 8052 microcontrollers have traditionally been programmed.

2. **In-System Programming.** Many modern microcontrollers, such as the Atmel AT89S8252 and Dallas DS89C420 used in the 8052.com SBC presented in chapter 14, include flash memory that can be programmed "in-system." This means that the program contained within the microcontroller's flash memory can be updated without having to remove the IC from the circuit and without the need for a device programmer as long as the circuit supports **in-system programming (ISP)**. In these cases the circuit must be designed to provide a connector to which a cable from the computer may be connected. The 8052 program is then downloaded through the cable directly to the microcontroller. This approach is quickly becoming the standard with modern microcontrollers and is gradually making EPROMs and device programmers less necessary than they were before.

3. **In-Circuit Emulator.** Another option that is useful during the debugging and testing phase is an **in-circuit emulator (ICE)**. An ICE is a device which plugs into the socket normally occupied by the microcontroller. The entire circuit sees the ICE as a real microcontroller but, in reality, the ICE is emulating the function of the microcontroller and is in constant communication with the developer's computer. The developer may use an application on the computer to stop, pause, and restart the application at any time or view the contents of memory, SFRs, etc. Breakpoints may be set and execution may be performed one instruction at a time. In many ways an ICE has the appearance of a simulator such as that contained within Pinnacle 52 but with the major difference that code that is executed by the ICE is actually executing within the environment of the circuit.

This chapter will examine each of these hardware tools that are important parts of the development process.

16.1 Device Programmers

A **device programmer** is a piece of hardware that is used to transfer programs or data from the computer to the microcontroller or to some type of non-volatile memory such as a flash or EPROM IC. In this manner the 8052 program that has been generated with the developer's toolset, such as the Pinnacle IDE, is transferred to the physical hardware that will be placed in the electronic circuit for execution.

Device programmers used to be the only way to transfer programs to the hardware. In recent years, a large number of derivatives have been produced that include integrated flash memory with in-system programming. These derivatives offer an alternative to using a device programmer which will be explained in the next section. But unless a microcontroller with on-chip code memory and in-system programming capabilities is being used and unless the circuit that includes the ISP-capable microcontroller has been designed with ISP in mind, a device programmer is still the most common method of transferring programs and data to non-volatile memory.

There are many companies that offer device programmers. It is up to the developer to research the capabilities of each product and decide which is most appropriate for his or her needs. When deciding upon a device programmer the following issues should be considered.

1. **Device Support**. Each device programmer has a list of the specific parts that it is capable of programming. Some programmers are made specifically for chips from a specific manufacturer while others are designed to be able to program a vast array of current and future parts. Be sure that the device programmer that is selected is able to program the parts that are to be used today and, preferably, will be capable of handling new parts that may need to be programmed in the future. Device programmers designed for a smaller number of parts are usually cheaper while programmers capable of programming a large number of parts are more expensive.

2. **Package Adapters**. Microcontrollers and memory devices come in many different "packages." The AT89S852 and DS89C420 used in the 8052.com SBC are DIP-40 which means they have two parallel lines of 20 pins for a total of 40 pins. The same microcontroller is also available in PLCC-44 which is a physically smaller, square IC with 11 pins on each side. Other derivatives come in many other packages. In order to use a device programmer with a given IC it must be possible to physically insert the IC into the device programmer. Normally this is done with optional adapters. These adapters allow an IC of one shape to be plugged into the device programmer that may accept ICs of a different shape. Even if the device programmer supports a specific IC, it may be necessary to purchase an adapter to make possible the physical connection between the IC and the programmer .

3. **PC Connection Cable**. Some device programmers may connect to the PC via a serial port, a parallel port, or a USB port. Make sure the device programmer being considered has a connection that is compatible with the PC it will be used with. Many modern PCs no longer have serial ports and it is probable that in the future even fewer PCs will have them, so give careful consideration before purchasing a device programmer that connects via the serial port.

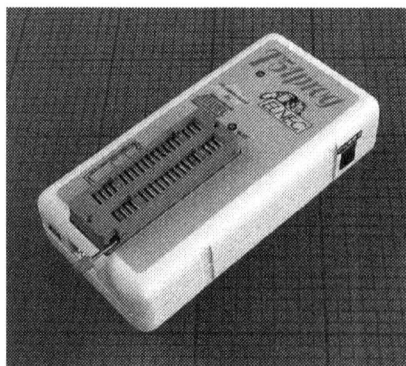

Operating System Support. Virtually all device programmers include software that will run under any recent version of Microsoft Windows. If an especially old version of Windows is being used it may be worthwhile to check whether the software supports it. Relatively few device programmers support Linux and Apple-based operating systems so if either of these platforms is to be used be sure to verify whether it is possible to use the device programmer on that platform.

The **T51Prog** from the Elnec company, pictured on the previous page, is an example of a part-specific device programmer. It is designed to program primarily 8052 derivatives with on-chip code memory that are programmed with normal TTL programming voltages but is also capable of programming popular EEPROM parts. Additionally an ISP port allows the T51Prog to double as an ISP programming cable and program parts without removing them from the circuitry as long as the circuit includes a connector compatible with the T51Prog's included ISP connector. Although the T51Prog is designed with 8052 derivatives in mind, it is capable of programming over 3000 different ICs.

The Phyton company offers the **ChipProg+** device programmer (pictured above). Like the T51Prog, the ChipProg+ is capable of programming virtually all modern microcontrollers but has the added advantage of being able to program and read traditional EPROMs which require higher programming voltages.

16.1.1 Using a Device Programmer

Regardless of which device programmer is used, the general process is basically the same.

1. Create the software with a toolset such as Pinnacle 52 (see chapter 15).
2. Run the software that came with the device programmer.
3. Use the device programmer's software to open or select the HEX file produced by the software toolset.
4. Instruct the device programmer's software to download the program into the microcontroller or EEPROM.

This section will go through this process with a ChipProg+ and will assume that a program called SBCMON.HEX has already been generated by a toolset such as Pinnacle 52.

Upon entering the ChipProg+ programming software the following window will appear:

The first step is to tell the application which type of IC is to be programmed by pressing the **Select Chip** button. This will bring up the following window in which Atmel is selected as the manufacturer. This brings up a list of all the ICs produced by Atmel that are supported by ChipProg+ (see figure below). The AT89S8252 is selected from the device list on the right side of the screen.

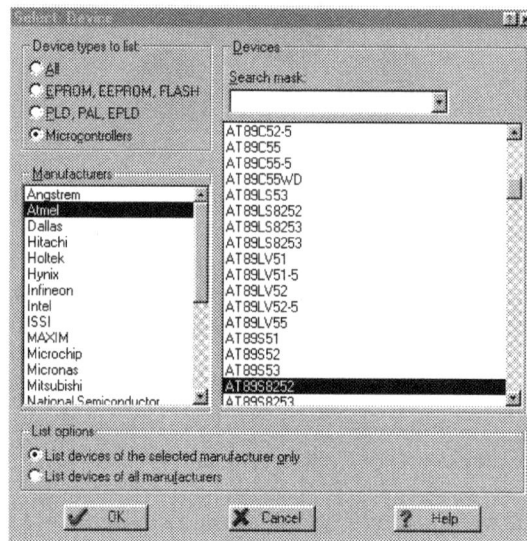

Once the Ok button is clicked, the AT89S8252 will be selected. From the main window the View drop-down may then be clicked and the Program option may be selected (or F6 may be pressed). This will bring up the following window:

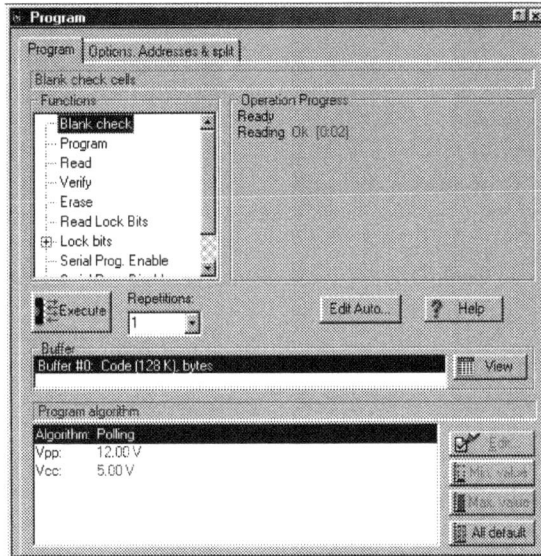

Each of the functions can be executed by double-clicking on the function in the listbox. For example, double-clicking on 'Blank' will verify that the device inserted into the ChipProg+ is blank. To program SBCMON.HEX, first click on the 'File' drop-down menu and select 'Load' (or press F2). This will open up a file selection window that allows the SBCMON.HEX file to be selected. Once the file is selected, insert the AT89S8252 in the ChipProg+ programmer and double-click on the 'Program' option in the listbox above.

The process for programming other parts is generally the same, although in some cases it may be necessary to double-click on 'Erase' prior to programming a new version into the part.

16.2 In-System Programming

In-system programming (ISP) is a relatively new approach to program storage in embedded microcontrollers. Rather than requiring an external EPROM or even requiring that the microcontroller be removed from the circuit and programmed by a device programmer, the programming of an ISP device is accomplished within the circuit. For example, the SBC described in chapter 14 specifically supports the Atmel AT89S8252 and Dallas 89C420. If either of these parts is used as the SBC's microcontroller then it is possible to download new versions of software to the SBC without removing any IC from the board. Instead, the SBC is connected to the PC, either via the serial or parallel port, and the new version of the program is sent to the SBC automatically.

In-system programming requires that the circuit be designed from the outset to support the ISP capability of the microcontroller that will be used. In the 8052.com SBC, the 74HC126 (U11) was included in the design to support ISP for the AT89S8252 while the 74LS125 (U11) was included to support ISP for the DS89C420. Had these parts not been included in the SBC's design it would be impossible to program the SBC using ISP. Therefore, it is very important to consider ISP from the very beginning of a design if ISP is a desired capability for the final product.

16.2.1 Using In-System Programming

The process of downloading a new program to a microcontroller via in-system programming is roughly the same regardless of the part being used.

1. Create the software with a toolset such as Pinnacle 52 (chapter 15).
2. Run the in-system programming software that is designed for the microcontroller and circuit being used.
3. Use the software to download the program into the microcontroller's on-chip memory.

The only aspect that changes in this process is using the correct software for the microcontroller being used. Since the 8052.com SBC supports both the Atmel AT89S8252 and the Dallas DS89C420, the ISP process for these two microcontrollers will be described in the following sections.

16.2.1.1 In-System Programming for the 8052.com SBC

The 8052.com SBC was designed to work with the **VisISP-52** in-system programming software. This software can be downloaded free of charge from **http://www.8052.com/visisp52** and allows the 8052.com SBC to be programmed regardless of whether it uses an Atmel 89S8252, 89S8253, or a Dallas DS89C420--the only aspect that changes is the microcontroller selected in the VisISP-52 listbox. The VisISP-52 program accomplishes ISP by using the PC's parallel port to communicate with the SBC—any standard 25-pin parallel cable will work; just connect one end of the cable to the PC and the other end to the SBC's J2 port.

The parallel port is used when performing ISP with an Atmel microcontroller. If a Dallas microcontroller is used, VisISP-52 uses a standard 9-pin RS-232 serial cable.

Once VisISP-52 has been downloaded and installed, the program may be executed which will result in the application window opening as pictured in the following graphic.

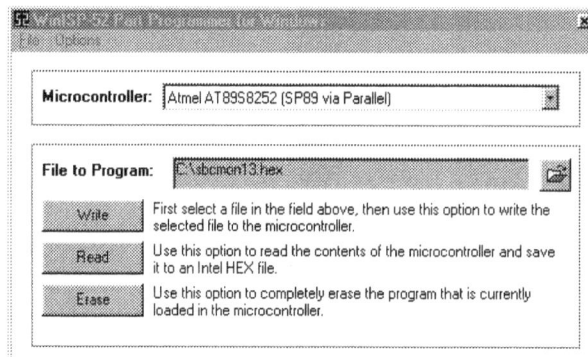

The interface is very straight-forward:

1. Make sure the right microcontroller is selected in the listbox.
2. Select the file to download to the microcontroller via ISP by pressing the folder button to the right of the "File to Program" field.
3. Click on the "Write" button to write the specified file to the microcontroller.
4. VisISP-52 will then display a progress dialog window that will show the download progress. Once complete, the microcontroller will contain the code from the specified file.

The VisISP-52 program was written for the Windows operating system and should work on all versions of Windows, including Windows 98, Windows 2000, and Windows XP. Additional information and troubleshooting information for the program may be found at the same link: **http://www.8052.com/visisp52.**

If you are using Linux, you may prefer to use the SP89 application program. This is an open source, native Linux application that is available at **http://www.8052.com/sp89** and is capable of programming the SBC using ISP as long as an Atmel 89S8252 is installed on the SBC. At the time this book was published, SP89 did not support the AT89S8253 nor will it support any non-Atmel microcontrollers.

16.3 In-Circuit Emulation

Another popular approach to executing code in a hardware environment is **in-circuit emulation** (ICE). Unlike device programmers and in-system programming, ICE is used exclusively for debugging and testing. ICE is never used to produce a final microcontroller product but rather to test the software in a real circuit with debug-level control.

Recall in section 15.5 it was explained that the Pinnacle 52 Simulator could load a HEX file into its memory and execute the program step by step within the PC, monitoring all the SFR registers, values of internal RAM, etc. Simulation provided a "look under the hood" so the developer could determine exactly what was happening at each step of the program's execution. Although the simulated program didn't actually execute within an 8052 microcontroller, the PC was capable of simulating most of the functions of an 8052 accurately. Although simulation is often sufficient, it can sometimes prove to be inadequate. If a program is interacting with specially connected external hardware it may be impossible to simulate the specialized design within a simulator. Likewise, some projects operate at high speed and may only be tested if they are executing at that specific speed—this is often true for analog to digital conversions and interaction with external high-speed parts.

ICE takes the concept of simulation one step further. ICE provides the same type of step by step program control, variable and memory monitoring as simulation, but the program is actually run on a special 8052 module which is inserted into the SBC's microcontroller socket (U1 of the SBC) and which is controlled and monitored by the PC. The program runs at full speed and has full access to all hardware components on the SBC. As far as the microcontroller program is concerned, it is running on real hardware and all hardware components are accessible at normal speeds; but from the developer perspective it is being controlled by a PC application that gives the developer the ability to look at individual SFRs, memory, and pause and continue program execution at full speed, or instruction by instruction.

16.3.1 Using ICE Within a Circuit

ICE is accomplished by inserting an ICE module (**PICE-52** from **Phyton** pictured below) into the socket that is normally occupied by the microcontroller. In this case, the ICE module is inserted into the U1 socket of the SBC. The microcontroller program is selected with a supporting Windows application which downloads it directly to the ICE module. The Windows application is then used to start, stop, pause, and otherwise manipulate the program even though the microcontroller program itself is running within the ICE module and in the actual circuit.

The Windows application is similar, in general concept and design, to the interface of Pinnacle 52 (chapter 15). The ICE application itself includes a host of "views" that allow the user to view internal RAM, bit memory, SFRs, and a number of other aspects of the microcontroller environment. Again, the functionality is very similar to that of a simulator but it has the added advantage of executing the program in the real hardware environment which allows the program to behave exactly as it would when executed by the microcontroller.

HARDWARE INTERFACE AND SOFTWARE EXAMPLES

CHAPTER 17: SBCMON MONITOR SOFTWARE

A "monitor program" is essentially an extremely small operating system that provides basic functionality and user interaction with a small system such as the 8052.com SBC described in chapter 14. To call a monitor program an operating system is actually to overstate its size and sophistication; a monitor should, instead, be considered a test environment to perform basic operations, run certain hardware tests, and confirm that the hardware itself is working as designed. A special monitor program for the SBC is discussed in this chapter since it makes it easier to verify that the SBC is functioning correctly and because the monitor program can be useful when testing the software examples contained in the rest of this book.

An original monitor program, called SBCMON, was written specifically for the 8052.com SBC. By monitor standards it is relatively sophisticated. It provides the ability to view and modify both external and internal RAM, either one byte at a time or in blocks, allows the user to download Intel HEX files into external RAM for subsequent execution, and provides a command to execute code anywhere in memory. SBCMON also allows complete access to all of the specialized hardware included on the SBC, including the DS1307 real time clock, the serial EEPROM IC, the 4x4 keypad, and the LCD connected either directly or as a memory-mapped device. In addition, SBCMON even includes a rudimentary mini-assembler which allows the user to create simple assembly language programs by typing them into the SBC with no need for an external assembler.

SBCMON is available free of charge as assembly language source code as well as a ready-to-run HEX file at **http://www.8052.com/sbc/sbcmon**.

Monitor programs such as SBCMON are useful for testing the hardware—often when a novice begins developing assembly language programs and runs into problems, it's not entirely clear whether the problem is due to an error in the program, in the process of downloading the program to the microcontroller, or even in the hardware itself. With so many potential points of failure it is easy to become frustrated.

SBCMON helps solve this problem. Once the SBC has been built the first thing the user should do is load SBCMON on the SBC as described in the next section; if the SBC was purchased prebuilt, SBCMON is mostly likely already installed in the SBC's microcontroller. Once installed on a functioning SBC and connected to a PC's serial port at 9600 baud, a welcome message from SBCMON should appear in the PC's terminal window when the SBC is powered on. At that point the user may begin to interact with the SBC using the monitor program—setting the real time clock, writing a string to the serial EEPROM and reading it back, writing a message to the LCD display, or assembling a small program using the mini-assembler.

When the user is ready to write his or her own programs, the assembly language source code of SBCMON may be referred to in order to better understand how the monitor accomplishes what it does. Additionally, the user will already have confirmed that the hardware does, in fact, work since it worked under SBCMON. Likewise, if some aspect of the hardware fails with SBCMON it's a very strong indication that there's a hardware problem of some sort.

17.1 Installing SBCMON

To install SBCMON, first download the HEX file from **http://www.8052.com/sbc/sbcmon**. The

SBCMON.HEX file is the assembled, ready-to-use version of the program that may be loaded directly to the SBC.

Follow the instructions in 16.1.1 to use a part programmer to program the SBCMON.HEX file into the microcontroller or an EPROM or follow the instructions in 16.2.1 to load SBCMON.HEX using in-system programming if the microcontroller supports it.

Once SBCMON has been installed on the SBC it may be used as described in the remainder of this chapter.

17.1.1 Configuring the PC Terminal Program

SBCMON itself is used by connecting the SBC to the PC via the serial port. The user then runs a terminal program—such as HyperTerminal under Windows—and configures it for 9600 baud. At that point everything that is typed in the terminal program is sent to SBCMON and everything SBCMON displays is sent to and displayed by the PC terminal program.

To use HyperTerminal in Windows with SBCMON follow these steps:

1. Launch the HyperTerminal application.
2. A window will open called "Connection Description: New Connection." Click on the Cancel button to close this window.
3. Click on the "File" drop-down menu and select "Properties" to open the "New Connection Properties" window.
4. Set the "Connect using" field to "Direct to Com1" (or to whichever communication port the SBC is connected on the PC).
5. Click the Configure button and set bits per second to 9600, data bits to 8, parity to none, stop bits to 1, and flow control to Hardware. Click "Ok" to close the window.
6. Click on the "Settings" tab at the top of the window and then press the "ASCII Setup" button which will open the "ASCII Setup" window.
7. Check the "Append line feeds to incoming line ends", then click on "Ok" to close the window.
8. Click on the "Ok" button to close the "New Connection Properties" window.

In order to avoid the HyperTerminal configuration process every time the application is started, the configuration may be saved as a new connection—perhaps named "SBCMON". This will add an entry to the HyperTerminal folder such that it will only be necessary to click on that icon in the future to launch HyperTerminal with all the right settings.

SBCMON may be used with any terminal program, computer, and operating system. The configuration process will be different on other platforms but the key configuration details are: 9600 baud, 8 data bits, 1 stop bit, no parity, and appending linefeeds to incoming carriage returns (or CRs as they are sometimes called).

17.2 Powering up SBCMON

When power is applied to the SBC loaded with SBCMON, the yellow LED (D6) will flash on and off three times. This serves two purposes:

1. It confirms that the SBCMON program is loaded correctly on the SBC and that the SBC is executing the monitor. This is useful since if there doesn't seem to be any communication between the PC and SBCMON, the flashing yellow LED will confirm that SBCMON is, in fact, executing. If the yellow LED flashes when the SBC is powered-up but nothing appears in the PC terminal program, the problem is almost certainly something to do with communications—either the wrong type of cable or the PC's terminal configuration is incorrect.

2. It provides a delay of approximately 600 milliseconds. This is useful because the Atmel AT89S8252 has a published errata which indicates that problems may occur if a program accesses the SPI pins (P1.5, P1.6, and P1.7) during the first 500 milliseconds following an ISP process. The 600ms delay eliminates any possibility that these pins will be manipulated within this period.

Once power has been applied and the yellow LED has flashed on and off three times, SBCMON will send the message:

```
SBC Initialized.
```

This informs the user that the SBC has powered-up or has been reset. The monitor then enters the main menu mode.

17.3 SBCMON Main Menu

When SBCMON is ready to receive a command from the user it will display the following message:

```
SBC Ready>
```

At this point the user may enter any valid command. A command consists of a single letter and zero or more additional parameters. The command is executed when the user presses ENTER.

When commands include a parameter, the first parameter is always entered immediately after the command letter with no space between them. For example:

```
X8000        - Correct
X 8000       - Incorrect, there should not be a space between the 'X' and '2000'
```

The previously executed instruction may be executed again by hitting the ESCAPE key alone on a line.

Commands are case-sensitive. This means the 'X' command is not the same as 'x'. All commands are upper-case except for one (the lower-case "w" command).

17.3.1 Command 'A': Mini-Assembler

Syntax: A[ASSEMBLE ADDRESS]

The 'A' command invokes SBCMON's mini-assembler. This is a very basic, no-frills assembler that is capable of taking standard 8052 assembly language instructions and converting them to the correct opcodes and storing them in memory. This may be used to make minor adjustments to code that has already been loaded into external RAM or may be used to actually create completely new code. While it is not likely that the mini-assembler will be used to create substantially long programs, it may certainly be used to create small test code segments and thereby save the user the hassle of creating the program in an assembler on the PC, generate the HEX file, and upload the HEX file to the SBC.

The 'A' command may be invoked with a single parameter which specifies the address to which the subsequently entered assembly language instructions should be compiled. For example, the command A8000 would assemble the first instruction at external RAM address 8000h.

If the 'A' command is issued without an assemble address then the assembler will default to the last address to which it assembled an instruction. Thus it is easy to enter a number of assembly language instructions, exit the assembler, issue other SBCMON commands, and then continue assembling the program by entering the 'A' command with no parameter.

Since the assembler stores the assembled code in RAM, it only makes sense to assemble code in the area of the address map that is allocated to the 62256 RAM IC—8000h through FFFFh. The assembler will allow the user to assemble to any address but assembling to addresses outside this range will produce useless and unpredictable results since the output of the assembler will be written to the LCD or the keypad depending on what assemble address is specified.

Once in the mini-assembler, the user will be presented with a prompt such as:

    8000:

This indicates the address to which the assembly language instruction provided by the user will be assembled. If, for example, the user enters the instruction MOV A,#25 then the value 74h will be stored at 8000h and the value 25h will be stored at 8001h since this is the machine language representation of that instruction. The user will then be prompted with the address of the next instruction which, in this case, would be:

    8002:

Since the MOV A,#25 instruction required two bytes of memory (addresses 8000h and 8001h), the next instruction will be stored at 8002h.

At any assembler prompt the user may press ENTER on a blank line to exit the assembler and return to SBCMON's main menu.

The user may also use the ORG directive to change the assemble address without leaving the assembler. For example, consider the sequence (bold-face text indicates user input):

    8000: **MOV A,#25**
    8002: **ORG 8005**
    8005: **RET**
    8006:

In this example assembly began at address 8000h. The MOV instruction required two bytes so the address for the next instruction was 8002h. Instead of assembling code at 8002h, the user entered the ORG 8005 instruction which set the assemble address to 8005h. The subsequent RET instruction was then assembled at 8005h thereby leaving addresses 8002h, 8003h, and 8004h unchanged.

Finally, the DB command may be used to insert a specific byte into the address of memory indicated by the assembler prompt. For example:

```
8000: MOV A,#25
8002: DB 15
8003: DB 99
8004:
```

In this case, as before, 74h and 25h are stored at addresses 8000h and 8001h, respectively. The DB 15 command instructs the assembler to insert the value 15h at address 8002h. The subsequent DB 99 command instructs the assembler to insert the value 99h at address 8003h. This provides an easy way to modify data that is contained within external RAM.

Compared to full-featured assemblers on the PC, such as Pinnacle 52, the mini-assembler within SBCMON is very limited. The following rules and restrictions must be kept in mind:

1. All numeric values are treated as hexadecimal. It is not necessary to include the traditional "h" at the end of a hexadecimal value. That is, MOV A,#25 is the same as MOV A,#25h.
2. No program labels are permitted. Instead of an instruction such as LCALL MyLabel, the instruction must be entered as LCALL 8153 where 8153h is the address that would normally correspond to MyLabel.
3. No standard 8052 SFR or bit names are recognized and must be entered as hexadecimal values. This means that MOV SBUF,#10 would have to be entered as MOV 99,#10 since the SFR address of SBUF is 99h.
4. No equations are permitted. Instructions such as MOV A,#10 + 25 will not be accepted by the mini-assembler.

Due to these restrictions the SBCMON mini-assembler is clearly not intended for extensive development. It is useful, however, for quickly modifying existing code that is loaded in external memory or for writing a very quick code segment. If, for instance, the user wanted to run a quick program that toggled P1.0 two times this could be done with the following command sequence:

```
SBC Ready> A8000
8000: CPL 90
8002: CPL 90
8004: RET
8005: (user presses return)
SBC Ready> R8000
```

The above sequence would compile CPL P1.0 (represented by bit address 90h) at 8000h and 8002h followed by a RET command. The code would then be executed with the R8000 command (more on this in section 17.3.9). This would be significantly faster than writing the same code on the PC, assembling it, and loading the HEX file into SBCMON.

17.3.2 Command 'C': Read/Set Real Time Clock

Syntax: C[ss nn hh w dd mm yy]

The 'C' command provides access to the Dallas DS1307 Real Time Clock (U13 on the schematic). When issued without any parameters this command will return the current time and date as read from the RTC. The information will be displayed in the following format:

```
SBC Ready> C
Secs:   49
Mins:   44
Hour:   09
Wday:   04
Day:    13
Month: 01
Year:   05
SBC Ready>
```

The above information indicates that the RTC reported a date of January 13, 2005 at 9:44:49 AM.

The 'C' command can also be used to set the time and date of the RTC. This is accomplished by issuing the 'C' command followed immediately by the seconds, then a space, then the current minute, space, hour, space, day of week (1-7), space, day of month, space, month of year, space, and finally the year. For example, to set the RTC to October 20, 2004 at 5:26:35 PM the command issued would be:

```
SBC Ready> C35 26 17 1 20 10 04
```

The day of the week ("1" in this example) is a value between 1 and 7 which represents the day of the week. When the clock rolls over from 23:59:59 to 0:00:00, it will increment the day of the week counter. The program may assign any day of the week to the value "1" and all subsequent values will be assigned to subsequent days. For example, if "1" is to represent Sunday then "2" would represent Monday and "7" would represent Saturday. The RTC, however, places no restriction on what day is assigned to the value "1".

When the RTC is set, the program *must* set the day of the week. The RTC does not calculate it based on the date provided. If the program does not set the day of the week value correctly then the value will be useless when read back from the RTC. If it's not important to the function of the program to know whether or not it is Monday, Tuesday, etc. then the RTC may simply write "1" to this field every time it sets the clock and ignore this field when it reads the clock.

Note that the RTC may not initially be active when the SBC is powered on—reading the clock immediately on power-up may show that the time is not changing. This is due to the fact that when the RTC powers up its internal registers are in a random state. One of the registers contains a flag that indicates whether the clock should be active or paused; if the RTC powers up and the register randomly starts with the clock disabled then the clock will appear to be stalled. The clock will automatically be activated as soon as the user sets the RTC date/time with the 'C' command.

17.3.3 Command 'E': Read/Write Serial EEPROM

Syntax: `E[text]`

The 'E' command allows easy read and write access to the AT25010A serial EEPROM (U14). When issued without a parameter, the command will read the serial EEPROM and display its results. When issued with a parameter, the command will write the rest of the command line to the serial EEPROM as an ASCII string.

For example, to write the string "HELLO WORLD" to the serial EEPROM the following command would be issued:

```
SBC Ready> EHELLO WORLD
```

To subsequently read the contents of the serial EEPROM, the command would be issued with no additional parameters:

```
SBC Ready> E
HELLO WORLD
SBC Ready>
```

17.3.4 Command 'I': Read/Write to Internal RAM

Syntax: `IstartAddress[,endAddress][=value,value,value...]`

The 'I' command allows the user to read and write internal RAM; that is, the 256 bytes of IRAM memory with addresses between 00h and FFh. This command has a number of different syntaxes that accomplish varying tasks.

In its simplest form, the 'I' command with a single hexadecimal parameter will display the contents of the specified internal RAM address. The `I50` command will display the contents of internal RAM address 50h:

```
SBC Ready> I50
IRAM[50]=25
```

In this case SBCMON is reporting that internal RAM address 50h contains the value 25h.

The command can also be used to display a block of internal RAM. In this case, the first address of the block must be followed immediately by a comma and the last address of the block. To view the contents of internal RAM from address 20h through 4F, the following command would be entered:

```
SBC Ready> I20,4F
IRAM[20]=00 01 02 03 04 05 06 07 08 09 0A 0B 0C 0D 0E 0F
IRAM[30]=10 11 12 13 14 15 16 17 18 19 1A 1B 1C 1D 1E 1F
IRAM[40]=20 21 22 23 24 25 26 27 28 29 2A 2B 2C 2D 2E 2F
SBC Ready>
```

As shown above, the command dumps all the internal RAM addresses between the two addresses specified in the command. The SBC automatically displays them in 16-byte groups for readability.

The 'I' command can also be used to modify the contents of internal RAM by specifying the address to modify followed immediately by an equals sign and the value to assign to that address. The command can modify a single internal RAM address or may modify a sequence of addresses in a single command.

To set the internal RAM address 60h to the value 27h the following command would be issued:

```
SBC Ready> I60=27
```

If it is necessary to change a consecutive sequence of addresses in internal RAM this may be accomplished by providing multiple values after the equal sign, separated by commas or spaces.

```
SBC Ready> I60=10,20,30,40
```

This command will set internal RAM address 60h to 10h, address 61h to 20h, address 62h to 30h, and address 63h to 40h. This is faster than setting the four address with four separate 'I' commands.

The syntax and features of the 'I' command are identical to that of the 'X' command (section 17.3.13). The only difference is that the 'I' command operates with internal RAM while the 'X' command operates with external RAM.

17.3.5 Command 'K': Keypad Test

Syntax: K[Disable Debounce]

The 'K' command provides an easy way to test the 4x4 keypad attached to J5, with and without key debounce. Issuing this command will cause SBCMON to enter an echo mode in which every key pressed on the keypad will be echoed to the PC's terminal program until such time as the user presses any key on the PC to terminate the test.

```
SBC Ready> K
Keypad test: Will echo keypresses to serial port
Press any key on PC keyboard to terminate
```

At this point every key pressed on the keypad will be echoed to the PC screen. Typing "123" on the keypad will result in the characters "123" being echoed to the PC. This verifies that the keypad is being read and properly debounced.

An additional feature of this command is to execute the same keypad test without key debounce. This is useful so the user may observe how the input from the keypad changes with and without debounce. To run the keypad test without debounce the command should be entered as KD.

```
SBC Ready> KD
Keypad test: Will echo keypresses to serial port
Press any key on PC keyboard to terminate
NOTE: DEBOUNCE DISABLED!
```

In this mode the user will observe that every time a key is pressed, the same key is echoed to the screen numerous times. This illustrates the importance of properly debouncing the keypad.

17.3.6 Command 'L': Load HEX File into SBC Memory

Syntax: L

The 'L' command allows the user to upload an Intel HEX file to the SBC's external RAM. Since the SBC is designed to allow code to be executed from external RAM, this feature provides a way to quickly upload programs into the SBC's memory for execution without having to load it into the microcontroller via ISP. This is useful since uploading the program to RAM is faster than programming via ISP. Additionally, the microcontroller may only be programmed a finite number of times—in the case of the AT89S8252, the part is guaranteed to be able to accept 1,000 program updates. Thereafter, the part may not be able to reliably store new programs. By uploading and executing programs in the SBC's RAM the microcontroller will not be subject to frequent program uploads that count against the 1,000 update limit.

Programs loaded with the 'L' command do not overwrite the program stored in the microcontroller's flash memory. That means that you will not overwrite SBCMON by uploading code with this command. Rather, SBCMON will receive the program you upload and store it in RAM (not flash memory). The program will then return to SBCMON so that the next command may be issued.

To load a program using this command, enter the 'L' command and press return. The SBC will send the following prompt:

```
SBC Ready> L
Send Intel HEX file in text format now, CTRL-X to abort
```

At this point the user should send the Intel HEX file to the SBC. There are two ways to accomplish this.

1. Copy/Paste. The user may select the text of the Intel HEX file in another editor or application, copy it, and paste it into the terminal window.

2. Send Text File. The user may use the terminal's "Send Text File" option to send the file directly without having to copy/paste it. In Windows' HyperTerminal this can be accomplished by clicking the Transfer drop-down menu, selecting the Send Text File option, and selecting the HEX file to send. The transfer will be invisible to the user but the SBC will indicate when the transfer is complete.

When the upload of the HEX file is complete the SBC will report:

```
Intel HEX file upload complete.
1512 bytes loaded.
SBC Ready>
```

Since HEX files normally store program code that is to be executed, the HEX file itself must contain a program that was assembled at an address that corresponds to the SBC's external RAM address range of 8000h – FFFFh. That means that all programs uploaded to the SBC using this command should be assembled such that they are located within this address range. This can be accomplished by starting all programs that are to be loaded with the 'L' command with `ORG 8000h` which will ensure that the program is located in the SBC's 32k of external RAM.

> The 'L' command will attempt to load any Intel HEX file it receives—even HEX files that instruct it to load data to other areas of memory where no RAM exists. If it receives a file that instructs it to load data at 4000h, it will obediently attempt to store the contents of the file at 4000h even though that corresponds to the address of the LCD. For this reason it doesn't make sense to load HEX files that contain data located outside the range of 8000h – FFFFh.

One a program has been uploaded to the SBC's external RAM, it may be executed with the 'R' command (see section 17.3.9).

17.3.7 Command 'M': Set LCD Access Mode

Syntax: `M[0|1|2][I]`

The 'M' command configures whether SBCMON will communicate with the LCD as a directly-connected device on J4 or as a memory-mapped device on J3. It will also optionally initialize the LCD so that it will be ready to display text sent with the 'W' command (see section 17.3.11).

The first parameter of the the command is either a '0', 1', or '2'. If the parameter is '0' then SBCMON will assume that the LCD is connected as a memory-mapped device on J3; if the parameter is '1' then the LCD will be accessed as a directly-connected device on J4 in 8-bit mode; if the parameter is '2' then the LCD will be accessed as a directly-connected device on J4 in 4-bit mode . The command will initialize the LCD if the letter 'I' immediately follows the '0', '1', or '2' parameter.

The initialization sequence consists of sending the LCD the commands 38h, 0Eh, 06h, 01h when in 8-bit mode (direct-connect or memory-mapped) or the commands 28h, 0Eh, 06h, 01h when in 4-bit mode (direct-connect only).

> The initialization sequence varies based on whether the LCD is being driven in 4-bit or 8-bit mode, not on whether the LCD is connected in memory-mapped or direct-connect mode. However, the LCD is always in 8-bit mode when connected in memory-mapped mode using connector J3.

Thus to configure the LCD for memory-mapped operation on J3 and initialize it so that it is ready to display data, the following command would be issued:

```
SBC Ready> MOI
LCD mode set to memory-mapped
Initializing LCD
```

> The 'M' command and the one or two characters of the parameter are all typed together. There are no spaces in this command.

17.3.8 Command 'Q': Quick HEX Load and Run

Syntax: Q

The 'Q' command is a shortcut that executes the 'L' command and, upon successful upload of a HEX file, automatically runs the 'R8000' command which executes the program at 8000h. Since most programs that are uploaded to the SBC's memory will be located at 8000h, this command provides a shortcut to quickly upload and run the program in a single step.

For this command to work the HEX file must contain code which begins at 8000h and terminates with a RET instruction.

To fully understand this command, sections 17.3.6 and 17.3.9 should be reviewed which provide full information about the two commands that are combined in the 'Q' instruction.

17.3.9 Command 'R': Run (Execute) Code at Specific Address

Syntax: RexecuteAddress

The 'R' command will cause the SBC to execute code at the specified hexadecimal address. This is useful primarily after loading a HEX file with the 'L' command (section 17.3.6) or assembling a small program in the mini-assembler with the 'A' command (section 17.3.1).

To execute a program that has been loaded with the 'L' command at external RAM address 8000h, the following command would be used:

```
SBC Ready> R8000
Executing code at 8000
Code execution complete
```

If the code that is executed generates any output it would be displayed after the "Executing code" line and before the "Execution complete" line.

> The 'R' command executes the code at the specified address by executing the equivalent of an LCALL, effectively treating the code as a subroutine. This means that the code at the address in question must end with a RET instruction to return control to SBCMON.

17.3.10 Command 'V': Verify External RAM

Syntax: V[F]

The 'V' command conducts a test of external RAM by writing to addresses 8000h through FFFFh and subsequently reading the memory to verify that the same data is read as was originally written. A quick-test will be executed if the V command is issued while the VF command will conduct a more exhaustive—and more time-consuming—memory test. Any discrepancies that could indicate a memory failure will be reported to the user.

The 'V' command will destroy any data or code that was previously stored in external RAM.

17.3.11 Command 'W': Write Text to LCD

Syntax: `W[text]`

The 'W' command is used to write text to the LCD. Any text that is included on the command line following the 'W' will be transmitted to the LCD as ASCII text and, when properly configured, be displayed on the LCD's screen.

For example, to display HELLO WORLD on the LCD's screen the following command would be used:

```
SBC Ready> WHELLO WORLD
```

Before writing text to the LCD it is necessary to initialize it with the 'M' command (see section 17.3.7).

17.3.12 Command 'w': Write Commands to LCD

Syntax: `w[text]`

The 'w' (lower-case 'w') command is used to send one or more commands to the LCD. The rest of the line is treated as one or more hexadecimal values which are sent one byte at a time to the LCD in command mode. The first hexadecimal value should immediately follow the 'w' character while subsequent values should be separated by spaces.

To send the initialization sequence to the LCD the following command could be used:

```
SBC Ready> w38 0E 06 01
```

The LCD initialization sequence can also be sent using the 'M' command (see section 17.3.7).

17.3.13 Command 'X': Read/Write to External RAM

Syntax: `IstartAddress[,endAddress][=value,value,value...]`

The 'X' command allows the user to read and write external RAM. This command has a number of different syntaxes that accomplish varying tasks.

In its simplest form, the 'X' command with a single hexadecimal parameter will display the contents of the specified external RAM address. The `X0050` command will display the contents of external RAM address 0050h:

```
SBC Ready> X8050
XRAM[8050]=25
```

In this case SBCMON is reporting that external RAM address 8050h contains the value 25h.

The command can also be used to display a block of external RAM. In this case, the first address of the block must be followed immediately by a comma and the last address of the block. To view the contents of external RAM from address 8020h through 804F, the following command would be entered:

```
SBC Ready> X8020,804F
XRAM[8020]=00 01 02 03 04 05 06 07 08 09 0A 0B 0C 0D 0E 0F
XRAM[8030]=10 11 12 13 14 15 16 17 18 19 1A 1B 1C 1D 1E 1F
XRAM[8040]=20 21 22 23 24 25 26 27 28 29 2A 2B 2C 2D 2E 2F
SBC Ready>
```

As shown above, the command dumps all the external RAM addresses between the two addresses specified in the command. The SBC automatically displays them in 16-byte groups for readability.

The 'X' command can also be used to modify the contents of external RAM by specifying the address to modify followed immediately by an equals sign and the value to assign to that address. The command can modify a single external RAM address or may modify a sequence of addresses in a single command.

To set the external RAM address 8060h to the value 27h the following command would be issued:

```
SBC Ready> X8060=27
```

If it is necessary to change a consecutive sequence of addresses in external RAM, this may be accomplished by providing multiple values after the equals sign, separated by commas or spaces.

```
SBC Ready> X8060=10,20,30,40
```

This command will set external RAM address 8060h to 10h, address 8061h to 20h, address 8062h to 30h, and address 8063h to 40h. This is faster than setting the four address with four separate 'X' commands.

The syntax and features of the 'X' command are identical to that of the 'I' command (see 17.3.4). The only difference is that the 'I' command operates with internal RAM while the 'X' command operates with external RAM.

17.4 Writing Programs that Execute in SBCMON

As explained earlier in this chapter, it is possible to load HEX files directly into SBCMON's external RAM and execute them without having to download the program to the microcontroller or ISP. This can make the development and testing process much easier and faster. The only restriction is that the HEX files that are loaded into SBCMON must be located between 8000h and FFFFh since that's the address corresponding to the SBC's external RAM.

A very simple example that could be coded in Pinnacle 52 might be:

```
ORG 8000h
CPL P1.0
RET
```

Once assembled in Pinnacle 52, the resulting HEX file could be uploaded to SBCMON using the 'L' command (see section 17.3.6). The code could then be executed with the R8000 command which would execute the program, toggle P1.0 (toggling the yellow LED on or off), and then return to SBCMON's main menu. Although this is as an extremely simple example, it illustrates the process of creating a program that can be loaded into SBCMON. The SBC's external RAM measures 32k so any program that requires less than 32k could be uploaded.

17.4.1 Using SBCMON Library Routines in External Programs

The SBCMON software includes a number of very useful subroutines. A "Send Serial" routine is available which will send a string of text to the serial port, a "Get Line" routine will read a string of text from the serial port into a buffer, a "Get Hex Value from Serial Port" will read two hexadecimal characters from the serial port and convert them to a single byte, and there are routines to send and receive data using SPI or I2C protocol, to name just a few.

It is likely that programs that are developed and subsequently loaded into SBCMON may have use for these same routines. The developer may certainly write his or her own subroutines that perform the same functions, but in many cases it may be easier to just call the SBCMON routines that are already available. Since the program is being executed from within SBCMON, these routines already exist in SBCMON's code memory—it's just a matter of executing an LCALL to the right address in code memory.

One problem, though, is that it's hard to predict the address of any given subroutine in SBCMON. If a single character is added to a text message in SBCMON, every subroutine that appears later in the code will be shifted by one byte which would require updating all references to that subroutine in external programs. Since the SBCMON source code is certain to evolve and improve over time, this is not a user-friendly design.

Instead, SBCMON provides a subroutine entry table. This is a sequence of LJMPs for each useful subroutine that begin at code memory address 0041h—in each case, the LJMP does nothing more than jump to the real address of the subroutine such that even if the address of the real subroutine changes, the LJMP itself is always in the same place.

For example, the instructions at code memory 0041h is:

```
0041h: LJMP SendSerial
```

This instruction references a subroutine that will send a null-terminated text string to the serial port. If a program that is to be loaded into SBCMON's external RAM needs to make use of this routine it may simply execute LCALL 0041h. This will call the above instruction at 0041h which will transfer program flow to wherever the SendSerial routine is really located within the SBCMON code memory. Even if the SBCMON program changes, the addresses in the subroutine entry table are guaranteed to remain valid.

A better way to use the SBCMON routines in external programs is to use the EQU directive (see section 11.6.2) in Pinnacle 52 to establish an equate for the subroutine name. Consider the following example:

```
1  SendSerial EQU 0041h
2  ORG 8000h
3  LCALL SendSerial
4  DB "Hello, world!",13,0
5  RET
```

Line 1 establishes an equate that tells Pinnacle 52 to use the value 0041h whenever the text SendSerial is encountered in the program. Line 2 sets the assemble address to 8000h. Line 3 then calls the SendSerial routine—Pinnacle 52 will translate this to LCALL 0041h and assemble it as such. The text in line 4 is the null-terminated text string that the SendSerial routine will send to the serial port. Line 5 uses the RET instruction to return to the SBCMON main menu.

Clearly this provides a very easy and legible way to develop external programs that will run under SBCMON. The developer simply uses the EQU directive to define the address of the routine as its entry in the subroutine entry point table and subsequently calls that subroutine name whenever it is needed. All that is needed is a list of all the subroutine entry point addresses, the parameters that need to be sent to the subroutine, and the values that are returned by the subroutine.

Since the examples in the remainder of this book will use these routines, the routines themselves will be covered in the following section.

17.4.1.1 List of SBCMON Subroutine Entry Points

The table on the next page is a list of all the SBCMON subroutines which have entry points. Using these subroutines is completely optional. External programs may use their own code to perform these same functions or may take advantage of the subroutines provided in the following table and described in the remainder of this chapter.

The source code for each of these routines may be found within SBCMON.ASM which may be obtained at **http://www.8052.com/sbc/sbcmon**. A Pinnacle 52 include file that contains a list of equates for all the entry point addresses may also be found at the same web page. This saves the developer the work of manually defining the entry point EQUs in each external program.

17.4.1.1.1 Subroutine: SendSerial

Subroutine:	**SendSerial**
Entry Point:	0041h
Purpose:	Send a null-terminated string to the serial port
Inputs Registers:	None
Output Registers:	None
Registers Destroyed:	DPTR

The SendSerial subroutine will send to serial port the null-terminated string that immediately follows the LCALL SendSerial instruction that called it. It will then return to the instruction following the terminating null value of the message.

SBCMON Entry Point Address Reference

Entry Point Address	Subroutine Name	Subroutine Description
0041	SendSerial	Sends null-terminated string in code memory to serial port
0044	SendSerialHexByte	Sends value in ACC to serial port as two hexadecimal digits
0047	SendSerialByte	Sends character in ACC to the serial port
004A	GetSerialByte	Waits for character on serial port, returns it in ACC
004D	GetSerialLine	Receives an enter-terminated line of input from serial port
0050	SendLCDText	Sends a character to the LCD as text
0053	SendLCDCommand	Sends byte to the LCD as an LCD command
0056	I2C_ReadByte	Reads one byte from the I2C bus
0059	I2C_SendByte	Sends one byte to the I2C bus
005C	I2C_Reset	Resets the I2C bus
005F	I2C_Start	Starts an I2C conversation
0062	I2C_Stop	Ends an I2C conversation
0065	SPI_ReadByte	Reads one byte from the SPI bus
0068	SPI_SendByte	Sends one byte to the SPI bus
006B	ByteTo2Hex	Converts accumulator to two hex ASCII digits
006E	HexToNibble	Converts ASCII value in accumulator to value 00h and 0Fh
0071	GetHexValue	Converts hex digits in input buffer to values
0074	GetSerialHex	Reads 2 hex digits from serial port, converts to value in ACC
0077	ToUpper	Converts the character in ACC to upper-case
007A	SendDecimalR4567	Sends value in R4-R7 as decimal value
007D	DivideBy10	Divides 32-bit value by 10
0080	Check4BytesForZero	Verifies that a 32-bit value is zero
0083	ShiftR765Left	Shifts registers R5 (MSB) through R7 (LSB) left1 bit
0086	AddR67to23	Adds 16-bit value in R6 and R7 to 16-bit value in R2 and R3
0089	AddR4567to0123	Adds 32-bit value R4 through R7 to R0 through R3
008C	InitializeLCD	Sends necessary commands to initialize LCD
008F	ReadLCDStatus	Reads the LCD status (cursor position and busy flag)
0092	ReadLCDText	Reads the LCD text/memory at current cursor position
0095	SetDptrMMKey1	Set DPTR to memory-mapped address for read keypad row 1
0098	SetDptrMMKey2	Set DPTR to memory-mapped address for read keypad row 2
009B	SetDptrMMKey3	Set DPTR to memory-mapped address for read keypad row 3
009E	SetDptrMMKey4	Set DPTR to memory-mapped address for read keypad row 4
00A1	SetDptrMMLCdWC	Set DPTR to memory-mapped address for write LCD cmd
00A4	SetDptrMMLcdWT	Set DPTR to memory-mapped address for write LCD text
00A7	SetDptrMMLcdRC	Set DPTR to memory-mapped address for read LCD cmd
00AA	SetDptrMMLcdRT	Set DPTR to memory-mapped address for read LCD text
00AD	MakeDecimal2Digit	Make 2 ASCII digits out of value in accumulator
00B0	GetKeyDebounced	Return keypad key with debounce
00B3	WaitKeyReleased	Wait for keypad key to be released

```
LCALL SendSerial
DB "Hello, World!", 13, 0
RET
```

The first instruction calls the `SendSerial` subroutine which will then send the contents of the string that follows ("Hello, World!') to the serial port. When the subroutine reaches the null character—the zero byte—it will return to the instruction immediately following; `RET` in this case.

It is absolutely imperative that the string be terminated with the null character (0). This is the marker that tells the `SendSerial` subroutine to stop sending data to the serial port. If the null character is missing the routine will continue to send data to the serial port until it reaches some 00 character later in memory. This usually results in garbage characters being sent to the serial port and the program behaving in an unexpected manner.

This subroutine is very useful for writing messages to the serial port. Since the text immediately follows the `LCALL`, the text being written to the serial port is embedded with the related code.

17.4.1.1.2 Subroutine: SendSerialHexByte

Subroutine:	**SendSerialHexByte**
Entry Point:	0044h
Purpose:	Sends the value in the accumulator to the serial port as two hexadecimal digits
Inputs Registers:	Accumulator: Value to send as two nibbles
Output Registers:	None
Registers Destroyed:	None

The `SendSerialHexByte` subroutine takes the value in the accumulator, converts it to two hexadecimal digits in ASCII format, and sends those two digits to the serial port. This is useful for displaying the contents of a register in a human-readable format. For example:

```
MOV A,#4Fh
LCALL SendSerialHexByte
```

The above code will send the digit '4' followed by the hexadecimal digit 'F' to the serial port. This differs from just sending the accumulator (4Fh) to the serial port which would result in the letter 'O' being sent since 4Fh is the ASCII value for 'O'.

17.4.1.1.3 Subroutine: SendSerialByte

Subroutine:	**SendSerialByte**
Entry Point:	0047h
Purpose:	Sends the value in the accumulator to the serial port
Inputs Registers:	Accumulator: Character to send to the serial port
Output Registers:	None
Registers Destroyed:	None

The `SendSerialByte` subroutine simply sends the character that is currently in the accumulator to the serial port and waits for it to be completely sent before returning.

```
MOV A,#41h
LCALL SendSerialByte
```

This code will send the letter 'A' to the serial port since 41h is the ASCII code for the letter 'A'.

17.4.1.1.4 Subroutine: GetSerialByte

Subroutine:	**GetSerialByte**
Entry Point:	004Ah
Purpose:	Waits for a character to be received on the serial port, then returns it
Inputs Registers:	None
Output Registers:	Accumulator: The character that was received by the serial port
Registers Destroyed:	None

The `GetSerialByte` subroutine waits indefinitely for a character to be received by the serial port. When the character is received, the subroutine reads the value and returns it in the accumulator.

```
LCALL GetSerialByte
```

This instruction will wait for a character to be received by the serial port. If the user presses the 'A' key on the terminal's keyboard then this routine will return 41h, the ASCII code for 'A', in the accumulator.

17.4.1.1.5 Subroutine: GetSerialLine

Subroutine:	**GetSerialLine**
Entry Point:	004Dh
Purpose:	Allows the user to enter a full line, waits for user to press RETURN
Inputs Registers:	None
Output Registers:	Accumulator: First character of the line entered by the user
	R0: Address of the first byte of internal RAM used for the buffer
Registers Destroyed:	PSW

The `GetSerialLine` subroutine allows the user to enter a full line of ASCII text into a buffer in internal RAM and terminates the string with a null character (00h) when the user presses the ENTER key. The subroutine correctly handles backspaces and ensures that the user does not enter more characters than there is room in the buffer.

The characters entered by the user are stored in internal RAM. When the subroutine returns, the accumulator will hold the first character of the line entered by the user; this is useful for quickly branching based on the first character of the line. The subroutine returns the internal RAM address of the first byte of the buffer in R0. The buffer may be assumed to continue until a null character is reached.

Assume that `GetSerialLine` is called and the user enters the following line:

```
TEST123 <ENTER>
```

When the subroutine returns the accumulator would hold the value 54h—which is the ASCII value for 'T'. To read the buffer the program would look at the value of R0 returned by the subroutine. If R0 is returned as B0h, that would indicate that:

```
IRAM[B0] = 54h (T)
IRAM[B1] = 45h (E)
IRAM[B2] = 53h (S)
IRAM[B3] = 54h (T)
IRAM[B4] = 31h (1)
IRAM[B5] = 32h (2)
IRAM[B6] = 33h (3)
IRAM[B7] = 00h (Null inserted at end of line)
```

Be sure to use the value of R0 returned by GetSerialLine to determine where the line buffer is in internal RAM. While it will always be the same as long as the program is run on the same version of SBCMON, if SBCMON is ever upgraded to a new version it is entirely possible that the buffer will be located somewhere else. If R0 is used to determine the location of the buffer then future upgrades to SBCMON will have no negative impact on user programs that use this library routine.

17.4.1.1.6 Subroutine: SendLCDText

Subroutine:	**SendLCDText**
Entry Point:	0050h
Purpose:	Sends the character in the accumulator to the LCD as a text character for display
Inputs Registers:	Accumulator: Text character to send to the LCD
Output Registers:	None
Registers Destroyed:	None

The SendLCDText subroutine sends the ASCII character held in the accumulator to the LCD as a text character. Text characters sent to the LCD are normally those that will be displayed on the LCD itself.

```
MOV A,#'H'
LCALL SendLCDText
MOV A,#'E'
LCALL SendLCDText
MOV A,#'L'
LCALL SendLCDText
MOV A,#'L'
LCALL SendLCDText
MOV A,#'O'
LCALL SendLCDText
```

The code above sends the word HELLO to the LCD's display, assuming the LCD has previously been properly initialized.

17.4.1.1.7 Subroutine: SendLCDCommand

Subroutine:	**SendLCDCommand**
Entry Point:	0053h
Purpose:	Sends the value in the accumulator to the LCD as an LCD command
Inputs Registers:	Accumulator: Value to send to the LCD as a command
Output Registers:	None
Registers Destroyed:	None

The SendLCDCommand subroutine sends the value contained in the accumulator to the LCD as an LCD command. LCD commands perform special functions such as initializing the LCD, clearing the screen, or positioning the cursor.

The following instructions will initialize the LCD and clear the display screen by sending the appropriate commands to the LCD.

```
MOV A,#38h
LCALL SendLCDCommand
MOV A,#0Eh
LCALL SendLCDCommand
MOV A,#06h
LCALL SendLCDCommand
MOV A,#01h
LCALL SendLCDCommand
```

17.4.1.1.8 Subroutine: I2C_ReadByte

Subroutine:	**I2C_ReadByte**
Entry Point:	0056h
Purpose:	Reads one byte from the I^{2}C bus
Inputs Registers:	Carry: 0=Send ACK, 1=Send NAK
Output Registers:	Accumulator: Byte received from I^{2}C bus
Registers Destroyed:	None

The I2C_ReadByte subroutine reads one byte from the I^{2}C bus and returns its value in the accumulator. If the carry bit is clear then the subroutine sends an ACK, otherwise it sends a NAK.

17.4.1.1.9 Subroutine: I2C_SendByte

Subroutine:	**I2C_SendByte**
Entry Point:	0059h
Purpose:	Sends the byte in the accumulator to the I^{2}C bus
Inputs Registers:	Accumulator: Byte to send to the I^{2}C bus
Output Registers:	Carry: 0=No error, 1=Error-no ACK received
Registers Destroyed:	None

The I2C_SendByte subroutine sends the byte in the accumulator to the I²C bus. When it returns, the carry bit indicates whether the byte was sent successfully or not. If the carry bit is clear, the send was successful. If the carry bit is set then no ACK was received from the receiving I²C device.

17.4.1.1.10 Subroutine: I2C_Reset

Subroutine:	**I2C_Reset**
Entry Point:	005Ch
Purpose:	Resets the I²C bus
Inputs Registers:	None
Output Registers:	Carry: 0=No error, 1=Error, couldn't reset bus
Registers Destroyed:	None

The I2C_Reset subroutine sends the I²C "stop" condition nine times to ensure that every device has aborted any ongoing transfers and released the I²C bus. The routine returns the carry cleared if the reset was successful and sets the carry if it appears that some device hasn't released the I²C bus.

17.4.1.1.11 Subroutine: I2C_Start

Subroutine:	**I2C_Start**
Entry Point:	005Fh
Purpose:	Starts an I²C conversation
Inputs Registers:	None
Output Registers:	None
Registers Destroyed:	None

The I2C_Start subroutine begins an I²C conversation by bringing the SDA line low while SCL is high. This should be executed before sending bytes to any I²C device.

17.4.1.1.12 Subroutine: I2C_Stop

Subroutine:	**I2C_Stop**
Entry Point:	0062h
Purpose:	Terminates an I²C conversation
Inputs Registers:	None
Output Registers:	None
Registers Destroyed:	None

The I2C_Stop subroutine ends an I²C conversation by bringing the SDA line high while SCL is high. This should be executed when I²C communication is complete.

17.4.1.1.13 Subroutine: SPI_ReadByte

Subroutine:	**SPI_ReadByte**
Entry Point:	0065h
Purpose:	Reads a single byte from the SPI bus into the accumulator
Inputs Registers:	None
Output Registers:	Accumulator: Byte received from the SPI bus
Registers Destroyed:	None

The SPI_ReadByte subroutine reads a single byte from the SPI bus and returns it in the accumulator.

17.4.1.1.14 Subroutine: SPI_SendByte

Subroutine:	**SPI_SendByte**
Entry Point:	0068h
Purpose:	Sends the byte in the accumulator to the SPI bus
Inputs Registers:	Accumulator: Byte to send to the SPI bus
Output Registers:	None
Registers Destroyed:	None

The SPI_SendByte subroutine sends the byte in the accumulator to the SPI bus.

17.4.1.1.15 Subroutine: ByteTo2Hex

Subroutine:	**ByteTo2Hex**
Entry Point:	006Bh
Purpose:	Converts the value in the accumulator to two ASCII hex digits
Inputs Registers:	Accumulator: Byte to convert to ASCII
Output Registers:	Accumulator: High nibble of hex pair (30h-39h, 40h-46h)
	R0: Low nibble of hex pair (30h-39h, 40h-46h)
Registers Destroyed:	None

The ByteTo2Hex subroutine converts the value in the accumulator to two ASCII values that represent the two hexadecimal digits of the original value. This is useful in converting a numeric value in the accumulator to two ASCII values which can be displayed on the LCD or sent to the serial port. For example:

```
MOV A,#41h
LCALL SendSerialByte        ;Sends letter 'A' to serial port
MOV A,#41h
LCALL ByteTo2Hex            ;Convert 41h to component hex digits
LCALL SendSerialByte        ;Sends 34h (digit '4') to the serial port
MOV A,R0
LCALL SendSerialByte        ;Sends 31h (digit '1') to the serial port
```

The above code illustrates the difference between sending the value 41h to the serial port—which will display the letter 'A' since 41h is its ASCII code—and converting it to two ASCII characters using the `ByteTo2Hex` routine. Upon returning from `ByteTo2Hex` the accumulator will hold 34h—which is the ASCII value for '4'—and R0 will hold 31h—which is the ASCII value for '1'.

17.4.1.1.16 Subroutine: HexToNibble

Subroutine:	**HexToNibble**
Entry Point:	006Eh
Purpose:	Convert ASCII value in accumulator to a value between 0 and 15 (00h and 0Fh)
Inputs Registers:	Accumulator: Hexadecimal ASCII character to convert to a value
Output Registers:	Accumulator: Converted value between 0 and 15 (00h and 0Fh)
	Carry: 0=No error, 1=ASCII value wasn't hexadecimal digit
Registers Destroyed:	None

The `HexToNibble` subroutine converts a hexadecimal ASCII character to a value between 0 and 15 (00h and 0Fh). This is useful when receiving hexadecimal data from the user and needing to convert it to a numeric value for subsequent processing.

Original Hex Digit	*Original ASCII Value*	*HexToNibble Result*	*Original Hex Digit*	*Original ASCII Value*	*HexToNibble Result*
'0'	30h	00h	'8'	38h	08h
'1'	31h	01h	'9'	39h	09h
'2'	32h	02h	'A', 'a'	41h, 61h	0Ah
'3'	33h	03h	'B', 'b'	42h, 62h	0Bh
'4'	34h	04h	'C', 'c'	43h, 63h	0Ch
'5'	35h	05h	'D', 'd'	44h, 64h	0Dh
'6'	36h	06h	'E', 'e'	45h, 65h	0Eh
'7'	37h	07h	'F', 'f'	46h, 66h	0Fh

The following code converts the ASCII value 'A' (41h) to 0Ah:

```
MOV A,#'A'
LCALL HexToNibble
```

17.4.1.1.17 Subroutine: GetHexValue

Subroutine:	**GetHexValue**
Entry Point:	0071h
Purpose:	Converts hexadecimal digits in input registers to numeric values
Inputs Registers:	R0: Points to first internal RAM address of hexadecimal digits
	R1: Points to first internal RAM address to store results
Output Registers:	Accumulator: The first non-hex character that terminated conversion
	R0: Points to the internal RAM address after the non-hex character
	Carry: 0=Successful conversion, 1=Nothing was converted
Registers Destroyed:	None

The `GetHexValue` subroutine converts between one and four hexadecimal ASCII digits contained within a buffer in internal RAM to their 16-bit numeric value. It is normally used to convert hexadecimal values that were received in a character buffer using the `GetSerialLine` subroutine (see section 17.4.1.1.5).

Before calling `GetHexValue`, the program must initialize R0 and R1. The R0 register must be set to point to where the first hexadecimal character to be decoded is in internal RAM. R1 must be set to point to where in internal RAM the decoded value will be placed. The decoded value is a 16-bit value which requires two bytes: The high byte will be stored at the address pointed to by R1 and the low byte will be stored in the byte immediately following that address. The subroutine will then scan the buffer and convert the hexadecimal values to the 16-bit value until it reaches a character that is not a valid hexadecimal digit. It will then load the accumulator with that character and set R0 to the address following that non-hexadecimal character.

For example:

```
LCALL GetSerialLine      ;Get line, R0 returns pointing to buffer
MOV R1,#50h              ;Set address for decoded 16-bit value
LCALL GetHexValue        ;Convert hex to 16-bit value in 50h & 51h
```

This code will first get a line of input from the serial port; for the sake of example, the user enters the ASCII string "681FXRS" and `GetSerialLine` returns with R0 set to B0h as the address used by SBCMON for the serial input buffer. The second line sets the decode address to 50h so that the high byte of the decoded value will be stored at internal RAM address 50h and the low byte of the decoded value will be stored at internal RAM address 51h. The program then calls `GetHexValue`.

Upon return, internal RAM address 50h will contain the value 68h (the high byte of the hexadecimal value entered by the user), address 51h will contain 1Fh (the low byte of the hexadecimal value entered by the user), the accumulator will contain the character 'X' (58h), and R0 will contain B5h which is the address of the character in the buffer following the letter that terminated the hexadecimal decoding.

The fact that R0 is always advanced by `GetHexValue` to point to the character after the non-hexadecimal character means that it can be called repeatedly to decode multiple hexadecimal values contained within a single line. For example, if the line entered by the user was "123A B981 C912" the following lines could decode it:

```
LCALL GetSerialLine      ;Get line, R0 returns pointing to buffer
MOV R1,#50h              ;Set address to decode 1st 16-bit value
LCALL GetHexValue        ;Convert 123A, store in 50h and 51h
MOV R1,#52h              ;Set address to decode 2nd 16-bit value
LCALL GetHexValue        ;Convert B981, store in 52h and 53h
MOV R1,#54h              ;Set address to decode 3rd 16-bit value
LCALL GetHexValue        ;Convert C912, store in 54h and 55h
```

17.4.1.1.18 Subroutine: GetSerialHex

Subroutine:	**GetSerialHex**
Entry Point:	0074h
Purpose:	Reads 2 hex digits from serial port, converts to value in accumulator
Inputs Registers:	None
Output Registers:	Accumulator: Converted value of the 2 hex digits received
	Carry: 0=No error, 1=Error, characters weren't hexadecimal digits
Registers Destroyed:	R1

The GetSerialHex subroutine waits for two characters on the serial port and treats them as hexadecimal digits. It converts each character to the numeric value it represents and combines them to form an 8-bit value that is returned in the accumulator. For example, if this subroutine is called and the serial report receives the character 'F' followed by the character '8', this routine will return the value F8h in the accumulator.

17.4.1.1.19 Subroutine: ToUpper

Subroutine:	**ToUpper**
Entry Point:	0077h
Purpose:	Converts the character in the accumulator to upper-case
Inputs Registers:	Accumulator: Character to be made upper-case
Output Registers:	Accumulator: Character converted to upper-case
Registers Destroyed:	None

This routine does nothing more than make a lower-case character into an upper-case character. If the character in the accumulator is not a letter or is already upper-case then the value is left unchanged.

17.4.1.1.20 Subroutine: SendDecimalR4567

Subroutine:	**SendDecimalR4567**
Entry Point:	007Ah
Purpose:	Sends the value in R4-R7 to the serial port as a decimal value
Inputs Registers:	R4 (high) – R7 (low): Value to be sent to serial port as decimal value
Output Registers:	None
Registers Destroyed:	R0, R1, Accumulator

This routine takes the 32-bit value contained in registers R4 (high) through R7 (low) and sends it to the serial port as a decimal value. For example, if R4=01h, R5=25h, R6=30h, and R7=15h, that represents the value 19,214,357. Thus the SendDecimalR4567 routine would send the string "19214357" to the serial port. This function is useful for displaying decimal values from within SBCMON programs.

17.4.1.1.21 Subroutine: DivideBy10

Subroutine:	**DivideBy10**
Entry Point:	007Dh
Purpose:	Divides the value pointed to by @R0 through @R0+3 by 10
Inputs Registers:	R0: Points to high byte of value to be divided by 10, remaining three bytes of value are contained in next three bytes of memory.
Output Registers:	Accumulator: Remainder of the division by 10.
Registers Destroyed:	None

This routine takes the 32-bit value that is contained in the internal RAM addresses pointed to by R0 (and the subsequent three bytes of memory) and divides it by 10 (decimal). The result is left in the same four bytes of memory and the remainder is left in the accumulator.

While the DIV AB instruction can be used to divide values less than 256 by 10, this routine is useful to divide large 32-bit values by 10.

17.4.1.1.22 Subroutine: Check4BytesForZero

Subroutine:	**Check4BytesForZero**
Entry Point:	0080h
Purpose:	Checks that value pointed to by @R0 through @R0+3 is zero.
Inputs Registers:	R0: Points to high byte of value to be checked for zero, remaining three bytes of value are contained in next three bytes of memory.
Output Registers:	Accumulator: 0=All four bytes are zero, Non-0: 1 or more bytes are non-zero
Registers Destroyed:	None

This routine analyzes the 32-bit value that is contained in the internal RAM addresses pointed to by R0 (and the subsequent three bytes of memory) and determines if all four bytes are zero. If all four bytes are zero, the routine returns zero in the accumulator. Otherwise, the accumulator is non-zero.

17.4.1.1.23 Subroutine: ShiftR765Left

Subroutine:	**ShiftR765Left**
Entry Point:	0083h
Purpose:	Shifts the value in R5 (high) through R7 (low) left one bit
Inputs Registers:	R5 (high) – R7 (low): Value to shift one bit to the left.
Output Registers:	R5 (high) – R7 (low): Shifted value.
Registers Destroyed:	Accumulator

This routine takes the value contained in R5 (high byte) through R7 (low byte) and shifts the entire value left by one bit, propagating through the three registers as necessary. This is effectively a RL A instruction that acts on R5 through R7 rather than the accumulator.

17.4.1.1.24 Subroutine: AddR67to23

Subroutine:	**AddR67to23**
Entry Point:	0086h
Purpose:	Adds R6 (high) and R7 (low) to R2 (high) and R3 (low)
Inputs Registers:	R6 (high), R7 (low): 16-bit value to add
Output Registers:	R2 (high), R3 (low): 16-bit value to be added to, and result
Registers Destroyed:	Accumulator

This routine takes the 16-bit value contained in R6 (high byte) and R7 (low byte) and adds it to the 16-bit value contained in R2 (high byte) and R3 (low byte). The result is stored in R2 and R3.

17.4.1.1.25 Subroutine: AddR4567to0123

Subroutine:	**AddR4567to0123**
Entry Point:	0089h
Purpose:	Adds R4 (high) through R7 (low) to R0 (high) through R3 (low)
Inputs Registers:	R4 (high) - R7 (low): 32-bit value to add
Output Registers:	R0 (high) - R3 (low): 32-bit value to be added to, and result
Registers Destroyed:	Accumulator

This routine takes the 32-bit value contained in R4 (high byte) through R7 (low byte) and adds it to the 32-bit value contained in R0 (high byte) through R3 (low byte). The result is stored in R0 through R3.

17.4.1.1.26 Subroutine: InitializeLCD

Subroutine:	**InitializeLCD**
Entry Point:	008Ch
Purpose:	Initializes the LCD
Inputs Registers:	Bit LCDDirect4: 0=8-bit mode, 1=4-bit mode
	Bit LCDMMMode: 1=Direct mode, 0=Memory-mapped mode
Output Registers:	None
Registers Destroyed:	Accumulator

This routine initializes the LCD by sending it one of the following sequence of commands: 38h, 0Eh, 06h (for 8-bit mode) or 28h, 0Eh, 06h (for 4-bit mode). The commands are automatically sent to the right bus (direct or memory-mapped) depending on the setting of the LCDMMMode bit.

The LCDMMMode bit should be set with the 'M' command prior to executing the user program.

17.4.1.1.27 Subroutine: ReadLCDStatus

Subroutine:	**ReadLCDStatus**
Entry Point:	008Fh
Purpose:	Read LCD status
Inputs Registers:	None
Output Registers:	Accumulator: Returned status of the LCD
Registers Destroyed:	None

This routine waits for the LCD to return its status register. This can be used for waiting for the LCD to become not busy and, upon becoming not busy, to return the status register which includes the LCD cursor position.

17.4.1.1.28 Subroutine: ReadLCDText

Subroutine:	**ReadLCDText**
Entry Point:	0092h
Purpose:	Read LCD text
Inputs Registers:	None
Output Registers:	Accumulator: Returned text from the LCD
Registers Destroyed:	None

This routine reads the value contained in the LCD at the current cursor position and returns it in the accumulator. This is useful for reading what the LCD is currently displaying which, in turn, is useful in many LCD scrolling applications.

17.4.1.1.29 Subroutine: SetDptrMMKey1

Subroutine:	**SetDptrMMKey1**
Entry Point:	0095h
Purpose:	Set DPTR to address of memory-mapped keypad row #1
Inputs Registers:	None
Output Registers:	DPTR: Set to the memory-mapped address of keypad row #1
Registers Destroyed:	None

This routine sets DPTR to the memory-mapped address of keypad row #1. By calling this routine instead of setting DPTR to a specific value, a program can operate under SBCMON secure in the knowledge that even if the underlying hardware that SBCMON runs on is changed, the correct address for keypad row #1 can always be obtained by calling this routine.

17.4.1.1.30 Subroutine: SetDptrMMKey2

Subroutine:	**SetDptrMMKey1**
Entry Point:	0098h
Purpose:	Set DPTR to address of memory-mapped keypad row #2
Inputs Registers:	None
Output Registers:	DPTR: Set to the memory-mapped address of keypad row #2
Registers Destroyed:	None

This routine sets DPTR to the memory-mapped address of keypad row #2. By calling this routine instead of setting DPTR to a specific value, a program can operate under SBCMON secure in the knowledge that even if the underlying hardware that SBCMON runs on is changed, the correct address for keypad row #2 can always be obtained by calling this routine.

17.4.1.1.31 Subroutine: SetDptrMMKey3

Subroutine:	**SetDptrMMKey1**
Entry Point:	009Bh
Purpose:	Set DPTR to address of memory-mapped keypad row #3
Inputs Registers:	None
Output Registers:	DPTR: Set to the memory-mapped address of keypad row #3
Registers Destroyed:	None

This routine sets DPTR to the memory-mapped address of keypad row #3. By calling this routine instead of setting DPTR to a specific value, a program can operate under SBCMON secure in the knowledge that even if the underlying hardware that SBCMON runs on is changed, the correct address for keypad row #3 can always be obtained by calling this routine.

17.4.1.1.32 Subroutine: SetDptrMMKey4

Subroutine:	**SetDptrMMKey4**
Entry Point:	009Eh
Purpose:	Set DPTR to address of memory-mapped keypad row #4
Inputs Registers:	None
Output Registers:	DPTR: Set to the memory-mapped address of keypad row #4
Registers Destroyed:	None

This routine sets DPTR to the memory-mapped address of keypad row #4. By calling this routine instead of setting DPTR to a specific value, a program can operate under SBCMON secure in the knowledge that even if the underlying hardware that SBCMON runs on is changed, the correct address for keypad row #4 can always be obtained by calling this routine.

17.4.1.1.33 Subroutine: SetDptrMMLcdWC

Subroutine:	**SetDptrMMLcdWC**
Entry Point:	00A1
Purpose:	Set DPTR to address of memory-mapped LCD for write command
Inputs Registers:	None
Output Registers:	DPTR: Set to the memory-mapped address of LCD for write command
Registers Destroyed:	None

This routine sets DPTR to the memory-mapped address of the LCD for writing a command. By calling this routine instead of setting DPTR to a specific value, a program can operate under SBCMON secure in the knowledge that even if the underlying hardware that SBCMON runs on is changed, the correct address for the LCD can always be obtained by calling this routine.

17.4.1.1.34 Subroutine: SetDptrMMLcdWT

Subroutine:	**SetDptrMMLcdWT**
Entry Point:	00A4
Purpose:	Set DPTR to address of memory-mapped LCD for write text
Inputs Registers:	None
Output Registers:	DPTR: Set to the memory-mapped address of LCD for write text
Registers Destroyed:	None

This routine sets DPTR to the memory-mapped address of the LCD for writing text. By calling this routine instead of setting DPTR to a specific value, a program can operate under SBCMON secure in the knowledge that even if the underlying hardware that SBCMON runs on is changed, the correct address for the LCD can always be obtained by calling this routine.

17.4.1.1.35 Subroutine: SetDptrMMLcdRC

Subroutine:	**SetDptrMMLcdRC**
Entry Point:	00A7
Purpose:	Set DPTR to address of memory-mapped LCD for read command
Inputs Registers:	None
Output Registers:	DPTR: Set to the memory-mapped address of LCD for read command
Registers Destroyed:	None

This routine sets DPTR to the memory-mapped address of the LCD for reading a command. By calling this routine instead of setting DPTR to a specific value, a program can operate under SBCMON secure in the knowledge that even if the underlying hardware that SBCMON runs on is changed, the correct address for the LCD can always be obtained by calling this routine.

17.4.1.1.36 Subroutine: SetDptrMMLcdRT

Subroutine:	**SetDptrMMLcdRT**
Entry Point:	00AA
Purpose:	Set DPTR to address of memory-mapped LCD for read text
Inputs Registers:	None
Output Registers:	DPTR: Set to the memory-mapped address of LCD for read text
Registers Destroyed:	None

This routine sets DPTR to the memory-mapped address of the LCD for reading text. By calling this routine instead of setting DPTR to a specific value, a program can operate under SBCMON secure in the knowledge that even if the underlying hardware that SBCMON runs on is changed, the correct address for the LCD can always be obtained by calling this routine.

17.4.1.1.37 Subroutine: MakeDecimal2Digit

Subroutine:	**MakeDecimal2Digit**
Entry Point:	00AD
Purpose:	Convert BCD value in accumulator and convert to two ASCII digits
Inputs Registers:	Accumulator: BCD value 00h through 99h
Output Registers:	Accumulator: High byte of result (0x30 – 0x39)
	B: Low byte of result (0x30 – 0x39)
Registers Destroyed:	None

This routine takes a BCD number stored in the accumulator and converts it two two ASCII digits in the accumulator (high byte) and B register (low byte). For example, if the routine is called with the value 78h, the routine would return accumulator as 0x37 and the B register as 0x38. These could then be sent immediately to the serial port or LCD for display purposes.

17.4.1.1.38 Subroutine: GetKeyDebounced

Subroutine:	**GetKeyDebounced**
Entry Point:	00B0
Purpose:	Reads one keypress from the keypad, debounces it, and returns it
Inputs Registers:	None
Output Registers:	Accumulator: The value of the key pressed
Registers Destroyed:	R3, R6, and R7

This routine waits for the user to press a key on the keypad. When a key is detected, it is properly debounced to protect against false readings. Once debounced, the value of the key is returned so that the calling program may process it. Once the key is processed by the main program, the main program should also call the WaitKeyReleased routine to verify that the key has been released so it doesn't automatically re-read the same keypress.

17.4.1.1.39 Subroutine: WaitKeyReleased

Subroutine:	**GetKeyDebounced**
Entry Point:	00B3
Purpose:	Waits for the key of the keypad to be released
Inputs Registers:	None
Output Registers:	None
Registers Destroyed:	R0, Accumulator

This routine is called after the GetKeyDebounced key is used to obtain a keypress. Once the key is obtained and processed, the program should call this routine to make sure the key is released so that the program doesn't re-interpret the keypress as a new keypress.

17.4.2 Using Interrupts in External Programs with SBCMON

SBCMON itself does not use any interrupts nor does it use interrupt service routines. This means it is possible for an external program that is running in SBCMON to implement interrupts. The problem that arises, however, is that when an interrupt occurs it will always call the interrupt service routine which is in the first few bytes of code memory. That is, external 0 interrupt, when enabled, will jump to address 0003h when it is triggered. In that case, how can an external program that is running in the address range of 8000h – FFFFh handle an interrupt that will branch to 0003h?

SBCMON solves this problem by including code at the interrupt service routine addresses which automatically jump to the same address, but in the 8000h memory area. Specifically, the following code exists in SBCMON:

```
ORG 0003h              ;External 0 Interrupt
LJMP 8003h
ORG 000Bh              ;Timer 0 Interrupt
LJMP 800Bh
ORG 0013h              ;External 1 Interrupt
LJMP 8013h
ORG 001Bh              ;Timer 1 Interrupt
LJMP 801Bh
ORG 0023h              ;Serial Port Interrupt
LJMP 8023h
ORG 002Bh              ;Timer 2 Interrupt
LJMP 802Bh
ORG 0033h              ;Unused in basic 8052
LJMP 8033h
ORG 003Bh              ;Unused in basic 8052
LJMP 803Bh
```

This means that if external 0 interrupt is enabled and is subsequently triggered, the interrupt will vector to 0003h as always, but SBCMON will immediately jump to 8003h. Thus if an external program needs to use interrupts, it simply places its interrupt code at the corresponding address offset by 8000h:

Interrupt Name	Real ISR Address	SBCMON External Program ISR Address
External 0	0003h	8003h
Timer 0	000Bh	800Bh
External 1	0013h	8013h
Timer 1	001Bh	801Bh
Serial Port	0023h	8023h
Timer 2	002Bh	802Bh
Unused 1	0033h	8033h
Unused 2	003Bh	803Bh

Therefore, if an external program needs to use external 0 interrupt, it would place its interrupt code at 8003h and enable the interrupt. The only difference between this and normal operation without SBCMON is that the LJMP from 0003h to 8003h takes two instruction cycles so any ISRs that are time-critical must take into account this additional execution time.

17.4.3 Using Timers in SBCMON

SBCMON uses only timer 1 in mode 2 in order to establish the baud rate for the serial port. External programs executing under SBCMON may use timer 0 and timer 2. When using timer 0, the external program must be careful not to change the mode of timer 1 when it sets TMOD or the serial port baud rate will be affected.

The external program may use or change timer 1 only if it specifically sets it to provide another baud rate, selects a different serial port mode such that the baud rate is determined by a source other than timer 1, or if it configures timer 2 to provide the baud rate clock instead.

CHAPTER 18: INTERFACING TO 4x4 KEYPAD

It is very common for an embedded system to include a device which allows the user to provide information to the microcontroller. This may be as simple as a push button or as complex as a full-fledged keyboard.

Somewhere between the two extremes is the "keypad" class of devices. Keypads come in many shapes and sizes that may be chosen based on the needs of the application. Some are flex-type keypads which can be mounted on a panel with adhesive tape, such as the keypad shown on the left; others are more rigid such as the one shown on the right. Both of these are 4x4 keypads which refer to the fact that they have four rows of four keys each. Other common keypads include 3x4 (usually used for numeric entry) and 4x5 (similar to the keypads shown but with an extra row of function keys on the right).

The 8052.com SBC uses the rigid keypad pictured to the right, but virtually any 4x4 keypad would work—though minor modifications may be necessary to the software depending on the layout of the keypad.

At first glance it would seem that a 4x4 keypad is nothing more than 16 push buttons. While this is technically true, the interface is slightly more complicated. If the keypad were nothing but a collection of 16 buttons, the keypad alone would require 16 I/O lines on the microcontroller—one for each button. This would be an extremely inefficient use of available I/O lines. Instead, a 4x4 keypad occupies just eight lines. This is accomplished by using four lines to select one of four rows of keys and using the remaining four lines to return the status for each key in that row.

As seen in the figure on the next page, there are four row select lines and four status lines. If row select 0 is activated by bringing the line low then status line 0 will return the state of the 'A' key, status line 1 will return the state of the 'B' key, status line 2 will return the state of the 'C' key, and status line 3 will return the state of the 'D' key. If row select 1 is activated instead then the status lines will hold the corresponding states of the '3', '6', '9', and '#' keys.

A low voltage level on the status line indicates the given key is pressed. So if, again, row 0 is selected and the 'A' key is pressed, status column 0 will return a low voltage. If the 'A' key is not pressed the status line will return a high voltage. This also allows the keypad to return multiple simultaneous keypresses—if row 0 is selected and the 'A' and 'B' keys are pressed at the same time then the keypad will simply return low voltages on both the column 0 and column 1 status lines. The microcontroller detects keypresses on the keypad by quickly scanning each of the four rows and checking the status of each column in each row.

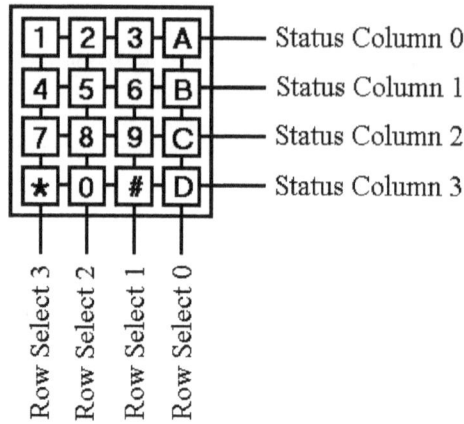

18.1 Direct Connection of Keypad

The simplest way to connect a keypad is to interface it directly to the I/O lines of the microcontroller, perhaps P1. Since P1 has eight I/O lines and the keypad needs eight lines, it is a perfect fit. This technique is pictured in the following schematic excerpt.

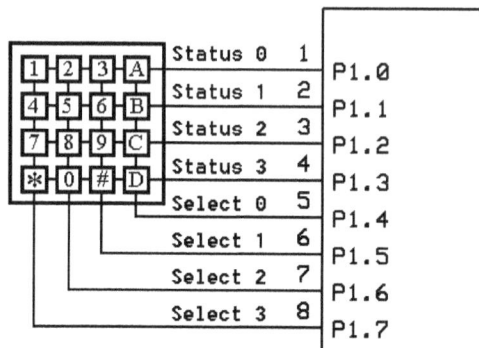

In such a design the row of the keypad is selected with P1.4, P1.5, P1.6, and P1.7 while the status of the keys is read using P1.0, P1.1, P1.2, and P1.3. In order to read the status of a port pin, that pin must be set to '1'—so the low four bits of P1 must be set in order to read the status of those lines. All four of the high bits must also be set to '1' except for the pin that corresponds to the row that is to be selected. For example, to read the value of the keys in row 3, the value 7Fh would be written to P1. This represents the highest bit being clear and the other 3 high bits being set ('7') and all four low bits being set ('F') so that the status of each line may be read.

232

The following table shows the value that must be written to P1 to read the status of the buttons in the corresponding row of the keypad.

Row to read	Value to write to P1
0	EFh
1	DFh
2	BFh
3	7Fh

Once the correct value has been written to P1 to select the row to be read, the value of P1 may be read back to get the status of the low four bits which will hold the status of the four keys in that row. When reading the value of P1, the program is only interested in the low four bits—that is, only P1.0, P1.1, P1.2 and P1.3. The value of the other four bits of P1 will correspond to the row that was selected.

Thus to read the status of row 3 of the keypad, the following assembly language instructions could be used:

```
MOV    P1,#7Fh     ;Select row 3 of the keypad
MOV    A,P1        ;Read P1 to get status of buttons in the row
ANL    A,#0Fh      ;Only the 4 low bits are of interest
```

These three instructions will select row three, read the current status of the keypad by reading P1, and zero out the high four bits which are not of interest. This will leave the four low bits that correspond to each of the status column lines intact. Each of these bits can then be analyzed to determine if the corresponding key was pressed. If the bit is set then the key isn't pressed; if the bit is clear then the key is pressed.

Thus the above code could be expanded:

```
JNB    ACC.0,Key1Pressed      ;The '1' key is pressed
JNB    ACC.1,Key4Pressed      ;The '4' key is pressed
JNB    ACC.2,Key7Pressed      ;The '7' key is pressed
JNB    ACC.3,KeyStarPressed   ;The star/asterisk key is pressed
```

This illustrates the logic behind how a keypad operates. It is very simple from both a hardware and software perspective: Four of the lines select the row of keys that are to be read while the other four lines return the state of each of the keys in that row.

However, connecting the keypad directly to the microcontroller is not a very efficient use of I/O lines. If a given application has plenty of I/O lines then the direct connection may be acceptable—if the application is such that there are eight extra I/O lines available then dedicating all of P1 to nothing but keypad input/output may be a valid solution. If, however, the application doesn't have eight I/O lines available for a keypad then the memory-mapped approach described in the next section should be considered.

18.2 Memory-Mapped Connection of Keypad

Many devices, such as the keypad, can be interfaced to the microcontroller using a memory-mapped approach. A memory-mapped design takes advantage of the fact that the microcontroller already has 16 address lines and an 8-bit data bus available. The address and data bus are normally used to access

EPROM and RAM but there is absolutely no reason why they can't be used to access other devices—such as the 4x4 keypad. The 8052.com SBC uses a memory-mapped approach to access the keypad so that the keypad alone doesn't occupy eight I/O lines.

Using a memory-mapped design only makes sense if the project already contains external memory (EPROM or RAM) or if multiple devices are to be memory-mapped. As a very minimum, a memory-mapped design will require the 74LS573 (U2) to create the address bus and the 74LS138 (U5) to select the memory-mapped device. If there is no external memory and the only device that is to be accessed is the keypad then it doesn't make sense to add two ICs to the hardware design to access the keypad when it could be wired directly to P1 with no additional ICs required. Memory-mapped designs allow existing external memory to be used to access other external devices and it only is useful if there are already other external memory or devices in the design that can share the address and data bus.

Recall from section 14.2.3 that the 74LS138 address decoder (U5) is used to select the EPROM or RAM IC based on the high four bits of the address that the microcontroller is accessing. Reviewing the schematic it can be seen that Y5 is labeled MM_KEYPAD. This line becomes the keypad enable line for the keypad while the low four address lines (A0, A1, A2, and A3) become the keypad row select line. Thus if the Y5 line on the 74LS138 is asserted, A0 is low, and A1, A2, and A3 are high, then keypad row 0 will be selected. Likewise, if A0, A2, and A3 are high and A1 is low, keypad row 1 will be selected.

This is depicted in the schematic excerpt below—it's actually not as complicated as it might look at first glance.

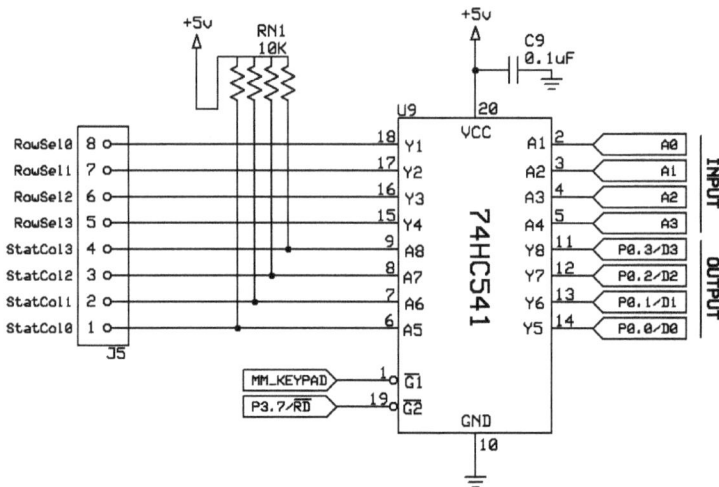

First, the 74HC541 is a buffer/drive IC which takes the inputs (A1-A8) and makes them available on the outputs (Y1-Y8) when both the G1 and G2 pins are low; if either G1 or G2 is high, the output pins are in high impedance mode. High impedance means, essentially, that the outputs are disconnected from the circuit and are not outputting a high nor a low—they are doing nothing and are not effecting the lines in any way. Thus when the keypad is selected by the MM_KEYPAD line from the '138, the 74HC541 will be active—but only when G2 is also brought low when the microcontroller is reading from external memory.

234

When the '541 is activated by a keypad row being selected, it takes the four low address lines A0 through A3 and outputs them on Y1-Y4. This is connected to the keypad row select lines which allows the selected keypad row to be active. The A5-A8 inputs of the '541 are attached to the status columns returned by the keypad and its outputs are attached to the the data bus (P0.0 – P0.3) such that the keypad's return values will be available in the low four bits of the value read by MOVX. The pull-up resistors (RN1) ensure that the value of each keypad status line is high when its corresponding key is not pressed.

The '541 serves two primary purposes:

1) The voltages of the "HC" line of ICs is more appropriate for driving external devices. Although the keypad will normally be attached to the SBC directly, the use of an HC IC makes the device more tolerant should a ribbon cable be used to attach the keypad.
2) It ensures that keypresses on the keypad will not interfere with the data bus when the keypad is not being addressed. If it were not for the '541, a user could interfere with an access to the static RAM (U4) by pressing a key on the keypad. The '541 isolates the keypad from the data bus except when it is specifically being addressed by a MOVX instruction.

Selecting a row of the keypad is a matter of accessing the correct external memory address. Since the low four address lines select the keypad row, the four least significant bits of the address will determine which row is selected—the address line that is low will be the row that is selected. For example, if all low four address lines were high except for A0, keypad row 0 would be selected. This could be accomplished by accessing external address 500Eh. The address 50xx would select the keypad while the low nibble, 0Eh, represents all low four address lines in a high state except for A0 which is low.

In reality, the only portions of the address that make a difference when accessing the keypad is the first digit of the address which must be 5, and the last digit of the address which selects the keypad row. This is demonstrated in the following table:

Address Range	Keypad row selected
5xxEh	Row 0
5xxDh	Row 1
5xxBh	Row 2
5xx7h	Row 3

This means that row 0 of the keypad could be accessed by accessing 500Eh, but it could also be accessed with the address 573Eh. The only address lines that matter are the ones that make up the first digit "5" and the last digit "E".

Now that the keypad's row has been selected by a memory address the only thing left to do is obtain the keypad's status. This is accomplished by connecting the four keypad status lines to the 8-bit data bus so that the MOVX instruction is able to read the value from the keypad just as if it were a memory IC. The keypad's column 0 status line is connected to bit 0 of the data bus, column 1 status is connected to bit 1 of the data bus, and so on. Reading the keypad then becomes a matter of reading any value in the specified address range. For example:

```
MOV     DPTR,#500Eh         ;Select row 0 keypad address
MOVX    A,@DPTR             ;Move value of keypad row 0 to accumulator
MOV     DPTR,#500Dh         ;Select row 1 keypad address
MOVX    A,@DPTR             ;Move value of keypad row 1 to accumulator
```

In this design, the 74LS138 and the low four address lines A0 through A3 do all the work of selecting the correct keypad row based on the memory address the program accesses. The program need only read any address in the address range for that row to successfully read the status of the keys in that keypad row.

18.2.1 Reading and Debouncing Keypad

Reading a 4x4 keypad efficiently requires more than just reading the keypad and branching based on the status of the bits read. There is the matter of decode efficiency and the matter of key debounce.

18.2.1.1 Brute Force Keypad Decoding

A straight-forward, brute-force, and terribly *inefficient* approach to decoding the keypad is shown in the following code.

```
1       MOV     R2,#00h             ;R2 will hold the key, 00=No key
2       MOV     DPTR,#500Eh         ;Select row 0 address range
3       MOVX    A,@DPTR             ;Get the key press
4       CJNE    A,#0Eh,NotA         ;It's not the 'A' character
5       MOV     R2,#'A'             ;It is 'A' so store it in R2
6       LJMP    DecodeDone          ;Character found so exit search mode
7     NotA:
8       CJNE    A,#0Dh,NotB         ;It's not the 'B' character
9       MOV     R2,#'B'             ;It is 'B' so store it in R2
10      LJMP    DecodeDone          ;Character found so exit search mode
11    NotB:
12      CJNE    A,#0Bh,NotC         ;It's not the 'C' character
13      MOV     R2,#'C'             ;It is 'C' so store it in R2
14      LJMP    DecodeDone          ;Character found so exit search mode
15    NotC:
16      CJNE    A,#07h,DecodeDone   ;It's not the 'D' character
17      MOV     R2,#'D'             ;It is 'D' so store it in R2
18    DecodeDone: ;R2 holds the value of the keypress, 0=no key
```

This will decode row 0 of the keypad. Lines 1-3 sets R2 to zero by default and reads the keypad row 0. Lines 4-6 check to see if the 'A' key was pressed and set R2 to 'A' if it was. Lines 7-10 check for the 'B' key, lines 11-14 check for the 'C' key, and lines 15-17 check for the 'D' key.

The above example, while functional, is terribly inefficient. It requires 2 bytes of memory for the initial MOV R0,#00h instruction and an additional 30 bytes for each row that is decoded. The above code would have to be inserted three more times, one for each row. Thus to fully decode the keypad into R2 would require a total of 122 bytes of program memory and approximately 69 lines of assembly language code.

18.2.1.2 Efficient Keypad Decoding

A better approach to decoding is to use a loop and a look-up table. Although the following code may look more complicated than the previous code, it is far easier to maintain since the memory addresses of the

236

keypad rows are all stored together in `KP_AddrTable` and the characters associated with each of the 16 keys are stored together in `KP_KeyTable`—this makes it very easy to modify if it is necessary to change the definition of a given key. Another major advantage of the following approach is that it only requires 40 lines of assembly language code and 61 bytes of memory to decode all four rows of the keypad—just half of what the previous approach would require.

This approach scans each of the keypad lines by obtaining the keypad address from the `KP_AddrTable`, determines if a key has been pressed, determines which key in the selected line has been sent, and then translates it to the ASCII value by consulting `KP_KeyTable`.

```
 1 KP_Loop:
 2     MOV R3,#03h                   ;Start with row 3 of the keypad
 3 KP_Cycle:
 4     MOV A,R3                      ;Get current row (0-3) in accumulator
 5     RL A                          ;Multiply row by 2 to get table offset
 6     MOV B,A                       ;Store offset temporarily in B
 7     MOV DPTR,#KP_AddrTable        ;Point to the address table
 8     MOVC A,@A+DPTR                ;Get the DPTR high byte from the table
 9     PUSH ACC                      ;Push the DPTR high byte on stack
10     INC DPTR                      ;Point to DPTR low byte in table
11     MOV A,B                       ;Restore offset into table from B
12     MOVC A,@A+DPTR                ;Get the DPTR low byte from the table
13     MOV DPL,A                     ;Move the low byte from table into DPL
14     POP DPH                       ;Pop high byte from table into DPH
15     MOVX A,@DPTR                  ;Read the status of the keypad row
16     ANL A,#0Fh                    ;Only interest in low 4 bits
17     XRL A,#0Fh                    ;Invert bits, keypress=1, 0=No key
18     JNZ KP_ProcessKey             ;If the value is not zero, process it
19     DEC R3                        ;Decrement row counter
20     CJNE R3,#0FFh,KP_Cycle        ;If not FF, process next row
21     SJMP   KP_Loop                ;After 4 rows, start with row 3 again
22 KP_ProcessKey:
23     CLR C                         ;Make sure carry starts clear
24     MOV R0,#00h                   ;R0 starts at zero
25 KP_FindKeyNum:
26     RRC A                         ;Rotate the value right into the carry
27     JC KP_KeyFound                ;If set, we found the key so exit
28     INC R0                        ;Increment the key number indicator
29     SJMP KP_FindKeyNum            ;Keep searching until we find the key
30 KP_KeyFound:
31     MOV A,R3                      ;Get row number
32     RL A                          ;Multiply by 2
33     RL A                          ;Multiply by 4
34     ADD A,R0                      ;Add the key offset
35     MOV DPTR,#KP_KeyTable         ;Get the start address of the key table
36     MOVC A,@A+DPTR                ;Get the keypress into accumulator
37     RET                           ;Return to calling program with keypress
38 KP_KeyTable:
39     DB "ABCD"                     ;Decode for keys in row 0
40     DB "369#"                     ;Decode for keys in row 1
41     DB "2580"                     ;Decode for keys in row 2
42     DB "147*"                     ;Decode for keys in row 3
43 KP_AddrTable:
44     DW 500Eh                      ;Address to access row 0 (500Eh)
45     DW 500Dh                      ;Address to access row 1 (500Dh)
46     DW 500Bh                      ;Address to access row 2 (500Bh)
47     DW 5007h                      ;Address to access row 3 (5007h)
```

Lines 1-2 initialize R3 to 3 so that the keypad scanning will start at row 3 and subsequently be decremented to scan rows 2, 1, and 0 before being reset again to 3.

Lines 3-14 take the current keypad row held in R3 and use it as an offset into the KP_AddrTable lookup table to get the address of the memory address to read for that row. For example, if R3 holds 3 then it will get the fourth entry from KP_AddrTable which is 5007h and use that as the the memory address to read using DPTR.

Lines 15-17 read the current status of the selected keypad row. The value returned by MOVX will be 0Fh if no key was pressed, 0Eh if the topmost key was pressed, 0Dh if the second highest key was pressed, and so on. Line 16 makes sure there is no garbage in the high bits that were read while line 17 inverts the bits—so instead of 0Eh being returned when the topmost key is pressed, the value 01h will be calculated; instead of 0Dh for the second highest key, 02h will be calculated, and so on.

Lines 18-21 check to see if a key was pressed that needs to be processed. If not, the R3 row variable is decremented. If it hasn't gone from 00h back to FFh then it's still a valid row between 0 and 3 so it branches back to KP_Cycle to process the next row. Once R3 reaches FFh, execution will fall through to line 21 which will jump back to KP_Loop which will re-initialize R3 to scan row 3.

Lines 22-29 are a tricky way of adjusting the value that was read from the keypad to a corresponding number between 0 and 3. The values returned by the keypad, after being inverted on line 17, will be either 1, 2, 4, or 8. Since the program will be using a lookup table, these values need to be converted to an offset between 0 and 3. These lines accomplish this by using R0 to hold the calculated offset—the value read from the keypad is then shifted into the carry bit one bit at a time. Each time the register is shifted, the R0 counter is incremented. When the bit is detected on line 27 the conversion will be complete such that keypad value 1 will have been converted to 0, keypad value 2 will have been converted to 1, keypad value 4 will have been converted to 2, and keypad value 8 will have been converted to 3.

Lines 30-34 calculate an offset into the KP_KeyTable. Lines 31-33 take the row number (held in R3) and multiply it by 4 (since there are four characters per row) by shifting the value to the left twice. Line 34 then adds the offset of the character that was just scanned to form a final offset for the keypress.

Lines 35-36 set DPTR to the address of KP_KeyTable and then read the value contained at that address added to the offset that was calculated in lines 30-34. The result is that the accumulator will hold the value of the key that was pressed.

Line 37 returns to the program that called the routine. At this time, the accumulator will hold the ASCII value of the keypress which can be displayed or handled as necessary.

This is a far more efficient algorithm than the brute force decoding method described in the previous section.

18.2.1.3 Debouncing the Keypad

While either of the solutions described in the previous two sections will be able to detect a keypress on the keypad, one additional important aspect is missing: key debounce.

238

The keys of a keypad are mechanical in nature. When a key is pressed, two electrical contacts are mechanically forced together which completes a circuit that is subsequently detected by the microcontroller; but the process of closing the contacts within the keypad is not perfectly clean. As the contacts are pushed together there will be a certain amount of "bouncing" as the connection closes momentarily, only to open again microseconds later. This can happen many times before the button reaches a steady "pressed" state. Since the microcontroller operates at nearly one instruction per microsecond it is probable that any bouncing that lasts for more than a few microseconds will be detected as if multiple key presses had been made when, in reality, the user only pressed the key once and the multiple keypresses are "phantom" presses caused by the mechanical and electrical limitations of the keypad.

For this reason it is necessary for software to validate whether a keypress is a real keypress or just one of the multiple bounces that may occur when a key is initially being pressed or released. If it is a phantom keypress then the software must disregard it. That is the purpose of the debounce routine.

The characteristic that distinguishes a real keypress from the phantom keypresses is that a real keypress lasts much longer. While a phantom keypress may last for microseconds or even tens of milliseconds, a real human keypress will normally last 50 or 100 milliseconds at the very least. Once a key is fully pressed, the output achieves a stable state for as long as the key remains pressed. Therefore the typical method of filtering out invalid bounces is to measure how long a keypress lasts—if the duration of the keypress is too short, it can be reasonably assumed that the keypress was actually just a phantom bounce caused by the closing or opening electrical connection and should be disregarded. If the duration of the keypress exceeds that limit then it may be reasonably assumed that it is a real keypress from the user.

The specific time interval that is used to determine whether a keypress is a bounce or genuine depends on the application as well as the keypad being used. Higher quality keypads may exhibit a shorter bouncing period than lower quality devices; if a low quality device is being used then the threshold to determine a valid keypress might have to be extended. The application also is an important factor. If the button that is being monitored is such that the user is not likely to press and release it quickly then the debounce time interval can be increased whereas a button that is only likely to be pressed very, very quickly will require a shorter debounce time.

SBCMON (see chapter 17) provides a command to test the keypad both with and without key debounce. To try the keypad with debounce use the 'K' command; to try the keypad without debounce use the 'KN' command. This will clearly illustrate the difference between the two and why debounce is so important.

In this example, a debounce period of 1/20th of a second will be assumed. Any keypress that lasts less than that period of time will be assumed to be electrical noise from the mechanical switching of the key. Anything that lasts longer than 1/20th of a second will be assumed to be a genuine keypress.

The debounce routine can be added to the previous program by inserting the following code at the KP_ProcessKey label on line 22 of the previous example.

```
22 KP_ProcessKey:
23      MOV R1,A                        ;Hold the keypress in R1
24      MOV R6,#23                      ;23*256=7680*8=approx. 46,080 cycles
25 KP_DebounceLoop2:
26      MOV R7,#00h                     ;Set R7 to 0 256 loops (00->FF->00)
27 KP_DebounceLoop:
28      MOVX A,@DPTR                    ;Re-read the same keypad row
29      ANL A,#0Fh                      ;Only interest in low 4 bits
30      XRL A,#0Fh                      ;Invert all the bits
31      XRL A,R1                        ;Exclusive OR with the original keypress
32      JNZ KP_Loop                     ;If not same as last value, debounce it!
33      DJNZ R7,KP_DebounceLoop         ;Loop R7 times 2
34      DJNZ R6,KP_DebounceLoop2        ;Loop R6 times
35      MOV A,R1                        ;Restore original value read from keypad
36      PUSH DPH                        ;Save high byte of keypad address
37 KP_DebounceDone:
```

Line 23 stores the original keypress in R1. Line 24-26 sets up the registers R6 and R7 such that the debounce loop will execute 23 * 256 = 5,888 times. Each iteration of the loop will require 8 instruction cycles so the entire debounce loop will execute for 5,888 * 8 = 47,104 instruction cycles. At 11.0592MHz and 12 cycles per instruction, the microcontroller will execute 11,059,200 / 12 = 921,600 instruction cycles per second—so the debounce delay of 47,104 instruction cycles is 47,104 / 921,600 = .051 seconds, or just slightly longer than 1/20th of a second.

This delay works with the Atmel AT89S8252 based on the assumption that the microcontroller executes one instruction cycle for every 12 oscillator cycles. The Dallas DS89C420, however, operates at one instruction cycle per oscillator cycle—effectively twelve times as fast given the same crystal speed. Thus if the DS89C420 is being used then the value of R6 in line 17 would need to be multiplied by 12 to compensate for the faster execution speed. Other microcontrollers may also operate faster than 12 oscillator cycles per instruction and the value of R6 would have to be similarly adjusted to compensate for this. Some operate at one instruction cycle for every 6 oscillator cycles (this would require R6 be doubled), some operate at one instruction cycle for every 4 oscillator cycles (this would require that R6 be tripled), and some operate at one instruction cycle for every 2 oscillator cycles (this would require that R6 be multiplied by 6). Check the datasheet for the microcontroller being used to determine the length of an instruction cycle and adjust R6 accordingly.

Lines 27-32 are the debounce loop itself. The loop repeatedly reads the same keypad row that was already read and compares the current value to the originally read value—if it changes then it is assumed that the keypress was not stable and, since 1/20th of a second has not yet passed, it jumps out of the loop back to KP_Loop which effectively discards the keypress. Line 28 reads the keypad row, line 29 zeros out the high four bits that we're not interested in, line 30 exclusive OR's the value with 0Fh to invert the bits so that a '1' bit represents a keypress and '0' represents a key not being pressed, and line 31 compares the current value to the original value that is now held in R1. If the current value is not the same as the original value, the result of the XRL instruction in line 24 will be non-zero so the program will branch back to KP_Loop and discard the keypress.

Lines 33-34 complete the loops that were initialized in lines 24-26 and which cause the debounce loop to repeat for approximately 1/20th of a second.

Line 35 restores the value of the keypress that was previously saved on line 23, and line 36 stores the high byte of the address of the keypad scan row on the stack.

This debounce logic detects an initial keypress and then continuously re-reads the same row to make sure that same keypress remains valid for at least 1/20th of a second. This is an effective and simple way to reject the phantom bounces caused by the mechanical limitations of the keys themselves.

18.2.1.3 Debouncing the Key Release

In addition to debouncing the initial keypress, it is also necessary to delay the key release. If a person presses the button for half a second, the revised program with debounce logic would report the key having been pressed nine or ten times—once for every 1/20th of a second that the key was pressed. This is normally undesirable since most users will not be able to press and release the key within 1/20th of a second. The 1/20th of a second debounce delay could be extended to prevent this, but then the response time of the debounce routine would become sluggish.

In order to avoid this it is necessary for the program to wait for the key to be released once a key has been detected and debounced. The process can be summarized as follows:

1. Detect a keypress.
2. Debounce the keypress by making sure it is a steady signal for at least 1/20th of a second.
3. Wait for the key to be released.

Steps one and two have already been handled in the previous two sections. The last step is to add code to the algorithm so that the program will pause until the user releases the key. This is accomplished by replacing the RET instruction on line 37 of the original program with the following code:

```
37        LCALL SendSerialByte        ;Echo keypress to the serial port
38        POP DPH                      ;Restore DPTR to access keypad
39        POP DPL                      ;Restore DPTR to access keypad
40 KP_Wait:
41        MOVX   A,@DPTR               ;Read the keypad
42        ANL A,#0Fh                   ;Only interested in low 4 bits
43        CJNE A,#0Fh,KP_Wait          ;If it's not 0Fh then keep waiting
44        LJMP KP_Loop                 ;Wait for another keypad press
```

Also, the following line must be inserted at line 31 after KP_KeyFound:

```
31 PUSH DPL   ;Save low byte of keypad address for later use
32 PUSH DPH   ;Save high byte of keypad address for later use
```

Line 31 and 32 saves the address that is held in DPTR and which points to the memory address of the keypad row that is currently being read.

Line 37 sends the key that was read to the serial port—this line could be replaced with any instruction or subroutine responsible for handling the key that was just read.

Lines 38 and 39 restore the value of DPTR so that it again points to the memory address that corresponds to the keypad row that was just scanned.

Lines 40-43 repeatedly read the keypad row until such time as the key is released by the user. When the user releases the key the value returned by the keypad will return to 0Fh which will cause the loop to exit.

Line 44 then jumps back to KP_Loop which restarts the process of scanning all the rows of the keypad for the next key.

18.2.1.4 Complete Keypad Decoder, Debouncer, and Key Release Delay

Combining the code from the previous three sections, the following code is derived for detecting a keypress, debouncing it, decoding it into a valid character, and pausing until the user releases the key.

```
1  KP_Loop:
2      MOV R3,#03h                    ;Start with row 3 of the keypad
3  KP_Cycle:
4      MOV A,R3                       ;Get current row (0-3) in accumulator
5      RL A                           ;Multiply row by 2 to get table offset
6      MOV B,A                        ;Store offset temporarily in B
7      MOV DPTR,#KP_AddrTable         ;Point to the address table
8      MOVC A,@A+DPTR                 ;Get the DPTR high byte from the table
9      PUSH ACC                       ;Push the DPTR high byte on stack
10     INC DPTR                       ;Point to DPTR low byte in table
11     MOV A,B                        ;Restore offset into table from B
12     MOVC A,@A+DPTR                 ;Get the DPTR low byte from the table
13     MOV DPL,A                      ;Move the low byte from table into DPL
14     POP DPH                        ;Pop high byte from table into DPH
15     MOVX A,@DPTR                   ;Read the status of the keypad row
16     ANL A,#0Fh                     ;Only interest in low 4 bits
17     XRL A,#0Fh                     ;Invert bits, keypress=1, 0=No key
18     JNZ KP_ProcessKey              ;If the value is not zero, process it
19     DEC R3                         ;Decrement row counter
20     CJNE R3,#0FFh,KP_Cycle         ;If not FF, process next row
21     SJMP KP_Loop                   ;After 4 rows, start with row 3 again
22 KP_ProcessKey:
23     MOV R1,A                       ;Hold the keypress in R1
24     MOV R6,#23                     ;23*256=7680*8=approx. 46,080 cycles
25 KP_DebounceLoop2:
26     MOV R7,#00h                    ;Set R7 to 0 256 loops (00->FF->00)
27 KP_DebounceLoop:
28     MOVX A,@DPTR                   ;Re-read the same keypad row
29     ANL A,#0Fh                     ;Only interest in low 4 bits
30     XRL A,#0Fh                     ;Invert all the bits
31     XRL A,R1                       ;Exclusive OR with original keypress
32     JNZ KP_Loop                    ;If not same as last value, debounce
33     DJNZ R7,KP_DebounceLoop        ;Loop R7 times 2
34     DJNZ R6,KP_DebounceLoop2       ;Loop R6 times
35     MOV A,R1                       ;Restore original value from keypad
36     PUSH DPH                       ;Save high byte of keypad address
37 KP_DebounceDone:
38     CLR C                          ;Make sure carry starts clear
39     MOV R0,#00h                    ;R0 starts at zero
40 KP_FindKeyNum:
41     RRC A                          ;Rotate the value right into the carry
42     JC KP_KeyFound                 ;If set, we found the key so exit
43     INC R0                         ;Increment the key number indicator
44     SJMP KP_FindKeyNum             ;Keep searching until we find the key
```

```
45 KP_KeyFound:
46     PUSH DPL                    ;Save low byte of keypad address
47     PUSH DPH                    ;Save high byte of keypad address
48     MOV A,R3                    ;Get row number
49     RL A                        ;Multiply by 2
50     RL A                        ;Multiply by 4
51     ADD A,R0                    ;Add the key offset
52     MOV DPTR,#KP_KeyTable       ;Get the start address of the key table
53     MOVC A,@A+DPTR              ;Get the keypress into accumulator
54     LCALL SendSerialByte        ;Echo keypress to the serial port
55     POP DPH                     ;Restore DPTR to access keypad
56     POP DPL                     ;Restore DPTR to access keypad
57 KP_Wait:
58     MOVX A,@DPTR                ;Read the keypad
59     ANL A,#0Fh                  ;Only interested in low 4 bits
60     CJNE A,#0Fh,KP_Wait         ;If it's not 0Fh then keep waiting
61     LJMP KP_Loop                ;Wait for another keypad press
62 KP_KeyTable:
63     DB "ABCD"                   ;Decode for keys in row 0
64     DB "369#"                   ;Decode for keys in row 1
65     DB "2580"                   ;Decode for keys in row 2
66     DB "147*"                   ;Decode for keys in row 3
67 KP_AddrTable:
68     DW 500Eh                    ;Address to access row 0 (500Eh)
69     DW 500Dh                    ;Address to access row 1 (500Dh)
70     DW 500Bh                    ;Address to access row 2 (500Bh)
71     DW 5007h                    ;Address to access row 3 (5007h)
```

Line 54 simply calls a routine called SendSerialByte which sends the character in the accumulator to the serial port.

This program may be downloaded as assembly language source code or as a ready-to-execute HEX file at **http://www.8052.com/sbc/keypad**.

CHAPTER 19: INTERFACING TO LCD

Liquid crystal displays (**LCDs**) are often a part of microcontroller projects. LCDs provide a convenient way to provide visual feedback to the user by allowing the program to display easy-to-read text messages that prompt the user to take some action or inform the user of some event or data. LCDs require significantly less supporting hardware than a full-fledged computer monitor and, obviously, a small LCD is far more suited for a typical microcontroller application where power consumption and size must be minimized.

LCDs come with a variety of different capabilities and features as shown in the figure below.

Most LCDs that are typically used in microcontroller projects use what is known as the **44780 standard**. This is a defacto standard that defines the electrical connections to the LCD as well as establishing a protocol that allows commands and text to be sent to the screen for subsequent display. Generally speaking, any program that utilizes the 44780 standard will work with any LCD that also uses that same standard.

The remainder of this chapter explains the process of communicating with a 44780-standard LCD in three different manners: 1) 8-bit direct connection, 2) 4-bit direct connection, 3) memory-mapped connection.

19.1 LCD Electrical Connections

The 44780 standard requires three control lines as well as either four or eight I/O lines for the data bus. The program may select whether it will communicate with the LCD with a 4-bit or 8-bit data bus. If 4-bit mode is used, the LCD will require a total of seven data lines (3 control lines plus the 4 lines for the data bus). If an 8-bit data bus is used, the LCD will require a total of eleven data lines (3 control lines plus the 8 lines for the data bus).

The three control lines are referred to as EN, RS, and RW.

EN is the "Enable" line. This control line is used to tell the LCD whether it is being sent data or a command. To send data to the LCD the program should first set this line low (0), then set the other two control lines and/or put data on the data bus, bring EN high (1), and then low again. The 1-0 transition tells the LCD to take the data currently found on the other control lines and on the data bus and to treat it as a command or data.

RS is the "Register Select" line. When RS is low (0), the data sent to the LCD is to be treated as a command or special instruction (such as clear screen, position cursor, etc.). When RS is high (1), the data being sent is text data which should be displayed on the screen. For example, to display the letter "T" on the LCD the RS line would be brought high.

RW is the "Read/Write" control line. When RW is low (0), the information on the data bus is being written to the LCD. When RW is high (1), the program is effectively querying (or reading) the LCD. Only one instruction ("Get LCD status") is a read command—all others are write commands. For this reason RW will almost always be low.

Finally, the data bus consists of four or eight lines, depending on the mode of operation selected by the user. In the case of an 8-bit data bus, the lines are referred to as DB0, DB1, DB2, DB3, DB4, DB5, DB6, and DB7.

The LCD generally has three other connections which provide power to the LCD, a ground for the LCD, and a variable voltage power supply to control the contrast of the LCD. Since these three lines do not have anything to do with the communication between the LCD and the microcontroller they will not be discussed further in this chapter. For details on these three lines please consult the datasheet for the LCD that is to be used.

19.2 Direct 8-bit Connection to LCD

The simplest design that allows an LCD to communicate with a microcontroller is to connect it in 8-bit mode directly to the pins of the microcontroller. This approach is also the easiest to understand so it will be used to explain the fundamentals of LCD communication and the LCD command set before going into more efficient—and complicated—methods of connecting the LCD.

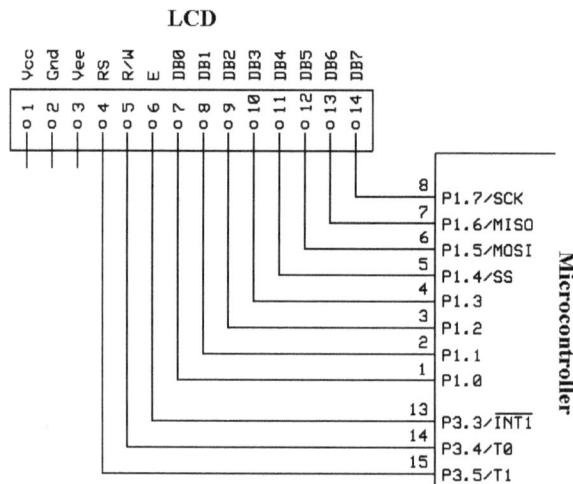

As can be seen in this schematic excerpt, there is a one-to-one relation between each line of the LCD and a pin of the 8052 (except for pins 1-3 on the LCD which are power and ground). To avoid accessing data lines by their port names (P3.3, P3.4, etc.) it is useful to assign a number of equates (see section 11.6.2) so that the program may refer to the lines by more meaningful names.

```
EN   EQU P3.3        ;LCD's ENable line is on P3.3
RW   EQU P3.4        ;LCD's RW line is on P3.4
RS   EQU P3.5        ;LCD's RS line is on P3.5
DATA EQU P1          ;Data will be written to P1
```

Having established the above equates it will be possible for the program to refer to the I/O lines by their 44780 name. For example, to set the RW line high the following instruction may be executed:

```
SETB RW      ;Set RW line high
```

19.2.1 Controlling the LCD's EN Line

As already mentioned, the EN line is used to inform the LCD that the program is ready for it to execute the instruction that has been put on the data bus and on the other two control lines. The EN line must be raised and brought low to execute the instruction —this is regardless of whether the instruction is read or write, text or instruction. In short, the EN line must *always* be manipulated when communicating with the LCD. EN is the LCD's way of knowing that it is being addressed. If EN isn't raised and lowered properly the LCD won't know it's receiving information on the other data lines.

Thus before a program interacts in any way with the LCD it will always bring the EN line low with the following instruction:

```
CLR EN
```

It will then configure the other data lines (RS, RW, and the data bus) and, once complete, it will bring the EN line high and back low with the following instruction:

```
SETB EN      ;Raise the EN line
CLR EN       ;Lower the EN line again
```

At the moment EN goes low the LCD will evaluate and process the command that was sent to it on the other two control lines and on the data bus. The EN line must stay high for a certain amount of time to give the LCD an opportunity to process the command—the exact time varies depending on the LCD but it is generally on the order of 250 nanoseconds or less (1/4th of a microsecond). An AT89S8252 operating at 11.0592MHz will execute one instruction every 1.08 microseconds. Thus by the time the next instruction executes sufficient time will have passed that the EN line can be brought low. If, however, a faster microcontroller is used, it may be necessary to insert a number of NOP instructions after raising the EN line to make sure there is sufficient time for the LCD to respond to the command. The DS89C420 operating at 11.0592 MHz executes one instruction every 0.09 microseconds. Since the LCD requires the EN line to be high at least 0.250 microseconds it is wise to add four NOP instructions between raising EN and subsequently lowering it. If the code is executed on a fast microcontroller the added NOPs will give the LCD enough time to work; if the code is executed on a typical, slower microcontroller then the added NOPs will have no harmful effect.

19.2.2 Checking the LCD's Busy Status

It takes a certain amount of time for each instruction to be executed by the LCD. The delay varies depending on the exact LCD model being used as well as the instruction which is being executed.

While it is possible to write code that waits for a specific amount of time to allow the LCD to execute instructions, this method of "waiting" is not very flexible. If the crystal frequency is changed the software will need to be modified to make the delay loop take longer. Additionally, if the LCD itself is changed for another LCD which, although 44780 compatible, requires more time to perform its operations, the program will not work until it is properly modified.

A more robust method of programming is to use the "Get LCD Status" command to determine whether the LCD is still busy executing the last instruction received.

The "Get LCD Status" command will return whether or not the LCD is busy on the DB7 line. When the LCD receives the "Get LCD Status" command it will immediately raise DB7 if it's still busy executing a previous command or lower DB7 to indicate that the LCD is ready for the next command. Thus the program may query the LCD status until DB7 goes low which indicates the LCD is no longer busy. At that point the program is free to send the next command.

Since the "wait" routine is used every time an instruction is sent to the LCD, it's a good candidate for a subroutine:

```
WAIT_LCD:
        CLR RS                  ;It's a command
        SETB RW                 ;It's a read command
        SETB EN                 ;Start LCD command
        MOV P1,#0FFh            ;Set all pins to FF initially
        MOV A,P1                ;Read the return value
        JB ACC.7,WAIT_LCD      ;If bit 7 high, LCD still busy
        CLR EN                  ;Finish the command
        CLR RW                  ;Turn off RW for future commands
        RET
```

Standard practice will be to send an instruction to the LCD and then call the WAIT_LCD routine to wait until the instruction is completely executed by the LCD. This will assure that the program gives the LCD the time it needs to execute instructions and also makes the program compatible with any LCD regardless of how fast or slow it is.

If the above routine were used in a real application a very definite improvement would need to be made. As written, if the LCD never becomes "not busy" the program will effectively "hang", waiting for DB7 to go low. If this never happens the program will freeze. Of course, this should never happen and won't happen when the hardware is working properly. But in a real application it would be wise to put some kind of time limit on the delay—for example, a maximum of 10,000 attempts to wait for the busy signal to go low. That would guarantee that even if the LCD hardware fails the program would not lock up.

19.2.3 Initializing the LCD

Before the LCD may be used for display purposes it must first be initialized and configured. This is accomplished by sending a number of commands to the LCD. Since multiple commands will be sent to the LCD, it makes sense to create a subroutine that will send the value in the accumulator to the LCD as a command:

```
LCD_COMMAND:
  CLR RS              ;RS low to indicate a command
  CLR RW              ;RW low to indicate write
  MOV DATA,A          ;Move command in accumulator to data   port
  SETB EN             ;Raise EN to initiate LCD command
  NOP                 ;Wait 4 cycles to give LCD time to process
  NOP
  NOP
  NOP
  CLR EN              ;Lower EN line
  LCALL WAIT_LCD      ;Wait for the previous command to execute
  RET                 ;Return to calling program code
```

19.2.3.1 LCD Command: Data Interface (20h)

The first command that is sent to the LCD indicates whether it will be operating with an 8-bit or 4-bit data bus and selects a 5x8 character font. These two options are selected by sending the value 38h to the LCD as a command. This is accomplished by loading the accumulator with the value 38h and calling the LCD_COMMAND routine:

```
MOV A,#38h
LCALL LCD_COMMAND
```

The 38h command is really the sum of a number of option bits. The instruction itself is the data interface instruction (20h) to which are added the values 10h to indicate an 8-bit data bus and 08h to indicate that the LCD is a two-line display. The options in the data interface command are:

Data Interface Command: 20h (then add the applicable values for options below)
    +10h: **Bus Size.** 8-bit data bus (otherwise 4-bit data bus)
    +08h: **Number of Lines.** 2-line LCD display (otherwise 1-line LCD)
    +04h: **Character Size.** Display has 5x10 characters (otherwise 5x8)

19.2.3.2 LCD Command: Data Interface (08h)

The second byte of the initialization sequence is the instruction 0Eh. Thus the above code is repeated but now with the 0Eh instruction:

```
MOV A,#0Eh
LCALL LCD_COMMAND
```

The 0Eh command is really the instruction 08h added to the value 04h to turn the LCD on and 02h to turn the cursor on. The options in the display cursor command are:

Display Cursor Command: 08h (then add the applicable values for options below)
+04h: Display on/off. Display on (otherwise display turned off)
+02h: Cursor on/off. Turns cursor on (otherwise no cursor displayed)
+01h: Blink mode. Cursor blinks (otherwise cursor is constant).

19.2.3.3 LCD Command: Cursor Move Direction (04h)

The third and final byte necessary to complete the initialization sequence is 06h:

```
MOV A,#06h
LCALL LCD_COMMAND
```

The command 06h is really the instruction 04h plus 02h to configure the LCD so that every time it displays a character the cursor position will automatically move to the right. The options in the cursor move direction command are:

Cursor Move Direction Command: 04h (then add the applicable values for options below)
+02h: Advance cursor. Advance cursor after write (otherwise don't)
+01h: Shift display. Shift display after each write(otherwise don't)

19.2.3.4 Complete LCD Initialization

So the entire initialization code for the LCD is:

```
INIT_LCD:
  MOV A,#38h       ;8-bit data bus, 5x8 character font
  LCALL LCD_COMMAND
  MOV A,#0Eh       ;Turn LCD on, turn cursor on
  LCALL LCD_COMMAND
  MOV A,#06h       ;Turn on cursor auto-advance
  LCALL LCD_COMMAND
  RET
LCD_COMMAND:
  CLR RS           ;RS low to indicate a command
  CLR RW           ;RW low to indicate write
  MOV DATA,A ;Move command in accumulator to data port
  SETB EN          ;Raise EN to initiate LCD command
  NOP              ;Wait 4 cycles to give LCD time to process
  NOP
  NOP
  NOP
  CLR EN           ;Lower EN line
  LCALL WAIT_LCD   ;Wait for the previous command to execute
  RET              ;Return to calling program code
```

Once the INIT_LCD subroutine is executed the the LCD will be fully initialized and ready to display data on the screen.

19.2.4 Clearing the LCD Screen

When the LCD is first initialized the screen should automatically be cleared by the 44780 controller. However, it's always a good idea to manually perform every function necessary to avoid depending on assumptions. For this reason it's not a bad idea to clear the screen as the very first operation after the LCD has been initialized.

An LCD command exists to accomplish this function: 01h. Since this is a command, just like the initialization commands in the previous section, the exact same LCD_COMMAND routine may be called to clear the screen:

```
CLEAR_LCD:
  MOV A,#01h               ;Set accumulator to clear screen command
  LCALL LCD_COMMAND        ;Execute command
  RET                      ;Return from subroutine
```

Clearing the LCD is now a simple matter of executing an LCALL CLEAR_LCD.

Sending the clear screen command LCD also positions the cursor in the upper left-hand corner as would be expected.

19.2.5 Writing Text to the LCD

Having initialized the LCD and cleared its screen, it is now possible to write text to the LCD that will be displayed on the screen.

Once again, writing text to the LCD is something that will be done often so it makes sense to make it a subroutine.

```
WRITE_LCD_TEXT:
  SETB RS            ;RS high to indicate text data
  CLR RW             ;RW low to indicate write
  MOV DATA,A         ;Move character in accumulator to data port
  SETB EN            ;Raise EN to initiate LCD command
  NOP                ;Wait 4 cycles to give LCD time to process
  NOP
  NOP
  NOP
  CLR EN             ;Lower EN line
  LCALL WAIT_LCD     ;Wait for the previous command to execute
  RET                ;Return to calling program code
```

The WRITE_LCD_TEXT subroutine above sends the character in the accumulator to the LCD which will, in turn, display it. To display text on the LCD all the program needs to do is load the value to display in the accumulator and call this routine.

19.2.6 An LCD "Hello World" Program

Now that all the component subroutines have been written, writing the classic "Hello World" program—which displays the text "Hello World" on the LCD—is a relatively trivial matter.

```
        LCALL INIT_LCD                      ;Initialize the LCD
        LCALL CLEAR_LCD                     ;Clear the LCD screen
        MOV A,#'H'
        LCALL WRITE_LCD_TEXT                ;Send the letter 'H' to the LCD
        MOV A,#'E'
        LCALL WRITE_LCD_TEXT                ;Send the letter 'E' to the LCD
        MOV A,#'L'
        LCALL WRITE_LCD_TEXT                ;Send the letter 'L' to the LCD
        MOV A,#'L'
        LCALL WRITE_LCD_TEXT                ;Send the letter 'L' to the LCD
        MOV A,#'O'
        LCALL WRITE_LCD_TEXT                ;Send the letter 'O' to the LCD
        MOV A,#' '
        LCALL WRITE_LCD_TEXT                ;Send a space to the LCD
        MOV A,#'W'
        LCALL WRITE_LCD_TEXT                ;Send the letter 'W' to the LCD
        MOV A,#'O'
        LCALL WRITE_LCD_TEXT                ;Send the letter 'O' to the LCD
        MOV A,#'R'
        LCALL WRITE_LCD_TEXT                ;Send the letter 'R' to the LCD
        MOV A,#'L'
        LCALL WRITE_LCD_TEXT                ;Send the letter 'L' to the LCD
        MOV A,#'D'
        LCALL WRITE_LCD_TEXT                ;Send the letter 'D' to the LCD
```

The above "Hello World" program will, when executed, initialize the LCD, clear the LCD screen, and display "Hello World" in the upper left-hand corner of the display.

19.2.7 LCD Cursor Positioning

The above "Hello World" program is simple in the sense that it prints its text in the upper left-hand corner of the screen. However, it is often necessary to display text in locations other than the upper left-hand corner. What if it were necessary to display the word "Hello" in the upper left-hand corner but to display the word "World" on the second line in the eleventh column? This sounds simple—and actually it *is* simple—but it does require a little additional understanding of the LCD.

The LCD contains a certain amount of memory which is assigned to the display. All text that is written to the LCD is stored in this memory which is subsequently used to display the text on the LCD itself. This memory can be represented with the following "memory map":

```
Display  00 01 02 03 04 05 06 07 08 09 10 11 12 13 14 15 16
Line 1   [00][01][02][03][04][05][06][07][08][09][0A][0B][0C][0D][0E][0F][10][11][12][13][14][15] ...
Line 2   [40][41][42][43][44][45][46][47][48][49][4A][4B][4C][4D][4E][4F][50][51][52][53][54][55] ...
```

In this map, the addresses with a white background are part of the visible display. This area measures 16 characters per line by 2 lines. The number in each box is the memory address that corresponds to that screen position. The first character in the upper left-hand corner is at address 00h, the following character position (character #2 on the first line) is address 01h, etc. This continues until the 16th character of the first line which is at address 0Fh. But the first character of line 2 is at address 40h. This means that if a character is written to the last position of the first line and then a second character is written, the second character will not appear on the second line. That is because the second character will effectively be written to address 10h which is not visible on the screen.

To display text on the second line it is necessary to send a command to the LCD that specifically positions the cursor at the desired location. The "Set Cursor Position" command is 80h to which the address of the location where the cursor is to be positioned must be added. In the 2-line "Hello World" example it was stated that we would like the word "World" to appear on the second line in the eleventh column.

The memory map indicates that the eleventh column of the second line is address 4Ah. Thus before writing "World" to the LCD it is necessary to send a "Set Cursor Position" command. The value of this command will be 80h (the command code to position the cursor) plus the address 4Ah, or 80h + 4Ah = CAh. Thus sending the command CAh to the LCD will position the cursor on the second line in the eleventh column.

```
MOV A,#0CAh              ;Set cursor to line 2, position 10
LCALL LCD_COMMAND        ;Send the command
```

This code will position the cursor on line 2, character 10. To display "Hello" in the upper left-hand corner with the word "World" on the second line at character position 10 just requires that these two lines of code be inserted into the existing "Hello World" program instead of the code that previously sent a space character to separate the two words. This results in the following:

```
LCALL INIT_LCD           ;Initialize the LCD
LCALL CLEAR_LCD          ;Clear the LCD screen
MOV A,#'H'
LCALL WRITE_LCD_TEXT     ;Send the letter 'H' to the LCD
MOV A,#'E'
LCALL WRITE_LCD_TEXT     ;Send the letter 'E' to the LCD
MOV A,#'L'
LCALL WRITE_LCD_TEXT     ;Send the letter 'L' to the LCD
MOV A,#'L'
LCALL WRITE_LCD_TEXT     ;Send the letter 'L' to the LCD
MOV A,#'O'
LCALL WRITE_LCD_TEXT     ;Send the letter 'O' to the LCD
MOV A,#0CAh              ;Set Cursor Position, line 2 char 10
LCALL LCD_COMMAND        ;Send command
MOV A,#'W'
LCALL WRITE_LCD_TEXT     ;Send the letter 'W' to the LCD
MOV A,#'O'
LCALL WRITE_LCD_TEXT     ;Send the letter 'O' to the LCD
MOV A,#'R'
LCALL WRITE_LCD_TEXT     ;Send the letter 'R' to the LCD
MOV A,#'L'
LCALL WRITE_LCD_TEXT     ;Send the letter 'L' to the LCD
MOV A,#'D'
LCALL WRITE_LCD_TEXT     ;Send the letter 'D' to the LCD
```

19.3 Direct 4-bit Connection to LCD

Although the 8-bit direct connection to the LCD described in the previous section is certainly the simplest, it can present problems when the application in question has insufficient I/O lines on the microcontroller such that it is not possible to dedicate the necessary 11 lines to the LCD. The LCD's 4-bit mode allows the exact same functionality to be achieved with just 7 I/O lines—4 data lines and the same 3 control lines. This is accomplished by using four of the eight data lines and sending and reading data to and from the LCD one nibble at a time—first the most significant four bits followed by the least significant four bits.

The only differences in 4-bit mode are:

1. The first command sent in the initialization sequence in 4-bit mode is 28h instead of 38h. This informs the LCD that all subsequent communication will be in 4-bit mode.
2. Each byte sent or read to or from the LCD requires that EN be raised and lowered twice—once for each nibble. This means that communication with the LCD in 4-bit mode is slightly slower, though the decreased speed would not be noticeable to a human user.

19.3.1 Writing a byte to LCD in 4-bit mode

Writing a byte to the LCD in 4-bit mode requires a few additional steps as compared to the 8-bit mode covered in the previous section. First, the subroutine will receive a single byte but it must be broken up into two component nibbles. Second, the subroutine must raise and lower EN for each nibble.

The following subroutine transmits the byte in the accumulator to the LCD in 4-bit mode. It is assumed that LCD lines DB4-DB7 are connected to P1.4 – P1.7 on the microcontroller.

```
 1  SendLCDByte4bit:
 2     PUSH ACC           ;Save accumulator, we'll need it again
 3     ORL A,#0Fh         ;Don't interfere with P0.0-P0.3
 4     ORL P1,#0F0h       ;Make sure P0.4-P0.7 start out set
 5     ANL P1,A           ;Send high nibble to LCD
 6     SETB EN            ;Raise EN for first nibble
 7     NOP                ;Give LCD time to work
 8     NOP
 9     NOP
10     NOP
11     CLR EN             ;Lower EN to prepare next nibble
12     POP ACC            ;Restore original value of accumulator
13     SWAP A             ;Move low nibble into high nibble of ACC
14     ORL A,#0Fh         ;Don't interfere with P0.0-P0.3
15     ORL P1,#0F0h       ;Make sure P0.4-P0.7 start out set
16     ANL LCDPORT,A      ;Send "A" to port
17     SETB EN            ;Raise EN for second nibble
18     NOP                ;Give LCD time to work
19     NOP
20     NOP
21     NOP
22     CLR EN             ;Lower EN since command is done
23     LCALL WaitForLCD   ;Wait for LCD to execute command
24     RET                ;Exit subroutine
```

Line 2 starts the subroutine by pushing a copy of the accumulator onto the stack. The subsequent instructions will be modifying the value of the accumulator but the original value will be needed again later in the routine so it is pushed on the stack for future use.

Lines 3-5 do the work of placing the high four bits of the accumulator on the high four I/O lines of P1 (P1.4 - P1.7)—but it does so without modifying the current value of lines P1.0 - P1.3 since these lines may be used by other external hardware. To accomplish this, line 3 first makes sure the low four bits of the accumulator are all set while line 4 makes sure the four LCD I/O lines start out all being set. It then ANDs the two values on line 5. This results in the LCD I/O lines assuming the value of the high four bits in the accumulator while leaving the four low bits of P1 unchanged.

Lines 6-11 bring EN high to clock the high nibble into the LCD, executes a short pause with NOP instructions to give the LCD time to process the data, and then brings the EN line low again.

Line 12 pops the accumulator off the stack so that the original value that was sent to the subroutine in the accumulator is restored. Line 13 uses the SWAP A instruction to swap the low nibble and the high nibble of the accumulator. Since the routine is now sending the low nibble but it must be sent out on the high nibble of P1, the nibbles in the accumulator are reversed before proceeding.

Lines 14-22 are exactly the same as lines 3-11 but now the code is clocking out the low nibble. As can be seen, the EN line is once again brought high, there is a pause, and EN is brought low.

Line 23 waits for the LCD to complete the command and line 24 returns from the subroutine.

The above routine is a generic "send byte" routine. It does not set or clear the RS line to instruct the LCD whether the data being sent is a command or text. If this routine were used in a functioning program the RS line would have to be set or cleared prior to calling the routine.

19.3.2 Reading a byte from LCD in 4-bit mode

The process of reading a byte from the LCD is similar to writing a byte, but somewhat simpler.

```
1   LCDRead2Nibbles:
2     SETB RW          ;Set RW to indicate reading from LCD
3     ORL P1,#0F0h     ;Set LCD lines high so they are readable
4     SETB EN          ;Bring EN line high to read first nibble
5     MOV A,P1         ;Read high nibble
6     CLR EN           ;Bring EN line low for end of first nibble
7     ANL A,#0F0h      ;Only interested in high nibble
8     PUSH ACC         ;Save high nibble on the stack
9     SETB EN          ;Bring EN line high to read second nibble
10    MOV A,P1         ;Read low nibble
11    CLR EN           ;Bring EN line low for end of second nibble
12    ANL A,#0F0h      ;Only interested in high nibble
13    SWAP A           ;This is low nibble, so swap it into place
14    MOV R7,A         ;Store low nibble temporarily in R7
15    POP ACC          ;Restore the high nibble saved on stack
16    ORL A,R7         ;Combine with low nibble
17    RET              ;Return from subroutine
```

Line 2 sets RW to indicate that the program is reading from the LCD. Line 3 then makes sure that the four high lines of P1 are set—the value of a port pin can only be read if it's high so this line makes sure the I/O lines will be readable.

Line 4 raises the EN line which causes the LCD to place the high nibble of the value to be read on the four I/O lines. Line 5 reads the value returned by the LCD and line 6 brings the EN line low again.

Line 7 zeros out the low four bits that were read and which are not involved in the LCD communication. The resulting value with the high nibble is then pushed on the stack by line 8.

Lines 9-12 repeat the function of lines 4-7, this time receiving the low nibble of the value. Since the value returned by the LCD is found in the high nibble of P1 but is actually the low nibble of the value being read from the LCD, the SWAP A instruction in line 13 puts the newly received nibble into the low nibble position. The low nibble is placed temporarily in R7 in line 14, the high nibble that was ready previous is restored from the stack on line 15, and the high and low nibbles are then combined on line 16.

The above routine is a generic "read byte" routine. It does not set or clear the RS line to instruct the LCD whether the data being read is the LCD status or text data. If this routine were used in a functioning program the RS line would have to be set or cleared prior to calling this routine.

19.4 Memory-Mapped Connection to LCD

Yet another approach to connecting the LCD to the microcontroller is a memory-mapped design. As explained in section 18.2, a memory-mapped design is one in which the device—in this case the LCD—is accessed by the microcontroller as if it were external memory. Reading and writing to the LCD is a simple matter of using the MOVX instruction to read and write from the appropriate address in memory.

This is a very attractive solution if the project design is already implementing some kind of external memory or other memory-mapped devices. In such cases, the LCD may be added to the design without requiring a single additional I/O pin on the microcontroller since it effectively shares the address and data bus with the other memory-mapped devices.

The 8052.com SBC described in chapter 14 uses a memory-mapped approach to access the LCD when it is connected to J3. The rest of this section will assume a similarly connected memory-mapped connection.

19.4.1 Electrical Connection of Memory-Mapped LCD

The electrical connections for the memory-mapped LCD are as follows:

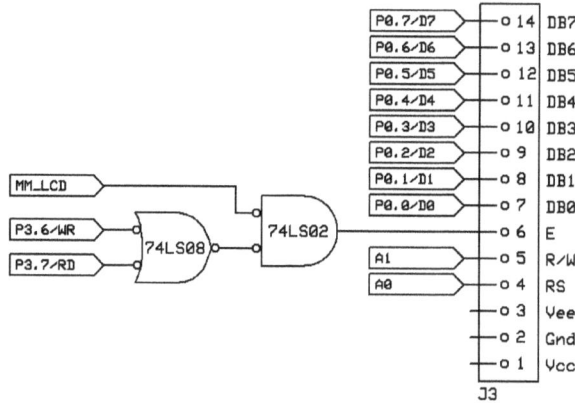

This design automatically controls the RW, RS, and EN lines such that the program does not have to drive them manually as was the case in the direct-connect approach described in the previous sections. Instead, these lines will be driven based on the memory address being accessed and whether or not the address is being read or written. This is handled by electrical logic circuits rather than program code.

Recall from section 14.2.5 that the P3.6/WR line is brought low when the microcontroller is writing to external RAM and that the P3.7/RD line is brought low when the microcontroller is reading from external RAM. If neither of these lines is low then the memory-mapped LCD is not being accessed. If either of these lines is low then the program *may* be accessing the LCD as long as it is referencing the right address. Thus the 74LS08 takes the WR and RD lines and asserts a low signal if either one is low.

Also recall from section 14.2.3 that MM_LCD is asserted by the 74LS138 IC (U5 on the SBC) whenever the program executes a MOVX instruction that references any address between 4000h and 4FFFh. When a memory location in this range of addresses is accessed, the MM_LCD line will be brought low which means the LCD is being accessed in some way—either written to or read from.

The 74LS02 then takes the MM_LCD signal and the output of the 74LS08 and will assert EN if both the inputs are low. Thus if either WR or RD is asserted *and* MM_LCD is also asserted, this indicates that the LCD is being accessed, the EN line will be brought high in order to clock the data that is on the data bus into the LCD.

The RW and RS lines are connected to A1 and A0, respectively. That means that whenever an address is referenced that has A0 set, the RS line of the LCD will be set. Likewise, any address that is referenced that has A1 set will cause the RW line to be set. Of course, the LCD will only be active when MM_LCD is asserted so the state of A0 and A1 will only matter to the LCD when the address in question is in the range of 4000h to 4FFFh.

The following table depicts the state of RW and RS based on the address that is being referenced. The first three columns show the binary state of each address line (the values of the 'x' bits don't matter). The fourth column shows the typical address used to generate this particular state, while the fifth and sixth column indicate the state of RW and RS when that address is referenced. The final column indicates what interaction will be made with the LCD based on the address being referenced.

A15-A2	A1	A0	Address	RW	RS	Function
010 xxxxx xxxxxx	0	0	4000h	0	0	Write command to LCD
010 xxxxx xxxxxx	0	1	4001h	0	1	Write text to LCD
010 xxxxx xxxxxx	1	0	4002h	1	0	Read status/data from LCD
010 xxxxx xxxxxx	1	1	4003h	1	1	Read text from LCD

When the microcontroller has finished reading or writing to the LCD address, it will stop asserting the RD or WR line. At that point, the 74LS08 OR gate will no longer be true so the output of the '08 will go high. At that point, since both signals aren't low, the 74LS02 AND gate will no longer be true so its output will go low—thereby bringing the EN line back low, exactly as required to write the data to the LCD.

19.4.2 Writing to the LCD in Memory-Mapped Mode

The table above shows that to write the first byte of the initialization sequence (38h) to the LCD, the following code would be used:

```
MOV DPTR,#4000h      ;Address 4000h is RW=0, RS=0, "write command"
MOV A,#38h           ;38h is the value of the command byte to send
MOVX @DPTR,A         ;Write the command to the LCD
```

These three lines of code are all that is necessary to send the command to the LCD. This is obviously significantly easier than the eleven lines of code necessary to send the same command to the LCD in 8-bit direct-connect mode or the twenty-three lines of code necessary in 4-bit mode. And, of course, this approach has the added advantage of not requiring a single dedicated I/O line beyond the address and data bus which are already being used for other purposes.

To write text to the LCD for display purposes, the exact same code would be used except the DPTR value would be 4001h instead of 4000h.

19.4.3 Reading from the LCD in Memory-Mapped Mode

Reading from the LCD is just as easy as writing to it in memory-mapped mode. The only difference is the address loaded into DPTR and changing the MOVX instruction so that the program transfers the data from the LCD to the accumulator.

To read the status of the LCD to check whether or not it is busy the following code may be used:

```
MOV DPTR,#4002h      ;Address 4002h is RW=1, RS=0, "read command"
MOVX @DPTR,A         ;Read the status from the LCD
```

To read text data from the LCD the address would simply be changed to 4003h. Once again, these two lines are significantly easier than the six lines necessary to accomplish the same thing in 8-bit direct-connect mode or the fifteen lines of code necessary to accomplish the same thing in 4-bit mode.

19.4.4 Additional Notes about Memory-Mapped LCD

There are a couple of aspects of memory-mapped LCD access that are worth noting.

The only address bits that matter when addressing the LCD are the high three and the low two bits. The high three bits must be 010 while the address lines A0 and A1 control RS and RW. The remaining bits of the address don't matter. This means that the program may write a command to the LCD by writing to the address 4000h, 4004h, 4008h, etc. Using any of these addresses will accomplish the same function. Likewise, writing text to the LCD may be accomplished by writing to the address 4001h, 4005h, 4009h, etc. This means that the LCD appears in the SBC memory map over and over throughout the 4000h – 4FFFh address range. Practically speaking, this is not a very efficient use of the memory map since only four addresses are necessary to access the LCD but it is occupying 4096 bytes of memory. A more efficient memory map, however, would require additional 74LS138 ICs to provide a more precise mapping of devices.

Another caution is that the program must be careful to read from and write to the correct address. For example, address 4001h is used for *writing* text to the LCD. If, instead, the program attempts to *read* from 4001h, garbage data will be sent to and possibly displayed by the LCD. This is due to the fact that 4001h specifically tells the LCD that the communication is a *write* operation but the the MOVX instruction will

258

try to *read* from the LCD. In essence, both the LCD and the microcontroller will be waiting for data. In tests with the 8052.com SBC, reading from address 4001h resulted in a black box/cursor being written to the next position on the LCD screen.

CHAPTER 20: INTERFACING TO SERIAL EEPROM (SPI)

The 8052.com SBC includes an AT25010A serial EEPROM (U14). Serial EEPROMs have the ability to retain information even after power is removed from the circuit. The data can subsequently be read when power to the system is restored.

In the case of the SBC, the AT25010A part was chosen due to the fact that the communication between the serial EEPROM and the microcontroller is accomplished using the Serial Peripheral Interface, or SPI. The SPI protocol was originally developed by Motorola and has since been adopted by many semiconductor firms for use in their own integrated circuits. Since such a large number of ICs use the SPI protocol, knowledge of how to communicate with these parts is very valuable.

The AT89S8252 microcontroller used with the 8052.com SBC has 2k of on-chip EEPROM. As such, it would actually make more sense for a program to store data in the on-chip EEPROM than on an external part such as the AT25010A which is an external IC and only has 128 bytes of EEPROM. However, the object of this chapter is to master SPI communication more than it is to focus on the usefulness of the serial EEPROM itself. Nevertheless, a serial EEPROM such as the AT25010A could be a very useful device if a microcontroller is being used that doesn't include on-chip EEPROM.

20.1 General Information about Serial Peripheral Interface (SPI)

The Serial Peripheral Interface is based on a master/slave relationship. The microcontroller—the master—has complete control over when communication is established with each SPI slave device. SPI slave devices don't speak unless first spoken to by the microcontroller.

The SPI bus consists of three communication lines plus a "chip select" line for each slave device. Multiple SPI-compatible devices may share the three communication lines while each device has its own chip select line. A device only responds to communication on the SPI bus when its chip select is asserted by the microcontroller.

The four lines of the SPI data bus are:

Master In/Slave Out (**MISO**): The MISO line is the input into the microcontroller (the master) which is also the output of the SPI slave—hence the name "Master In Slave Out". The microcontroller receives data from SPI devices on this line.

Master Out/Slave In (**MOSI**): The MOSI line is the output from the microcontroller (the master) which is also the input to the SPI slave—hence the name "Master Out Slave In". The microcontroller sends data to the SPI devices on this line.

Serial Clock (**SCK**): The SCK line is the serial clock line. The clock line is used to indicate when the data on the MISO and MOSI lines are valid.

Slave Select (SS): The SS line is the "chip select" line. There is one for each SPI device. The microcontroller asserts the SS line for the SPI device with which it wishes to communicate by bringing it low. When an SPI device's SS line is high, the device will not respond to any SPI communication on the other three lines of the bus.

A typical SPI byte transfer is depicted in the following figure:

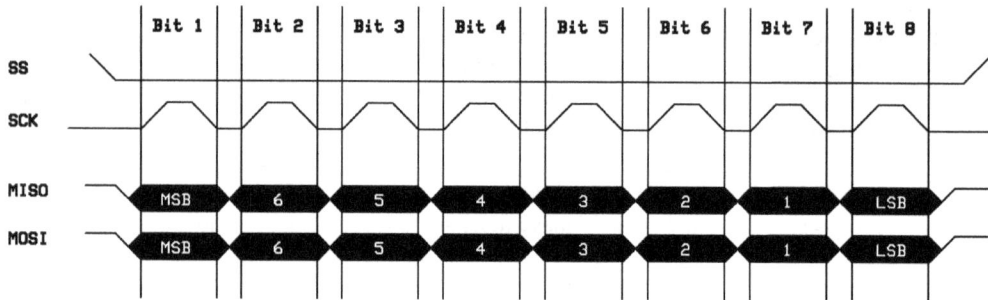

As this diagram indicates, the microcontroller first drives SS low to select the SPI slave device. As soon as SS goes low, the SPI device puts its first bit on the MISO line while the microcontroller puts its first bit on the MOSI line. The microcontroller then raises the SCK line to indicate that it has placed valid data on MOSI and also reads the value of MISO. It then lowers the SCK line to prepare for the next bit. This process is repeated eight times—once for each bit. After eight SCK cycles one full byte will have been transmitted to the SPI device from the microcontroller and one full byte will have also been received by the microcontroller from the SPI device.

SPI is a full duplex synchronous protocol. This means that data is sent and received at the same time time. In fact, reading a byte from an SPI device actually results in a dummy byte being sent to the device. It is up to the slave device to discard this dummy byte.

Beyond these signals, the actual communication between the microcontroller and SPI slave device in terms of the data actually sent and received is not defined. Each SPI device may use its own protocol. It is up to the microcontroller software to send and receive data as defined by the SPI slave device.

20.1.1 Coding for SPI Communication

The previous section illustrated how SPI communication is accomplished by the microcontroller driving three lines (SS, SCK, and MOSI) and treating the remaining line (MISO) as input. The steps a program must take to accomplish this are:

1. Set MISO pin high so that it will be an input pin.
2. Drive SCK low so that the initial state of the clock line is low.
3. Drive SS low to select SPI slave device.
4. Put the first bit on the MOSI line.

5. Raise the SCK line to clock MOSI out to the SPI device and cause the SPI device to clock out its first bit.
6. Read the MISO line to get the first bit returned by the SPI device.
7. Lower SCK to signal the end of the first bit.
8. Repeat steps 4-7 for each of the eight bits of the byte
9. Raise SS to deselect the SPI device.

This is accomplished in the following code:

```
1  SPI_SendByte:
2      SETB MISO              ;MISO line configured as input
3      CLR SCLK               ;SCLK starts low
4      MOV R7,#08h            ;8 bits to clock in/out
5  SPI_SendLoop:
6      MOV C,ACC.7            ;Get the MSB to clock out
7      MOV MOSI,C             ;Set MOSI to bit to clock out
8      SETB SCLK              ;Raise SCLK to clock data out
9      MOV C,MISO             ;Get the bit to clock in
10     CLR SCLK               ;Lower SCK to end cycle
11     MOV ACC.7,C            ;Store it at the current position
12     RL A                   ;Rotate left in accumulator
13     DJNZ R7,SPI_SendLoop   ;Send next bit out
14     RET
```

Lines 2 and 3 start by setting the MISO pin for input and clearing the SCK line. Line 4 sets R7 to 8 so that the routine will clock out each of the eight bits in the accumulator.

Lines 5 through 13 constitute the actual loop which clocks out the bits in the accumulator to the SPI device and at the same time receives bits from the SPI device and stores them in the accumulator. Line 6 and 7 start by getting the most significant bit and sending it out the MOSI line. SCK is then brought high on line 8.

The data sent to the microcontroller is read in line 9 after which SCK is immediately brought low on line 10. Line 11 takes the bit that was received from the SPI device and stores it in the high bit of the accumulator. Line 12 then rotates the accumulator to the left such that the most significant bit becomes the least significant bit. Line 13 causes lines 5 through 12 to be executed eight times—once for each bit of the byte.

This subroutine receives a value in the accumulator and sends it to the SPI device. At the same time, it receives a value from the SPI device and returns it in the accumulator.

It is common to also implement an SPI "receive" routine even though the send routine above effectively receives data from the SPI device. Such a receive routine usually consists of nothing more than:

```
SPI_ReadByte:
   MOV A,#0FFh            ;Dummy value to be sent to SPI device
   LCALL SPI_SendByte     ;Read byte from SPI device
   RET                    ;Return
```

This illustrates the fact that data cannot be received without data also being sent. Whenever the SPI_ReadByte routine is called it will simply load the accumulator with a "dummy value" of FFh.

Presumably the SPI device will ignore this data. However, attempting to "read" a byte from the SPI device when the SPI device is itself awaiting input will cause the SPI device to receive FFh which may or may not trigger a valid response.

20.1.2 Additional Information about SPI Communication

SPI is rather non-specific as communication standards go. Not only is the actual byte-level communication exchange between the microcontroller and the SPI device dependent on the SPI device being used, the actual signal synchronization in the figure in section 20.1 can vary from device to device.

One variable, normally called clock polarity (CPOL) defines whether SCK is active low or active high. In the SPI communication diagram, clock polarity is "0"—this means that SCK is idle low and active high. If clock polarity were "1" then the SCK signal would be reversed meaning that SCK is idle high and active low. In this case all the CLR SCK instructions would be exchanged for SETB SCK, and vice versa.

Another variable is known as clock phase (CPHA). This variable refers to where in the SCK cycle the data on the SPI bus is valid. If clock phase is "0" then the data on the SPI bus is valid when the clock is high; this is what is described in the figure above and in the source code provided. If clock phase is "1" then the data on the SPI bus is valid when the clock is low. This means that data should be read while SCK is low rather than high.

The clock phase and clock polarity of SPI devices should be indicated in the part's datasheet.

20.2 Software Communication with AT25010A

Communication with the AT25010A is achieved using the SPI protocol described in the previous section. Using this protocol, a higher-level command/response protocol is implemented such that the AT25010A responds to specific commands that it receives over the SPI bus.

Once the AT25010A's chip select line has been asserted by the microcontroller, the first byte sent to the AT25010A is a command followed, in some cases, by additional data. The AT25010A responds once the command byte and any necessary parameters are received.

The AT25010A commands are:

Command Byte	Command Description
02h	WR: Write Byte
03h	RD: Read Byte
05h	RDSR: Read Status Register
06h	WREN: Write Enable

20.2.1 AT25010A Command: Write Byte (WR)

The Write Byte (WR) command is used to write a byte of data to the serial EEPROM. The write command byte (02h) is followed immediately by the address to write to (in the range of 00h and 7Fh since the

AT25010A has 128 bytes of EEPROM memory), followed by the byte to write to that address. For example, the sequence of bytes sent to the AT25010A to write the value 47h to address 54h of its memory would be:

```
02h 54h 47h
```

The first byte, 02h, is the write command itself. The second byte, 54h, tells the AT25010A in which memory address the data will be stored, and 47h is the value to be stored in memory. The code to accomplish this would be:

```
CLR AT25010A_SS         ;Select the AT25010A as the SPI device
MOV A,#02h              ;Command byte: 02h-Write
LCALL SPI_SendByte      ;Send the command byte
MOV A,#54h              ;Address byte: Write to 54h
LCALL SPI_SendByte      ;Send the address byte
MOV A,#47h              ;Data Byte: Value to be written is 47h
LCALL SPI_SendByte      ;Send the data byte
SETB AT25010_SS         ;Deselect the AT25010A
```

Before sending the Write Byte command, it is necessary to first send the Write Enable (WREN) command. The WP line (pin 3) of the AT25010A must also be held high.

20.2.2 AT25010A Command: Read Byte (RD)

The Read Byte (RD) command is used to read a byte (or bytes) of data from the serial EEPROM. The read command byte (03h) is followed immediately by the address to read from (in the range of 00h and 7Fh since the AT25010A has 128 bytes of EEPROM memory). After the address has been sent to the AT25010A, the microcontroller reads the value returned in the next SPI operation. This requires that a dummy byte be sent out. The AT25010A will not process the dummy byte but, instead, will send the contents of the requested EEPROM address to the microcontroller on the MISO line as the dummy by is received.

For example, the sequence of bytes that need to be sent to the AT25010A to read the value from address 54h of its memory would be:

```
03h 54h FFh
```

The first byte, 03h, is the write command itself. The second byte, 54h, tells the AT25010A from which memory address the data will be read. The A25010A then clocks out the contents of the specified memory address while the FFh dummy byte is sent by the microcontroller.

The code to accomplish this would be:

```
CLR AT25010A_SS         ;Select the AT25010A as the SPI device
MOV A,#03h              ;Command byte: 03h-Read
LCALL SPI_SendByte      ;Send the command byte
MOV A,#54h              ;Address byte: Read from 54h
LCALL SPI_SendByte      ;Send the address byte
LCALL SPI_ReadByte      ;Read the response byte from AT25010A
SETB AT25010_SS         ;Deselect the AT25010A
```

20.2.3 AT25010A Command: Set Write Enable Latch (WREN)

The Set Write Enable Latch (WREN) command is used to to put the EEPROM into a write-enabled mode. When the AT25010A is powered-up it begins in a write-protect mode in which all write commands are ignored. In order to write information to the EEPROM, the Set Write Enable command must be issued prior to executing Write Byte commands.

This command is issued by simply sending the command byte 06h. There are no parameters.

```
CLR AT25010A_SS          ;Select the AT25010A as the SPI device
MOV A,#06h               ;Command byte: 06h-Set Write Enable
LCALL SPI_SendByte       ;Send the command byte
SETB AT25010_SS          ;Deselect the AT25010A
```

20.2.4 AT25010A Command: Read Status Register (RDSR)

The Read Status Register (RDSR) command is used to to read the status of the AT25010A. For the purposes of the 8052.com SBC this is used exclusively to check whether or not the AT25010A is busy. After a byte is written to the AT25010A with the write command, a certain amount of time is required for the actual write to the EEPROM to take place. During this time all commands to the AT25010A are ignored except for the Read Status Register command. The Read Status Register command can be called repeatedly to wait for a previous write operation to complete.

The command consists of the command byte (05h) with no parameters. However, a dummy byte FFh is sent after the command byte in order to clock in the response. If bit 0 of the byte that is returned is set then the AT25010A is still busy. The following code can be used to repeatedly poll the AT25010A until it is no longer busy:

```
AT25010_Wait:
    CLR        AT250X0_CS        ;Select AT250X0 as the SPI device
    MOV        A,#05h            ;Set Read Status Register command
    LCALL      SPI_SendByte      ;Send command
    LCALL      SPI_ReadByte      ;Read response from AT25010A
    SETB       AT250X0_CS        ;Deselect the AT250X0
    JB         ACC.0,AT25010_Wait ;Loop while still busy
    RET                          ;Return from routine
```

CHAPTER 21: INTERFACING TO REAL TIME CLOCK (I²C)

A Dallas DS1307 Real Time Clock (RTC) is included on the 8052.com SBC. An RTC is, as the name suggests, a clock. As long as power is applied, the RTC will provide a highly accurate and user-friendly measure of time. Specifically, the RTC will track the hour, minute, second, day of week, day of month, month, and year.

This part was chosen primarily as an example of Inter-IC (I²C) communication protocol. I²C is similar in function to the Serial Peripheral Interface protocol described in chapter 20, but I²C accomplishes communication with just two bus lines. No chip select line is necessary and the data is sent between the microcontroller and the I²C devices on a single line. Whereas SPI is capable of sending and receiving data simultaneously, I²C devices are either sending or receiving—they can't do both at the same time.

The I²C protocol was originally developed by Philips Semiconductor and has since been adopted by many semiconductor firms for use in their own integrated circuits.

21.1 General Information about Inter-IC (I²C) Protocol

The I²C protocol is based on a master/slave relationship. The microcontroller—the master—has complete control over when communication is established with each I²C slave device. I²C slave devices don't speak unless first spoken to by the microcontroller.

The I²C bus consists of just two communication lines to which all I²C devices in the circuit are connected. They are:

> **Serial Data (SDA)**: The SDA line is the serial data line. This is a bidirectional line that is used to write data from the microcontroller to the I²C slave device or to read data from the I²C slave back to the microcontroller.

> **Serial Clock (SCL)**: The SCL line is the serial clock line. The clock line is used to indicate when the data on SDA is valid and to clock data out of the microcontroller or out of the I²C device, depending on whether data is being written or read.

Typical I²C read and write sequences are depicted in the following diagram in which the components with a dark background represent data sent by the microcontroller and the components with a white background represent data sent by the I²C slave device back to the microcontroller.

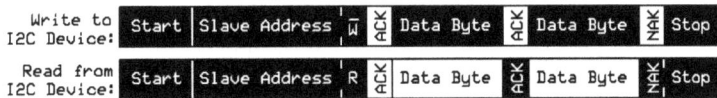

The microcontroller initiates all communication by sending a start condition, a slave address byte which identifies the I²C device to be addressed, and a read/write flag. The addressed device then responds with an ACK signal to indicate it received the command. A data exchange sequence consisting of one or more bytes then takes place depending on whether it is a read or write operation.

In the case of a write operation, the microcontroller continues sending bytes to the I²C device one byte at a time, with the slave acknowledging each byte with an ACK.

In the case of a read operation, the microcontroller waits for the initial acknowledge from the slave to the slave address byte that was initially sent, and then clocks in data on the SDA line. Thus the SDA line becomes an input line rather than an output line. Each data byte is acknowledged by the microcontroller with an ACK, except for the last byte which is answered with a NAK. All I²C communication is completed with the microcontroller sending a stop condition to the slave.

There are three conditions that an I²C bus handles:

Start condition: This condition proceeds all communication. It is characterized by the SDA line going from a high condition to a low condition while the SCL line is high.

Stop condition: This condition terminates an I²C transfer. It is characterized by the SDA line going from a low condition to a high condition while the SCL line is high.

Data transfer: After a start condition, a bit of data is considered valid when it remains stable for as long as SCL is in its high condition. Changes to the SDA state must be made while SCL is low.

These three conditions are depicted in the following figure which illustrates the difference between the conditions: the SDA line changes while the SCL line is high for start and stop conditions while the SDA changes while SCL line is low for data bits. The three condition diagrams are then followed by a figure that shows the start condition, eight data bits, and the stop condition combined to start an I²C transaction, send the value 34h (00110100 in binary), and terminate the transaction.

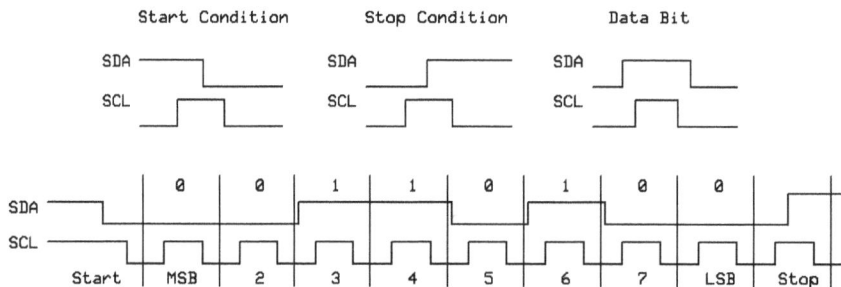

21.1.1 Coding for I²C Communication

The previous section illustrated how I2C communication is accomplished by the microcontroller driving two lines (SDA, SCL). A program that wishes to communicate via I2C must include code for the three possible data conditions:

1. Transmit start condition.
2. Transmit the eight bits that make up a byte.
3. Transmit stop condition.

21.1.1.2 I²C Communication: Transmitting Start Condition

As described in section 21.1, the start condition consists of SDA going from high to low while SCL remains high. This is accomplished with the following code:

```
1  I2C_START:
2    SETB SDA              ;Make sure SDA starts high
3    SETB SCL              ;Make sure SCL starts high
4    LCALL WAIT47us        ;Wait 47us for I2C to timeout
5    CLR SDA               ;Bring SDA high->low while SCL high
6    LCALL WAIT4us         ;Wait 47us for I2C timeout
7    CLR SCL               ;Bring SCL low at end of cycle
8    RET                   ;Return
```

Lines 2 and 3 set the initial state of the SDA and SCL lines as high. In theory these lines are redundant since the I2C specification assumes that an idle bus has both SDA and SCL high—but these two lines guarantee that the I2C bus is as at a known state. Line 4 then waits for 4.7 microseconds to give the bus time to settle since a start condition must not occur for at least 4.0 microseconds after a stop condition and at least 4.7 microseconds immediately following a previous start condition.

Line 5 brings SDA low while SDA remains high, thus creating the start condition. Line 6 waits for 4 microseconds to ensure that the minimum SCL high period time has transpired and then line 7 finishes the condition by bringing SCL low.

21.1.1.3 I²C Communication: Transmitting Stop Condition

The stop condition consists of SDA going from low to high while SCL remains high. This is accomplished with the following code:

```
1  I2C_STOP:
2    CLR SDA               ;SDA starts condition low
3    LCALL WAIT4us         ;Wait for 4us timeout
4    SETB SCL              ;Bring SCL high for transaction
5    LCALL WAIT47us        ;Wait 4.7us timeout
6    ORL C,/SCL            ;Verify that SCL went high correctly
7    SETB SDA          ;Bring SDA low->high for stop condition
8    LCALL WAIT47us ;Wait 4.7us timeout
9    ORL C, /SDA           ;Verify that SDA went high correctly
10   RET                   ;Return
```

Line 2 ensures that the SDA line starts low while SCL is still low. Line 3 then waits for 4 microseconds which is the minimum time SCL must be low between cycles. Line 4 then raises the SCL line to initiate the cycle and waits 4.7 microseconds to make sure the slave has adequate time to recognize the high level on SCL.

Line 6 checks the status of SCL and ORs its inverted value into the carry bit. After line 4 brings SCL high, the SCL line should actually *be* high. If it remains low then there is some kind of error condition or the bus didn't respond correctly. Thus SCL remaining low is an error condition. Line 6 inverts the condition (i.e. a high value flags an error condition) and stores that in the carry bit.

Line 7 then brings the SDA line high while SCL remains high, thus creating the stop condition. Line 8 waits another 4.7 microseconds before checking to make sure that SDA actually *went* high. As is the case with SCL, if SDA remains low then it is an error condition and that is combined with the previous error condition in the carry bit.

When the routine returns, the carry bit will indicate whether the stop condition was successful. A clear carry indicates that the stop condition was successful while a set carry indicates some kind of error occurred.

21.1.1.4 I²C Communication: Transmitting a Byte of Data

Transmitting a byte of data to the I2C bus actually consists of sending 9 bits rather than 8: The eight data bits plus an ACK or NAK bit. The following subroutine sends the byte in the accumulator to the I²C bus and then sends either an ACK or a NAK depending on the status of the carry bit. A cleared carry sends an ACK while a set carry sends a NAK.

```
 1  I2C_Send:
 2     MOV R2,#9            ;8 data bits + ACK = 9 cycles
 3  I2C_S2:
 4     RLC A                ;Send MSB first
 5     MOV SDA,C            ;Send bit to SDA while SCL low
 6     LCALL WAIT47us       ;Wait 4.7 microseconds for SDA to settle
 7     SETB SCL             ;Raise SCL once SDA is settled
 8     LCALL WAIT4us        ;Wait 4.7 microseconds to give device time
 9     MOV C,SDA            ;Read the bit status of SDA
10     CLR SCL              ;Bring SCL low for end of cycle
11     DJNZ R2,I2C_S2       ;Repeat for each of 9 cycles
12     RET                  ;Return
```

Line 2 initializes R2 to 9 which represents the nine bits of data that will be clocked in and out (8 data bits plus the ACK or NAK bit).

Lines 3 through 11 make up the loop that transmits each of the nine bits. Line 4 rotates the accumulator and carry right such that the first bit to be transmitted is the most significant bit of the accumulator and the last bit to be transmitted is the value that was originally in the carry bit. Line 5 sets the SDA line to the value of the bit that is currently being clocked out. Line 6 then waits for 4.7 microseconds to make sure the SDA line has a chance to settle before line 7 raises the SCL line to initiate the cycle. Line 8 waits another 4 microseconds to give the slave device a chance to clock the data in while like 9 gets the current status of SDA into the current bit position. Line 10 then brings SCL low to end the cycle. Line 11 repeats the loop for each of the nine bits that are to be transmitted.

When line 12 returns from the subroutine, the carry bit will contain the ACK or NAK returned by the I²C slave device. A clear carry means that an ACK was received from the I²C device and that the byte transfer was successful; if the carry bit is set then it means the I²C device didn't respond or responded with a NAK.

The exact same routine is used to receive data from an I²C device—the only difference is that the byte to be sent must be FFh. The value returned by this routine in the accumulator is the value of the byte that was received from the I²C slave device. This may be accomplished with the code:

```
I2C_Read:
    MOV A,#0FFh        ;Set FFh as the dummy byte to output
    LCALL I2C_Send     ;Send dummy byte out, read byte from I2C
    RET                ;Return with byte read in accumulator
```

21.2 I²C Communication with the DS1307

Communicating with the DS1307 via I²C requires that a slave address byte be sent along with a read/write bit. The DS1307's I2C slave address is D0h (11010000 binary) which is the value used to address the device when writing to it. If the DS1307 is to be read then the read bit must be set so the slave address becomes D1h (11010001). The data exchange then depends on whether the microcontroller is reading or writing to the RTC.

21.2.1 Sending Data to DS1307

When writing to the DS1307, the sequence of events is as follows:

1. Send I2C start condition.
2. Send DS1307 slave address and write flag (D0h).
3. Send address of register to write to or point to, one bit at a time.
4. Send data bytes, if any, one bit at a time.
5. Send I2C stop condition.

The DS1307 contains 64 registers numbered 00h through 3Fh. Eight of these registers hold the date and time while the other 56 bytes can be used as general purpose RAM. The register address sent to the DS1307 in step 3 indicates which register will be read or written to in the rest of the operation.

Writing to the DS1307 serves one of two purposes: 1) Setting the clock or writing to the RAM, or 2) Setting the register pointer for a subsequent read. Since the read command has no parameters, the data returned by a read command will begin at the address last pointed to by a write command. Thus a read sequence normally sends a write command to set the register pointer and then subsequently issues the read command itself.

21.2.2 Receiving Data from the DS1307

When reading the time and date from the DS1307, the sequence of events is as follows:

1. Send a write instruction and set register address to 00h. That is, send the byte sequence D0h 00h as described in the previous section. The register address can be set to any other value between 00h and 3Fh if the desired information is not the first byte of the clock registers.
2. Send I²C start condition.
3. Send DS1307 slave address and read flag (D1h).
4. Read 8 I²C bytes from the slave device which correspond to the eight DS1307 registers which contain the date/time. Any number of bytes may be read—it need not necessarily be eight. Eight is simply the number of bytes that hold clock data.
5. Send I²C stop condition.

Since the register address is retained between read and write operations, reading the DS1307 will return data starting at the next register address. For this reason it is generally appropriate to first set the register pointer as described in step 1.

21.2.3 DS1307 Registers

The DS1307 itself contains 64 bytes of memory, the first 8 of which have specific clock-related functions. The remaining memory locations can be used for any purpose the program requires—or, as is often the case, simply left unused.

The registers in the DS1307 are as follows:

Addr	B7	B6	B5	B4	B3	B2	B1	B0
00h	CH	10 Seconds			Seconds			
01h	0	10 Minutes			Minutes			
02h	0	SEL12	10HR/AP	10HR	Hours			
03h	0	0	0	0	0	Day Of Week		
04h	0	0	10 Date		Date			
05h	0	0	0	10 Mth	Month			
06h	10 Year				Year			
07h	OUT	0	0	SQWE	0	0	RS1	RS0
08h-3Fh	USER RAM 56 bytes – May be used for anything the user wishes							

The clock-related registers are stored as binary coded decimal (BCD). That means if the time is, for example, 11:37:25, seconds (register 00h) will contain the value 25h, minutes (register 01h) will contain the value 37h, and hours (register 02h) will contain the value 11h.

> The internal representation of the time and date may seem completely logical, but when manipulating the data read from the clock it is important to remember that the data is stored as BCD and not binary. For example, when the seconds register goes from :09 to :10 the seconds counter will go from 09h to 10h—which means its value actually went from 9 (decimal) to 16 (decimal). If binary calculations—such as adding seconds—are required, it may be necessary to first convert the BCD values returned by the clock to binary.

A number of bits contained in the first eight registers have special functions.

Clock Hold (CH): Bit 7 of register 00h is the "Clock Hold" flag. When this bit is set the clock will be paused; when this bit is clear the clock will be enabled. When power is first applied to the DS1307 the contents of the DS1307's memory are undetermined so it is important to clear this bit initially to start the clock. This normally happens automatically when the program sets the "seconds" value of the clock since any value for this register between 00h and 59h will clear bit 7.

SEL12 (Select 12-Hour Mode): Bit 6 of register 02h is the "12 hour select" flag. When this bit is set the RTC will operate in 12-hour mode; when this bit is clear it will operate in 24-hour mode. This affects the behavior of the 10HR/AP bit and also controls whether the hour register will roll over from 12:59:59 pm to 01:00:00 pm or 13:00:00, and whether 11:59:59 pm will roll over to 0:00:00 or to 12:00:00 am.

10HR/AP (10 Hour/AM/PM Flag): Bit 5 of register 02h is either the PM flag in 12-hour mode or the second digit of the high nibble of the hour in 24-hour mode. If the RTC is in 12-hour mode then this bit will be clear when the time is between midnight and 11:59:59am and will be set between noon and 11:59:59pm. If the RTC is in 24-hour mode then this bit will be set between 20:00:00 and 23:59:59 in order to represent the value "2" of the hour.

Day of Week: The day of the week is a value between 1 and 7 that can be used to represent the day of the week such as Sunday, Monday, etc. This value must be specifically programmed when the clock is set—the DS1307 will not automatically calculate this value. The RTC will simply increment this value each time the clock rolls past midnight and reset it back to 1 when it increments from 7. It is up to the program to choose what value equals which day (1=Sunday, 2=Monday, etc.).

Register address 07h is the clock control register. Once this is written it is not normally necessary to update it even if the date and time in the clock are reset. The control register has the following control bits:

Output Control (OUT): Bit 7 controls the state of pin 7 of the DS1307 when the square wave is disabled. When disabled, pin 7 will be high if this bit is set and will be low if this bit is clear.

Square Wave Enable (SQWE): Bit 4 controls whether or not the square wave output is enabled on pin 7. If this bit is set, pin 7 will be toggled at a frequency selected by the two rate select bits described below. If this bit is clear then pin 7 will be set to the value of the Output Control bit (bit 7).

Rate Select (RS0 and RS1): Bits 0 and 1 select the frequency of the square wave that can be enabled on pin 7. The rate for specific values of RS0 and RS1 are illustrated in the following table.

RS1	RS0	SQW (Pin 7) Output Frequency
0	0	1Hz
0	1	4,096 Hz
1	0	8,192 Hz
1	1	32,768 Hz

The optional square wave on pin 7 of the DS1307 can be connected to an interrupt pin on the microcontroller in order to trigger an interrupt at regular intervals.

21.3 Software Communication with DS1307

Based on the I²C routines presented in section 21.1.1 and the the DS1307 presented in the previous section, setting and reading the time from the RTC becomes a relatively simple exercise in calling the I²C routines.

21.3.1 Setting the DS1307 Clock

Setting the DS1307 clock is a write operation. It begins with the DS1307's slave address for writes (D0h) followed by the initial address to write to (00h) which is in turn followed by the seven bytes of data that make up the date and time.

The following code sets the DS1307's clock to Wednesday, January 26, 2005, 15:50:27.

```
LCALL I2C_START      ;Start I2C transaction
MOV A,#0D0h          ;DS1307's write address
LCALL I2C_Send       ;Send the slave address
MOV A,#00h           ;Set register address to 0
LCALL I2C_Send       ;Send the register byte
MOV A,#27h           ;Seconds = 27h in BCD
LCALL I2C_Send       ;Send the seconds to the RTC
MOV A,#50h           ;Minutes = 50h in BCD
LCALL I2C_Send       ;Send the minutes to the RTC
MOV A,#15h           ;Hours = 15h in BCD, 24-hour mode
LCALL I2C_Send       ;Send the hours to the RTC
MOV A,#03h           ;Day of week Wednesday, with 1=Monday
LCALL I2C_Send       ;Send the day of week to the RTC
MOV A,#26h           ;Day of month = 26h in BCD
LCALL I2C_Send       ;Send the day of month to the RTC
MOV A,#01h           ;Month = 01h (January) in BCD
LCALL I2C_Send       ;Send the month to the RTC
MOV A,#05h           ;Year = 05h (2005) in BCD
LCALL I2C_Send       ;Send the year to the RTC
LCALL I2C_STOP       ;Finish the I2C transaction
```

In a practical implementation the date and time would not be hard-coded into the program but would rather be entered by some type of input. Refer to the source code for SBCMON for an example of setting the clock based on user input.

The source code to SBCMON may be found at **http://www.8052.com/sbc/sbcmon**.

21.3.1 Reading the DS1307 Clock

Reading the current time and date from the DS1307 is a two-step process that requires a write operation followed by a read operation.

The program first sends the DS1307's slave address for writes (D0h) followed by the initial address that will subsequently be read from (00h). An I²C stop condition is then sent followed immediately by an I²C start condition. The DS1307's slave address for reads (D1h) is then sent; the seven bytes of the date and time registers are then read from the DS1307 into the microcontroller.

The following code reads the DS1307's date and time into internal RAM address 60h (seconds) through 66h (year).

```
                   ;Set the initial register address to 00h with write command

LCALL I2C_START    ;Start I2C transaction
MOV A,#0D0h        ;DS1307's write address
LCALL I2C_Send     ;Send the slave address
MOV A,#00h         ;Set register address to 0
LCALL I2C_Send     ;Send the register byte
LCALL I2C_STOP     ;Terminate the I2C write transaction

                   ;Start new I2C read transaction to read the 7 bytes of the
                   ;date and time from the DS1307.

LCALL I2C_START    ;Start a new I2C transaction
MOV A,#0D1h        ;DS1307's read address
LCALL I2C_Send     ;Send the slave address
LCALL I2C_Read     ;Read register 0 - Seconds
MOV 60h,A          ;Store seconds in internal RAM 60h
LCALL I2C_Read     ;Read register 1 - Minutes
MOV 61h,A          ;Store minutes in internal RAM 61h
LCALL I2C_Read     ;Read register 2 - Hours
MOV 62h,A          ;Store hours in internal RAM 62h
LCALL I2C_Read     ;Read register 3 - Day of week
MOV 63h,A          ;Store day of week in internal RAM 63h
LCALL I2C_Read     ;Read register 4 - Day of month
MOV 64h,A          ;Store day of month in internal RAM 64h
LCALL I2C_Read     ;Read register 5 - Month
MOV 65h,A          ;Store month in internal RAM 65h
LCALL I2C_Read     ;Read register 6 - Year
MOV 66h,A          ;Store year in internal RAM 66h
LCALL I2C_STOP     ;Finish the I2C transaction
```

Once the data is in internal RAM the program may process the data as necessary. It is worth mentioning, again, that the data will be stored in internal RAM in BCD representation, not binary. This means that if it is 5:37pm, the minutes register will hold the value 37h (hexadecimal) and not the value 37 (decimal).

Additionally, if the program is only interested in a specific DS1307 register, it's not necessary to read them all. For example, if the program is only interested in obtaining the current month from the DS1307 the above code could be shortened to:

```
                   ;Set the initial register address to 05h with write command
LCALL I2C_START    ;Start I2C transaction
MOV A,#0D0h        ;DS1307's write address
LCALL I2C_Send     ;Send the slave address
MOV A,#05h         ;Set register address to 5 (month)
LCALL I2C_Send     ;Send the register byte
LCALL I2C_STOP     ;Terminate the I2C write transaction

                   ;Start new I2C read transaction to read the month only
LCALL I2C_START    ;Start a new I2C transaction
MOV A,#0D1h        ;DS1307's read address
LCALL I2C_Send     ;Send the slave address
LCALL I2C_Read     ;Read register 5 - Month
MOV 65h,A          ;Store month in internal RAM 65h
LCALL I2C_STOP     ;Finish the I2C transaction
```

CHAPTER 22: ADDITIONAL SOFTWARE TOPICS

This chapter will provide additional information and touch on a number of other software concepts that may be useful to the developer.

22.1 Reading the Value of the Program Counter

As mentioned in section 4.4, there is no direct way to read the value of the program counter. It is, however, possible to read the program counter with a small subroutine which copies the value of the program counter into DPTR by way of popping the return address off the stack into DPTR.

```
            LCALL ReadProgramCounter
                  .
                  .
                  .
ReadProgramCounter:
        POP     DPH             ;Get high byte of return address
        POP     DPL             ;Get low byte of return address
        CLR     A               ;Clear accumulator for zero offset
        JMP     @A+DPTR         ;Execute the equivalent of a RET instruction
```

This code takes advantage of the fact that when an LCALL is executed, the address of the instruction following the LCALL is pushed on the stack, low byte first, high byte second. In this case, the ReadProgramCounter pops the two bytes of the return address into the corresponding bytes of DPTR. It then clears the accumulator so that there's no offset for the subsequent JMP @A+DPTR instruction which jumps to the address contained in DPTR—which is the return address.

This is an unusual subroutine in that it doesn't end with a RET instruction. This is due to the fact that the two POP instructions remove the return address from the stack and put it in DPTR. Since the return address is no longer on the stack, the RET instruction would not return properly. But since the return address is in DPTR, the JMP @A+DPTR will accomplish the same function as a RET normally would.

This routine returns in DPTR the address of the instruction following the LCALL that called the subroutine. It thereby provides an indirect method of reading the program counter.

22.2 Power Saving Modes

The 8052 provides two power saving modes that may be very useful in projects that operate with power constraints; typically those that operate on battery power. In these cases it is often desirable to execute a function and then cause the 8052 to go into a low-power mode to minimize power consumption when the microcontroller doesn't need to operate at full capacity.

There are two power-saving modes: 1) Idle mode and 2) Power-down mode. These modes are entered by setting corresponding bits in the Power Control (PCON) SFR at address 87h.

In both cases, a reset condition on the RST (pin 9) line will always reset the microcontroller and take it out of any power-saving mode it may be in. Additionally, other conditions may cancel the power saving mode and resume normal operation depending on which of the two modes is enabled.

22.2.1 Idle Mode

The idle mode is entered by setting the IDL bit (bit 0) of PCON. When an instruction sets PCON.0, program execution will immediately stop. Although program execution will stop, the timers, serial port, and interrupts will remain active. While in idle mode, the microcontroller will consume approximately $1/3^{rd}$ to $1/4^{th}$ of the current it normally requires to operate, depending on the derivative chip in use.

The idle mode will be terminated and program execution will resume with the next instruction when any enabled interrupt occurs. This allows the program to "go to sleep" during periods of time when nothing is happening and thereby save energy rather than executing code that does nothing but wait for an event.

Consider a typical serial "echo loop" which does nothing but wait for serial input and echo it back to the serial port:

```
Loop:
        JNB     RI,$            ;Wait for serial input
        MOV     A,SBUF          ;Get the byte that was received
        CLR     RI              ;Clear the RI flag
        MOV     SBUF,A          ;Write the byte back to serial port
        JNB     TI,$            ;Wait for serial output to be sent
        CLR     TI              ;Clear the TI flag
        SJMP    Loop            ;Repeat to wait for next character
```

The above code could be rewritten using idle mode as follows:

```
        SETB    EA      ;Enable global interrupts
        SETB    ES      ;Enable serial interrupt
Loop:
        MOV PCON,#01h           ;Go to sleep until serial byte received
        MOV A,SBUF              ;Get the byte that was received
        CLR RI                  ;Clear the RI flag
        MOV SBUF,A              ;Write the byte back to the serial port
        MOV PCON,#01h           ;Wait for serial output to be sent
        CLR TI                  ;Clear the TI flag
        SJMP Loop               ;Repeat to wait for next character

        ORG 0023h               ;Serial interrupt vector address
        RETI                    ;Do nothing but return from interrupt
```

The function of these two programs is identical but the second program will consume 66% to 75% less power since the microcontroller will be in idle mode when it is waiting to receive or send a character. Using idle mode in situations such as these can greatly reduce the amount of power needed to run the circuit in battery-powered environments.

Note that interrupts must be enabled for the microcontroller to be able to leave idle mode. If, in the example above, the serial interrupt were not enabled, the microcontroller would never leave idle mode even if data is received on the serial port. It is the triggering of the interrupt which causes the microcontroller to "wake up."

When the interrupt is triggered, execution will resume immediately with the interrupt routine that corresponds to the interrupt that was triggered. In this case, a serial interrupt is waking the microcontroller up, so as soon as the microcontroller wakes up it will branch to the serial interrupt service routine at 0023h. The serial interrupt routine in this example does nothing except return which allows the program to continue at the instruction that followed the `MOV PCON,#01h` command which originally put the microcontroller in idle mode.

22.2.2 Power-Down Mode

The second power-saving mode is known as power-down mode. This functions in generally the same way as idle mode but powers down even more aspects of the microcontroller, including the oscillator. This means that power consumption is reduced to a small fraction of the normal mode—along the lines of $1/500^{th}$. Needless to say, a device that can achieve the same functionality on $1/500^{th}$ of the power is certainly attractive if it is powered by a battery or, perhaps, solar power.

The functionality of power-down mode is identical to idle mode except that 1) The instruction used to put the microcontroller in power-down mode is `MOV PCON,#02h` (rather than #01h), 2) Only external interrupts can wake-up a microcontroller that is in power-down mode. Further, the external interrupt must be configured to be level-active so that it is triggered by a low on the corresponding port pin; it may not be edge triggered. 3) The microcontroller's timers and serial port are not active in this mode.

Consider the following code:

```
            LCALL  GoToSleep      ;Go to sleep until interrupt 1 occurs
            .
            .
            .
GoToSleep:
            SETB   EA             ;Enable global interrupts
            SETB   EX1            ;Enable external 1 interrupt
            MOV PCON,#02h         ;Go to sleep until external 1 interrupt
            RET                   ;Ext. 1 interrupt occurred, return

            ORG 0013h             ;External 1 interrupt vector address
            RETI                  ;Do nothing but return from interrupt
```

This code will cause the microcontroller to go to sleep until an external 1 interrupt occurs. If P3.3 is connected to a momentary push-button, pressing that button will wake-up the microcontroller. Similarly, an external RTC with an alarm interrupt signal (such as the DS1337) can be used such that the microcontroller first sets an alarm in the external RTC and then goes into power-down mode until the external RTC wakes it up by triggering a low level on the interrupt pin.

22.3 Software-Based Real-Time Clock

Although the 8052.com SBC has a hardware-based RTC (see chapter 21), it is often useful to be able to keep track of time in software without depending on external hardware. This may be because the time doesn't have to be maintained when the power is lost or because an existing device needs to keep track of time without new hardware being retrofitted to the existing design. This concept can also be used in conjunction with a hardware RTC (see chapter 21) by reading the time and date from the hardware RTC at power-up and then subsequently updating the time using this software-based RTC.

This section will cover the development of a simple software-base RTC solution using timer 0. Software may implement this in order to obtain the benefits of an RTC without requiring external hardware.

22.3.1 RTC Variables

Before developing the code, a few variables should be established. These variables will be used frequently within interrupts so it is a good idea to put them in internal RAM. These variables will be placed arbitrarily at 5Ch through 5Fh, but they could be placed anywhere in internal RAM.

```
HOURS      EQU   07Ch
MINUTES    EQU   07Dh
SECONDS    EQU   07Eh
TICKS      EQU   07Fh
```

The timer interrupt will use these four variables to keep track of time. Additionally, the main program may access these variables whenever it needs to determine the "current time" from the RTC.

22.3.1 Crystal Frequency

The first thing that needs to be taken into account is the speed of the crystal being used. Keep in mind that with a crystal of 11.0592 MHz, timer 0 will increment 11,059,200 / 12=921,600 times per second.

A standard 8052 timer increments every 12 crystal cycles. However, some derivatives increment their timers after a different number of crystal cycles: For example, Dallas microcontrollers can be programmed to increment every 4 cycles. If a derivative that uses some value other than 12 is used in the circuit, the corresponding changes will need to be made to this code.

Additional equates will be established to make the code more portable between different derivatives:

```
CRYSTAL EQU 11059200          ;The crystal speed
TMRCYCLE EQU 12               ;Crystal cycles per timer increment
TMR_SEC EQU CRYSTAL/TMRCYCLE  ;The # of timer increments per second
```

Thus, should the crystal frequency change or should an alternative derivative be used that increments its timers at some value other than every once every 12 crystal cycles, it will be necessary to do nothing more than modify these equates.

22.3.2 Calculating Timer 0 Overflow Frequency

Recall from chapter 8 that a 16-bit timer will count from 0 to 65,535 before resetting. Obviously it will overflow it's 65,535 maximum value multiple times in the course of one second—to be exact, it will overflow 921,600 / 65,536=14.0625 times per second. If the timer were in 8-bit or auto-reload mode, the timer would end up overflowing 921,600 / 256 = 3600 times per second which is even more difficult to keep track of.

Since it's easier and more efficient to count approximately 14 overflows per second than 3,600, timer 0 will be placed in 16-bit mode (mode 1). However, timer 0 will actually overflow 14.0625 times per second, not an even 14. This means that the software can't simply count the number of overflows from it counting from 0 to 65,536 since this would introduce approximately 0.44% of unnecessary inaccuracy.

For this reason it is necessary to configure timer 0 to overflow at some frequency that adds up nicely to one second intervals. For example, 65,536 timer 1 cycles is 65,536/921,600 = .071 seconds which means it would take the timer 0.071 seconds to count from 0 up to 65,536. The problem is that 1.00 seconds divided by .071 seconds does not produce an integer result, thus we have inaccuracy. The goal is to have timer 0 overflow at a frequency that can be multiplied by an integer to arrive at 1.00 seconds.

For example, if timer 0 were to overflow every .05 seconds, the software would known that after exactly 20 overflows, one second had passed. How long is .05 seconds in terms of timer cycles? Simple: 921,600 * .05 = 46,080. In other words, after timer 0 has been incremented 46,080 times, 1/20th of a second (.05 seconds) have passed.

So the trick is to have the timer overflow every .05 seconds instead of every .071 seconds. Remember that the timer overflows when it reaches 65,535 and is incremented to 0. It was just calculated that the timer needs to overflow every 46,080 cycles. To do that the counter needs to start counting up from some value other than 0. Specifically, it should be initialized to 65,536 - 46,080 = 19,456. In other words, if timer 0 is initialized to 19,456, it will take 46,080 cycles for it to reset to 0. When it resets to 0, it needs to be reset to 19,456.

Since the code should be as portable and as easy to maintain as possible, a number of additional equates will be defined that represent how many timer cycles will pass in .05 seconds. The TMR_SEC equate, which indicates how many timer cycles pass in a second, was already defined, so to determine how many cycles make up 1/20th of a second simply requires multiplying the first value by .05.

```
F20TH_OF_SECOND EQU TMR_SEC * .05
```

Thus, F20TH_OF_SECOND indicates how many cycles the timer will count in 1/20th of a second. However, an "initialization value" is needed for the timer. The initialization value, as just discussed above, is actually the number 65,536 less the constant that was just calculated:

```
RESET_VALUE EQU 65536 - F20TH_OF_SECOND
```

22.3.3 Starting Timer 0

Timer 0 will be configured for 16-bit mode. The first step is to initialize timer 0 to the reset value that was calculated using the equates above.

```
MOV TH0,#HIGH RESET_VALUE   ;Initialize timer high-byte
MOV TL0,#LOW RESET_VALUE    ;Initialize timer low-byte
```

Now that the timer has been initialized with a reset value, timer 0 must be configured for 16-bit mode and subsequently set to run. The following code sets timer 0 to mode 1 without modifying the configuration for timer 1.

```
        ANL TMOD,#0F0h      ;Make sure timer 0 bits are off
                            ;don't affect timer 1 configuration
        ORL TMOD,#01h       ;Set timer 0 to mode 1 (16-bit mode)
                            ;without affecting timer 1 configuration
        SETB TR0            ;Start timer 0 running
```

After executing this code, timer 0 will overflow in 46,080 instruction cycles. But then what? The 8052's interrupt facility needs to be used so that whenever timer 0 overflows, special RTC clock code is executed to handle the event.

Of course, it is very important the clock variables be initialized with the following instructions.

```
        MOV HOURS,#00       ;Initialize to 0 hours
        MOV MINUTES,#00     ;Initialize to 0 minutes
        MOV SECONDS,#00     ;Initialize to 0 seconds
        MOV TICKS,#20       ;Initialize countdown tick counter to 20
```

This initializes the clock to 0 hours, 0 minutes, and 0 seconds. The tick counter is initialized to 20—more on that later.

22.3.4 Configure Timer 0 Interrupt

Configuring the timer 0 interrupt is very easy. It is simply necessary to enable the global interrupts (set the EA bit) and also enable timer 0 interrupt itself (set the ET0 bit). This is accomplished with the following two lines of code.

```
        SETB EA    ;Initialize interrupts
        SETB ET0   ;Initialize timer 0 interrupt
```

Once enabled, whenever timer 0 overflows (i.e., is incremented from 65,535 to 0), a timer 0 interrupt will be triggered and the interrupt service routine at 000Bh will be executed. So the task at hand is to write the ISR that will be executed each time 1/20th of a second has passed.

22.3.5 Writing the Timer 0 Interrupt Service Routine

Before writing the code, consider what needs to be done every 1/20th of a second:

1. Reset timer 0 to the reset value of 19,456.
2. Decrement the TICKS variable.
3. If TICKS has been decremented from 20 all the way down to 0, it means a second has passed and the SECONDS variable must be incremented.
4. If SECONDS is equal to 60, it means an entire minute has passed and the MINUTES variable must be incremented.
5. If MINUTES is equal to 60, it means an entire hour has passed and the HOURS variable must be incremented.
6. Exit the interrupt routine.

The code for each of these steps will be explained one step at a time.

22.3.5.1 Reset Timer 0

The first thing that needs to be done is reset timer 0 to the reset value. Otherwise, timer 0 will take the necessary .05 seconds to overflow the first time, but subsequent overflows will occur every .071 seconds as the timer counts from 0 up to 65,535.

Thus whenever the interrupt is triggered the timer must be set to the RESET_VALUE that was calculated earlier. Also remember that the interrupt must leaves the main working variables in the same state they were in when the interrupt started, so the routine starts by pushing the registers that will be used by the interrupt onto the stack so that they can be restored at the end of the interrupt.

The interrupt service routine starts with:

```
ORG 000Bh                    ;Timer 0 interrupt routine starts here
PUSH ACC                     ;We'll use the accumulator, so protect it
PUSH PSW                     ;We may modify PSW flags, so protect it
CLR TR0                      ;Turn off timer 0 as we reset the value
MOV TH0,#HIGH RESET_VALUE    ;Set the high byte of the reset value
MOV TL0,#LOW RESET_VALUE     ;Set the low byte of the reset value
SETB TR0                     ;Restart timer 0
```

22.3.5.2 Countdown TICKS variable

Now that timer 0 has been reset, the interrupt must "do what needs to be done"—the guts of the interrupt.

First, the interrupt must count this interrupt as a "tick." When 20 ticks have passed the interrupt knows that a full second has transpired. If 20 ticks have not yet passed it is not necessary to do anything else: the code simply exists the interrupt service routine. This is accomplished with the following code:

```
DJNZ TICKS,EXIT_RTC ;Decrement TICKS, if not yet zero, exit immediately
```

This will decrement the TICKS countdown timer and, if it hasn't reached zero yet, will exit. Recall from step #3 above that TICKS was initialized to 20. Thus each time our interrupt is triggered, TICKS will be decremented. If it hasn't reached 0, a second has not yet passed so the routine simply skips ahead to the EXIT_RTC label which will terminate the interrupt.

22.3.5.3 One Second Has Passed

Once TICKS is decremented to 0, the DJNZ instruction above will fail and execution will continue with the next section of code—meaning a full second has passed.

The TICKS variable must first be reset to 20 so that the countdown variable is ready for another second to pass. The SECONDS variable must then be incremented.

```
MOV TICKS,#20      ;Reset the ticks variable
INC SECONDS        ;Increment the second variable
```

22.3.5.4 Have 60 Seconds Passed?

After incrementing SECONDS, the next step is to make sure that the number of seconds has not overflowed—it would not make sense to have a count of 85 seconds. Rather, our RTC should indicate 1 minute and 25 seconds. Thus the code must check to see if the value of SECONDS is equal to 60. If it isn't, that means the clock has not yet counted 60 seconds and may simply exit the interrupt routine.

```
MOV A,SECONDS       ;Move the seconds variable into the accumulator
CJNE A,#60,EXIT_RTC ;If we haven't counted 60 seconds, we're done.
```

22.3.5.5 Have 60 Minutes Passed?

If the above test fails, it means the program has counted 60 seconds. In this case, the SECONDS variable must be reset to zero and the MINUTES variable must be incremented—and if 60 minutes have passed, the program needs to reset MINUTES to 0 and increment the HOURS variable.

```
MOV SECONDS,#0      ;Reset the seconds variable
INC MINUTES         ;Increment the number of minutes
MOV A,MINUTES       ;Move the minutes variable into the accumulator
CJNE A,#60,EXIT_RTC ;If we haven't counted 60 minutes, we're done
MOV MINUTES,#0      ;Reset the minutes variable
INC HOURS           ;Increment the hour variable
```

22.3.5.6 Exit the Interrupt Routine

Finally, the standard housekeeping of any interrupt service routine has to be completed: the values that were protected by being pushed onto the stack must be restored by popping them off the stack in reverse order. The routine then terminates with the RETI instruction.

```
EXIT_RTC:
        POP PSW ;Restore the PSW register
        POP ACC ;Restore the accumulator
        RETI    ;Exit the interrupt routine
```

22.3.5.7 Putting it all Together

All the individual code fragments have been coded in the previous sections. All that's left is to put it all together in a single program.

```
HOURS           EQU 07Ch      ;Our HOURS variable
MINUTES         EQU 07Dh      ;Our MINUTES variable
SECONDS         EQU 07Eh      ;Our SECONDS variable
TICKS           EQU 07Fh      ;Our 20th of a second countdown timer
CRYSTAL         EQU 11059200  ;The crystal speed
TMRCYCLE        EQU 12        ;Number of crystal cycles per timer increment
TMR_SEC         EQU CRYSTAL/TMRCYCLE        ;# of cycles per second
F20TH_OF_SECOND EQU TMR_SEC * .05           ;# of cycles per 1/20 second
RESET_VALUE     EQU 65536-F20TH_OF_SECOND ;Timer reload value
```

```
        ORG 0000h                    ;Start assembly at 0000h
        LJMP MAIN                    ;Jump to the main routine

        ORG 000Bh                    ;Timer 0 Interrupt Routine starts here
        PUSH ACC                     ;We'll use the accumulator, protect it
        PUSH PSW                     ;Protect PSW flags
        CLR TR0                      ;Turn off timer 1 as we reset the value
        MOV TH0,#HIGH RESET_VALUE    ;Set the high byte of the reset value
        MOV TL0,#LOW RESET_VALUE     ;Set the low byte of the reset value
        SETB TR0                     ;Restart timer 0
        DJNZ TICKS,EXIT_RTC          ;Decrement TICKS, if not zero, exit
        MOV TICKS,#20                ;Reset the ticks variable
        INC SECONDS                  ;Increment the second variable
        MOV A,SECONDS                ;Move seconds variable into the accumulator
        CJNE A,#60,EXIT_RTC          ;If we haven't counted 60 seconds, we're done.
        MOV SECONDS,#0               ;Reset the seconds variable
        INC MINUTES                  ;Increment the number of minutes
        MOV A,MINUTES                ;Move minutes variable into the accumulator
        CJNE A,#60,EXIT_RTC          ;If we haven't counted 60 minutes, we're done
        MOV MINUTES,#0               ;Reset the minutes variable
        INC HOURS                    ;Increment the hour variable
EXIT_RTC:
        POP PSW                      ;Restore the PSW register
        POP ACC                      ;Restore the accumulator
        RETI                         ;Exit the interrupt routine
MAIN:
        MOV TH0,#HIGH RESET_VALUE    ;Initialize timer high-byte
        MOV TL0,#LOW RESET_VALUE     ;Initialize timer low-byte
        ANL TMOD,#0F0h               ;Make sure timer 0 bits are off
                                     ;don't affect timer 1 configuration
        ORL TMOD,#01h                ;Set timer 0 to mode 1 (16-bit mode)
                                     ;without affecting timer 1 configuration
        MOV HOURS,#00                ;Initialize to 0 hours
        MOV MINUTES,#00              ;Initialize to 0 minutes
        MOV SECONDS,#00              ;Initialize to 0 seconds
        MOV TICKS,#20                ;Initialize countdown tick counter to 20
        SETB TR0                     ;Start timer 0 running
        SETB EA                      ;Initialize interrupts
        SETB ET1                     ;Initialize timer 1 interrupt
        .... Main program continues here ...
```

22.3.5.8 Using the RTC

Once the code above has been included in the program, the application will have its own RTC. The program may set the RTC by setting the HOUR, MINUTE, and SECONDS variables, or may obtain the current time by reading them. Other than that, the RTC can be pretty much ignored since it will be running by itself in the background using timer 0 interrupt.

22.3.6 Additional Comments about the RTC

It is worthwhile to point out a few shortcomings and observations about the above solution.

First, there is a slight error introduced in the ISR. As can be seen in the code, the ISR turns off timer 0 while it resets TH0 and TL0. In all, the timer is turned off for three instructions: It is turned off for the two MOV instructions, and it is turned off until the end of the SETB instruction. On a standard 8052, each

MOV instruction requires 2 clock cycles to operate and the SETB instruction requires 1. Thus the clock effectively loses five cycles due to the ISR implementation. This error may be compensated for by taking into account the five "lost" cycles by using the following code instead of the code used in the interrupt above.

```
CLR TR0
MOV TH0,#HIGH (RESET_VALUE-5)
MOV TL0,#LOW (RESET_VALUE-5)
SETB TR0
```

Second, this solution is based on interrupts. If other interrupts are used in the program, timer 0 interrupt may not necessarily execute right away. If another interrupt of the same priority is executing when timer 0 overflows, the RTC interrupt will not execute until the other interrupt has finished. This will introduce inaccuracy.

One way way to guarantee that the RTC interrupt will always execute immediately is to give it a high interrupt priority and give all other interrupts a low priority. This can be done with the following instruction:

```
MOV IP,#2 ;Timer 0 Priority=1, all others = 0
```

Another solution to the problem of other interrupts delaying the timer 0 interrupt is to subtract the reload value in the interrupt rather than set it. This is illustrated in the following code:

```
CLR TR0
CLR C
MOV A,TL0
SUBB A,#LOW (F20TH_OF_SECOND-8)
MOV TL0,A
MOV A,TH0
SUBB A,#HIGH (F20TH_OF_SECOND-8)
MOV TH0,A
SETB TR0
```

The above implementation assumes that the timer 0 reset value is properly initialized by the main program before the interrupt was enabled. Once enabled, it won't matter if the interrupt executes immediately or not because the subtraction is calculating a new value based on the *current* timer value at the point at which the interrupt is executed. For example, assuming that the 1/20th of a second requires 46,080 instruction cycles, the reset value would be 65,536 – 46,080 = 19,456. But if the timer overflows while another interrupt is executing and the RTC interrupt isn't executed for an additional 1,000 instruction cycles, the code above would calculate a new value for the timer as 1,000 – 46,080 = -45,080. Since the timer is unsigned, the number -45,080 would be interpreted as 20,456—so the reload value would be calculated as 20,456 instead of 19,456—so it would take 1,000 instruction cycles less time for the timer to overflow again which is consistent with the fact that 1,000 instruction cycles already had passed when the RTC interrupt was finally executed. For this solution to work, the timer 0 interrupt must be delayed no more than 46,080 instruction cycles.

The final disadvantage is that this solution requires the dedicated use of timer 0. The main program isn't allowed to change the value of the timer since doing so will cause the RTC to become completely inaccurate. The main program can *read* the value of the timer, but not change it.

REFERENCE & APPENDIXES

Appendix A: 8052 Instruction Set Quick-Reference

00	NOP	40	JC *relAddr*	80	SJMP *relAddr*	C0	PUSH *direct*
01	AJMP *pg0Addr*	41	AJMP *pg2Addr*	81	AJMP *pg4Addr*	C1	AJMP *pg6Addr*
02	LJMP *addr16*	42	ORL *direct,A*	82	ANL C,*bitAddr*	C2	CLR *bitAddr*
03	RR A	43	ORL *direct,#data8*	83	MOVC A,@A+PC	C3	CLR C
04	INC A	44	ORL A,#data8	84	DIV AB	C4	SWAP A
05	INC *direct*	45	ORL A,*direct*	85	MOV *direct,direct*	C5	XCH A,*direct*
06	INC @R0	46	ORL A,@R0	86	MOV *direct,*@R0	C6	XCH A,@R0
07	INC @R1	47	ORL A,@R1	87	MOV *direct,*@R1	C7	XCH A,@R1
08	INC R0	48	ORL A,R0	88	MOV *direct,*R0	C8	XCH A,R0
09	INC R1	49	ORL A,R1	89	MOV *direct,*R1	C9	XCH A,R1
0A	INC R2	4A	ORL A,R2	8A	MOV *direct,*R2	CA	XCH A,R2
0B	INC R3	4B	ORL A,R3	8B	MOV *direct,*R3	CB	XCH A,R3
0C	INC R4	4C	ORL A,R4	8C	MOV *direct,*R4	CC	XCH A,R4
0D	INC R5	4D	ORL A,R5	8D	MOV *direct,*R5	CD	XCH A,R5
0E	INC R6	4E	ORL A,R6	8E	MOV *direct,*R6	CE	XCH A,R6
0F	INC R7	4F	ORL A,R7	8F	MOV *direct,*R7	CF	XCH A,R7
10	JBC *bitAddr,relAddr*	50	JNC *relAddr*	90	MOV DPTR,#*data16*	D0	POP *direct*
11	ACALL *pg0Addr*	51	ACALL *pg2Addr*	91	ACALL *pg4Addr*	D1	ACALL *pg6Addr*
12	LCALL *address16*	52	ANL *direct,A*	92	MOV *bitAddr*,C	D2	SETB *bitAddr*
13	RRC A	53	ANL *direct,#data8*	93	MOVC A,@A+DPTR	D3	SETB C
14	DEC A	54	ANL A,#data8	94	SUBB A,#*data8*	D4	DA A
15	DEC *direct*	55	ANL A,*direct*	95	SUBB A,*direct*	D5	DJNZ *direct,relAddr*
16	DEC @R0	56	ANL A,@R0	96	SUBB A,@R0	D6	XCHD A,@R0
17	DEC @R1	57	ANL A,@R1	97	SUBB A,@R1	D7	XCHD A,@R1
18	DEC R0	58	ANL A,R0	98	SUBB A,R0	D8	DJNZ A,R0
19	DEC R1	59	ANL A,R1	99	SUBB A,R1	D9	DJNZ A,R1
1A	DEC R2	5A	ANL A,R2	9A	SUBB A,R2	DA	DJNZ A,R2
1B	DEC R3	5B	ANL A,R3	9B	SUBB A,R3	DB	DJNZ A,R3
1C	DEC R4	5C	ANL A,R4	9C	SUBB A,R4	DC	DJNZ A,R4
1D	DEC R5	5D	ANL A,R5	9D	SUBB A,R5	DD	DJNZ A,R5
1E	DEC R6	5E	ANL A,R6	9E	SUBB A,R6	DE	DJNZ A,R6
1F	DEC R7	5F	ANL A,R7	9F	SUBB A,R7	DF	DJNZ A,R7
20	JB *bitAddr,relAddr*	60	JZ *relAddr*	A0	ORL C,*/bitAddr*	E0	MOVX A,@DPTR
21	AJMP *pg1Addr*	61	AJMP *pg3Addr*	A1	AJMP *pg5Addr*	E1	AJMP *pg7Addr*
22	RET	62	XRL *direct,A*	A2	MOV C,*bitAddr*	E2	MOVX A,@R0
23	RL A	63	XRL *direct,#data8*	A3	INC DPTR	E3	MOVX A,@R1
24	ADD A,#data8	64	XRL A,#data8	A4	MUL AB	E4	CLR A
25	ADD A,*direct*	65	XRL A,*direct*	A5		E5	MOV A,*direct*
26	ADD A,@R0	66	XRL A,@R0	A6	MOV @R0,*direct*	E6	MOV A,@R0
27	ADD A,@R1	67	XRL A,@R1	A7	MOV @R1,*direct*	E7	MOV A,@R1
28	ADD A,R0	68	XRL A,R0	A8	MOV R0,*direct*	E8	MOV A,R0
29	ADD A,R1	69	XRL A,R1	A9	MOV R1,*direct*	E9	MOV A,R1
2A	ADD A,R2	6A	XRL A,R2	AA	MOV R2,*direct*	EA	MOV A,R2
2B	ADD A,R3	6B	XRL A,R3	AB	MOV R3,*direct*	EB	MOV A,R3
2C	ADD A,R4	6C	XRL A,R4	AC	MOV R4,*direct*	EC	MOV A,R4
2D	ADD A,R5	6D	XRL A,R5	AD	MOV R5,*direct*	ED	MOV A,R5
2E	ADD A,R6	6E	XRL A,R6	AE	MOV R6,*direct*	EE	MOV A,R6
2F	ADD A,R7	6F	XRL A,R7	AF	MOV R7,*direct*	EF	MOV A,R7
30	JNB *bitAddr,relAddr*	70	JNZ *relAddr*	B0	ANL C,*/bitAddr*	F0	MOVX @DPTR,A
31	ACALL *pg1Addr*	71	ACALL *pg3Addr*	B1	ACALL *pg5Addr*	F1	ACALL *pg7Addr*
32	RETI	72	ORL C,*bitAddr*	B2	CPL *bitAddr*	F2	MOVX @R0,A
33	RLC A	73	JMP @A+DPTR	B3	CPL C	F3	MOVX @R1,A
34	ADDC A,#data	74	MOV A,#data8	B4	CJNE A,#data8,relAddr	F4	CPL A
35	ADDC A,*direct*	75	MOV *direct,#data8*	B5	CJNE A,*direct,relAddr*	F5	MOV *direct,A*
36	ADDC A,@R0	76	MOV @R0,#data8	B6	CJNE @R0,#data8,relAddr	F6	MOV @R0,A
37	ADDC A,@R1	77	MOV @R1,#data8	B7	CJNE @R1,#data8,relAddr	F7	MOV @R1,A
38	ADDC A,R0	78	MOV R0,#data8	B8	CJNE R0,#data8,relAddr	F8	MOV R0,A
39	ADDC A,R1	79	MOV R1,#data8	B9	CJNE R1,#data8,relAddr	F9	MOV R1,A
3A	ADDC A,R2	7A	MOV R2,#data8	BA	CJNE R2,#data8,relAddr	FA	MOV R2,A
3B	ADDC A,R3	7B	MOV R3,#data8	BB	CJNE R3,#data8,relAddr	FB	MOV R3,A
3C	ADDC A,R4	7C	MOV R4,#data8	BC	CJNE R4,#data8,relAddr	FC	MOV R4,A
3D	ADDC A,R5	7D	MOV R5,#data8	BD	CJNE R5,#data8,relAddr	FD	MOV R5,A
3E	ADDC A,R6	7E	MOV R6,#data8	BE	CJNE R6,#data8,relAddr	FE	MOV R6,A
3F	ADDC A,R7	7F	MOV R7,#data8	BF	CJNE R7,#data8,relAddr	FF	MOV R7,A

Appendix B: 8052 Instruction Set

This appendix is a reference for all instructions in the 8052 instruction set. For each instruction, the following information is provided:

- **Instruction**: Indicates the correct syntax for the given opcode.
- **OpCode**: The operation code, in the range of 0x00 through 0xFF, that represents the given instruction in machine code.
- **Bytes**: The total number of bytes (including the opcode byte) that make up the instruction.
- **Cycles**: The number of machine cycles required to execute the instruction.
- **Flags**: The flags that are modified by the instruction, if any.

When listing instruction syntax, the following terms will be used:

- *bitAddr*: Bit address value (00-FF)
- *pgXAddr*: Absolute 2k (11-bit) Address
- *data8*: Immediate 8-bit data value
- *data16*: Immediate 16-bit data value
- *address16*: 16-bit code address
- *direct*: Direct address (IRAM 00-7F, SFR 80-FF)
- *relAddr*: Relative address (-128 to +127 bytes)

Calculating Relative Addresses

Relative addresses, indicated in this appendix with the term *relAddr*, use a single byte to express an address that is 128 bytes before or 127 bytes after the address of the instruction that follows the instruction with the relative address. The new value for the program counter is calculated by treating the relative address byte as a signed value represented as a 2's complement and adding it to the address of the instruction that follows the instruction.

For example, if the instruction SJMP 0000h is located at address 0002h, the opcodes would be 80 FC. The value 80h is the instruction code for SJMP. Since SJMP is a 2-byte instruction, the next instruction would start at 0004h. Thus to obtain the address 0000h, the value -4 must be added to 0004h. FC is the 2's complement representation of -4.

Calculating Absolute Addresses

Absolute addresses, indicated in this appendix with the term *pgXAddr*, use part of the instruction operation code plus a second byte to express an address that is within the same 2k block as the address of the instruction that follows the absolute address instruction. The new value for the program counter is calculated by replacing the least-significant-byte of the program counter with the second byte of the instruction, and replacing bits 0-2 of the most-significant-byte of the program counter with bits 5-7 of the instruction code value. Bits 3-7 of the most-significant-byte of the program counter remain unchanged. Since only 11 bits of the program counter are affected by these instructions, only addresses that are within the same 2k block as the first byte that follows the instruction may be referenced.

For example, if the instruction AJMP 0325h is located at address 0000h, the opcodes would be `61 25`. The value 61h corresponds to the instruction AJMP *pg3Addr* (corresponding to the 03 of the target address) while 25h is the low byte of the target address.

ACALL – Absolute Call within 2k Block
Syntax: ACALL *codeAddress*

Instructions	OpCode	Bytes	Cycles	Flags
ACALL *pg0Addr*	0x11	2	2	None
ACALL *pg1Addr*	0x31	2	2	None
ACALL *pg2Addr*	0x51	2	2	None
ACALL *pg3Addr*	0x71	2	2	None
ACALL *pg4Addr*	0x91	2	2	None
ACALL *pg5Addr*	0xB1	2	2	None
ACALL *pg6Addr*	0xD1	2	2	None
ACALL *pg7Addr*	0xF1	2	2	None

ACALL unconditionally calls a subroutine at the indicated code address. ACALL pushes the address of the instruction that follows ACALL onto the stack, least-significant-byte first, most-significant-byte second. The program counter is then updated so that program execution continues at the indicated address.

Please see "Calculating Absolute Addresses" at the beginning of this appendix for information regarding the calculation of each byte of this instruction.

See Also: LCALL, RET

ADD, ADDC – Add Value, Add Value with Carry
Syntax: ADD A,*operand*
Syntax: ADDC A,*operand*

Instructions	OpCode	Bytes	Cycles	Flags
ADD A,#*data8*	0x24	2	1	C, AC, OV
ADD A,*direct*	0x25	2	1	C, AC, OV
ADD A,@R0	0x26	1	1	C, AC, OV
ADD A,@R1	0x27	1	1	C, AC, OV
ADD A,R0	0x28	1	1	C, AC, OV
ADD A,R1	0x29	1	1	C, AC, OV
ADD A,R2	0x2A	1	1	C, AC, OV
ADD A,R3	0x2B	1	1	C, AC, OV
ADD A,R4	0x2C	1	1	C, AC, OV
ADD A,R5	0x2D	1	1	C, AC, OV
ADD A,R6	0x2E	1	1	C, AC, OV
ADD A,R7	0x2F	1	1	C, AC, OV
ADDC A,#*data8*	0x34	2	1	C, AC, OV

Instructions	OpCode	Bytes	Cycles	Flags
ADDC A,*direct*	0x35	2	1	C, AC, OV
ADDC A,@R0	0x36	1	1	C, AC, OV
ADDC A,@R1	0x37	1	1	C, AC, OV
ADDC A,R0	0x38	1	1	C, AC, OV
ADDC A,R1	0x39	1	1	C, AC, OV
ADDC A,R2	0x3A	1	1	C, AC, OV
ADDC A,R3	0x3B	1	1	C, AC, OV
ADDC A,R4	0x3C	1	1	C, AC, OV
ADDC A,R5	0x3D	1	1	C, AC, OV
ADDC A,R6	0x3E	1	1	C, AC, OV
ADDC A,R7	0x3F	1	1	C, AC, OV

ADD and ADDC both add the value operand to the value of the accumulator, leaving the resulting value in the accumulator. The value operand is not affected. ADD and ADDC function identically except that ADDC adds the value of operand as well as the value of the carry flag whereas ADD does not add the Carry flag to the result.

The **Carry bit** (**C**) is set if there is a carry-out of bit 7. In other words, if the unsigned summed value of the accumulator, operand and (in the case of ADDC) the carry flag exceeds 255 Carry is set. Otherwise, the Carry bit is cleared.

The **Auxiliary Carry** (**AC**) bit is set if there is a carry-out of bit 3. In other words, if the unsigned summed value of the low nibble of the accumulator, operand and (in the case of ADDC) the Carry flag exceeds 15 the Auxiliary Carry flag is set. Otherwise, the Auxiliary Carry flag is cleared.

The **Overflow** (**OV**) bit is set if there is a carry-out of bit 6 or out of bit 7, but not both. In other words, if the addition of the accumulator, operand and (in the case of ADDC) the Carry flag treated as signed values results in a value that is out of the range of a signed byte (-128 through +127) the Overflow flag is set. Otherwise, the Overflow flag is cleared.

See Also: SUBB, DA, INC, DEC

AJMP – Absolute Jump within 2k Block
Syntax: AJMP *codeAddress*

Instructions	OpCode	Bytes	Cycles	Flags
AJMP *pg0Addr*	0x01	2	2	None
AJMP *pg1Addr*	0x21	2	2	None
AJMP *pg2Addr*	0x41	2	2	None
AJMP *pg3Addr*	0x61	2	2	None
AJMP *pg4Addr*	0x81	2	2	None
AJMP *pg5Addr*	0xA1	2	2	None
AJMP *pg6Addr*	0xC1	2	2	None
AJMP *pg7Addr*	0xE1	2	2	None

AJMP unconditionally jumps to the indicated *codeAddress*. Please see "Calculating Absolute Addresses" at the beginning of this appendix for information regarding the calculation of each byte of this instruction.

See Also: LJMP, SJMP

ANL– Bitwise AND
Syntax: ANL *operand1,operand2*

Instructions	OpCode	Bytes	Cycles	Flags
ANL *direct*,A	0x52	2	1	None
ANL *direct*,#*data8*	0x53	3	2	None
ANL A,#*data8*	0x54	2	1	None
ANL A,*direct*	0x55	2	1	None
ANL A,@R0	0x56	1	1	None
ANL A,@R1	0x57	1	1	None
ANL A,R0	0x58	1	1	None
ANL A,R1	0x59	1	1	None
ANL A,R2	0x5A	1	1	None
ANL A,R3	0x5B	1	1	None
ANL A,R4	0x5C	1	1	None
ANL A,R5	0x5D	1	1	None
ANL A,R6	0x5E	1	1	None
ANL A,R7	0x5F	1	1	None
ANL C,*bitAddr*	0x82	2	2	C
ANL C,/*bitAddr*	0xB0	2	2	C

ANL does a bitwise "AND" operation between *operand1* and *operand2*, leaving the resulting value in *operand1*. The value of *operand2* is not affected. A logical "AND" compares the bits of each operand and sets the corresponding bit in the resulting byte only if the bit was set in both of the original operands, otherwise the resulting bit is cleared.

See Also: ORL, XRL

CJNE – Compare and Jump if Not Equal
Syntax: CJNE *operand1,operand2,relAddr*

Instructions	OpCode	Bytes	Cycles	Flags
CJNE A,#*data8,relAddr*	0xB4	3	2	C
CJNE A,*direct,relAddr*	0xB5	3	2	C
CJNE @R0,#*data8,relAddr*	0xB6	3	2	C
CJNE @R1,#*data8,relAddr*	0xB7	3	2	C
CJNE R0,#*data8,relAddr*	0xB8	3	2	C
CJNE R1,#*data8,relAddr*	0xB9	3	2	C
CJNE R2,#*data8,relAddr*	0xBA	3	2	C
CJNE R3,#*data8,relAddr*	0xBB	3	2	C
CJNE R4,#*data8,relAddr*	0xBC	3	2	C

Instructions	OpCode	Bytes	Cycles	Flags
CJNE R5,#*data8*,*relAddr*	0xBD	3	2	C
CJNE R6,#*data8*,*relAddr*	0xBE	3	2	C
CJNE R7,#*data8*,*relAddr*	0xBF	3	2	C

CJNE compares the value of *operand1* and *operand2* and branches to the indicated relative address if the two operands are not equal. If the two operands are equal program flow continues with the instruction following the CJNE instruction.

The **Carry bit** (**C**) is set if *operand1* is less than *operand2*, otherwise it is cleared.

Please see "Calculating Relative Addresses" at the beginning of this appendix for information regarding the calculation of the relative address byte of this instruction.

See Also: DJNZ

CLR – Clear Register
Syntax: CLR *register*

Instructions	OpCode	Bytes	Cycles	Flags
CLR *bitAddr*	0xC2	2	1	None
CLR C	0xC3	1	1	C
CLR A	0xE4	1	1	None

CLR clears (sets to 0) all the bit(s) of the indicated register. If the register is a bit (including the carry bit), only the specified bit is affected. Clearing the accumulator sets the accumulator's value to 0.

See Also: SETB

CPL – Complement Register
Syntax: CPL *operand*

Instructions	OpCode	Bytes	Cycles	Flags
CPL A	0xF4	1	1	None
CPL C	0xB3	1	1	C
CPL *bitAddr*	0xB2	2	1	None

CPL complements *operand*, leaving the result in *operand*. If *operand* is a single bit then the state of the bit will be reversed. If *operand* is the accumulator then all the bits in the accumulator will be reversed. This can be thought of as "accumulator Logical Exclusive OR 255" or as "255-accumulator." If *operand* refers to a bit of an output port, the value that will be complemented is based on the last value written to that bit, not the last value read from it.

See Also: CLR, SETB

DA– Decimal Adjust Accumulator

Syntax: DA A

Instructions	OpCode	Bytes	Cycles	Flags
DA A	0xD4	1	1	C

DA adjusts the contents of the accumulator to correspond to a BCD (Binary Coded Decimal) number after two BCD numbers have been added by the ADD or ADDC instruction.

If the carry bit is set or if the value of bits 0-3 exceed 9, 0x06 is added to the accumulator. If the carry bit was set when the instruction began, or if 0x06 was added to the accumulator in the first step, 0x60 is added to the accumulator.

The **Carry bit** (**C**) is set if the resulting value is greater than 0x99, otherwise it is cleared.

See Also: ADD, ADDC

DEC – Decrement Register

Syntax: DEC *register*

Instructions	OpCode	Bytes	Cycles	Flags
DEC A	0x14	1	1	None
DEC *direct*	0x15	2	1	None
DEC @R0	0x16	1	1	None
DEC @R1	0x17	1	1	None
DEC R0	0x18	1	1	None
DEC R1	0x19	1	1	None
DEC R2	0x1A	1	1	None
DEC R3	0x1B	1	1	None
DEC R4	0x1C	1	1	None
DEC R5	0x1D	1	1	None
DEC R6	0x1E	1	1	None
DEC R7	0x1F	1	1	None

DEC decrements the value of *register* by 1. If the initial value of *register* is 0, decrementing the value will cause it to reset to 255 (0xFF Hex). Note: The carry flag is *not* set when the value "rolls over" from 0 to 255.

See Also: INC, SUBB

DIV – Divide Accumulator by B

Syntax: DIV AB

Instructions	OpCode	Bytes	Cycles	Flags
DIV AB	0x84	1	4	C, OV

Divides the unsigned value of the accumulator by the unsigned value of the "B" register. The resulting quotient is placed in the accumulator and the remainder is placed in the "B" register.

The **Carry flag** (**C**) is always cleared. The **Overflow flag** (**OV**) is set if division by 0 was attempted, otherwise it is cleared.

See Also: MUL AB

DJNZ – Decrement and Jump if Not Zero
Syntax: DJNZ *register,relAddr*

Instructions	OpCode	Bytes	Cycles	Flags
DJNZ *direct,relAddr*	0xD5	3	2	None
DJNZ R0,*relAddr*	0xD8	2	2	None
DJNZ R1,*relAddr*	0xD9	2	2	None
DJNZ R2,*relAddr*	0xDA	2	2	None
DJNZ R3,*relAddr*	0xDB	2	2	None
DJNZ R4,*relAddr*	0xDC	2	2	None
DJNZ R5,*relAddr*	0xDD	2	2	None
DJNZ R6,*relAddr*	0xDE	2	2	None
DJNZ R7,*relAddr*	0xDF	2	2	None

DJNZ decrements the value of *register* by 1. If the initial value of *register* is 0, decrementing the value will cause it to reset to 255 (0xFF Hex). If the new value of register is *not* 0 the program will branch to the address indicated by *relAddr*. If the new value of *register* is 0, program flow continues with the instruction following the DJNZ instruction.

Please see "Calculating Relative Addresses" at the beginning of this appendix for information regarding the calculation of the relative address byte of this instruction.

See Also: DEC, JZ, JNZ

INC – Increment Register
Syntax: INC *register*

Instructions	OpCode	Bytes	Cycles	Flags
INC A	0x04	1	1	None
INC *direct*	0x05	2	1	None
INC @R0	0x06	1	1	None
INC @R1	0x07	1	1	None
INC R0	0x08	1	1	None
INC R1	0x09	1	1	None
INC R2	0x0A	1	1	None
INC R3	0x0B	1	1	None
INC R4	0x0C	1	1	None
INC R5	0x0D	1	1	None

Instructions	OpCode	Bytes	Cycles	Flags
INC R6	0x0E	1	1	None
INC R7	0x0F	1	1	None
INC DPTR	0xA3	1	2	None

INC increments the value of *register* by 1. If the initial value of register is 255 (0xFF Hex), incrementing the value will cause it to reset to 0. Note: The carry flag is *not* set when the value "rolls over" from 255 to 0.

In the case of "INC DPTR", the two-byte value of DPTR is incremented as an unsigned integer. If the initial value of DPTR is 65,535 (FFFF Hex), incrementing the value will cause it to reset to 0. Again, the Carry Flag is *not* set when the value of DPTR "rolls over" from 65,535 to 0.

See Also: ADD, ADDC, DEC

JB – Jump if Bit Set
Syntax: JB *bitAddr,relAddr*

Instructions	OpCode	Bytes	Cycles	Flags
JB *bitAddr,relAddr*	0x20	3	2	None

JB branches to the address indicated by *relAddr* if the bit indicated by *bitAddr* is set. If the bit is not set program execution continues with the instruction following the JB instruction.

Please see "Calculating Relative Addresses" at the beginning of this appendix for information regarding the calculation of the relative address byte of this instruction.

See Also: JBC, JNB

JBC – Jump if Bit Set and Clear Bit
Syntax: JBC *bitAddr,relAddr*

Instructions	OpCode	Bytes	Cycles	Flags
JBC *bitAddr,relAddr*	0x10	3	2	None

JBC will branch to the address indicated by *relAddr* if the bit indicated by *bitAddr* is set. Before branching to *relAddr* the instruction will clear the indicated bit. If the bit is not set program execution continues with the instruction following the JBC instruction and the value of the bit is not changed.

Please see "Calculating Relative Addresses" at the beginning of this appendix for information regarding the calculation of the relative address byte of this instruction.

See Also: JB, JNB

<u>**JC**</u> – Jump if Carry Set
Syntax: JC *relAddr*

Instructions	OpCode	Bytes	Cycles	Flags
JC *relAddr*	0x40	2	2	None

JC will branch to the address indicated by *relAddr* if the carry flag is set. If the carry flag is not set, program execution continues with the instruction following the JC instruction.

Please see "Calculating Relative Addresses" at the beginning of this appendix for information regarding the calculation of the relative address byte of this instruction.

See Also: JNC

<u>**JMP**</u> – Jump to Data Pointer + Accumulator
Syntax: JMP @A+DPTR

Instructions	OpCode	Bytes	Cycles	Flags
JMP @A+DPTR	0x73	1	2	None

JMP jumps unconditionally to the address represented by the sum of the value of DPTR and the value of the accumulator.

See Also: LJMP, AJMP, SJMP

<u>**JNB**</u> – Jump if Bit Not Set
Syntax: JNB *bitAddr,relAddr*

Instructions	OpCode	Bytes	Cycles	Flags
JNB *bitAddr,relAddr*	0x30	3	2	None

JNB will branch to the address indicated by *relAddr* if the indicated bit is not set. If the bit is set program execution continues with the instruction following the JNB instruction.

Please see "Calculating Relative Addresses" at the beginning of this appendix for information regarding the calculation of the relative address byte of this instruction.

See Also: JB, JBC

<u>**JNC**</u> – Jump if Carry Not Set
Syntax: JNC *relAddr*

Instructions	OpCode	Bytes	Cycles	Flags
JNC *relAddr*	0x50	2	2	None

JNC branches to the address indicated by *relAddr* if the carry bit is not set. If the carry bit is set program execution continues with the instruction following the JNB instruction.

Please see "Calculating Relative Addresses" at the beginning of this appendix for information regarding the calculation of the relative address byte of this instruction.

See Also: JC

JNZ – Jump if Accumulator Not Zero
Syntax: JNZ *relAddr*

Instructions	OpCode	Bytes	Cycles	Flags
JNZ *relAddr*	0x70	2	2	None

JNZ will branch to the address indicated by *relAddr* if the accumulator contains any value except 0. If the value of the accumulator is zero, program execution continues with the instruction following the JNZ instruction.

Please see "Calculating Relative Addresses" at the beginning of this appendix for information regarding the calculation of the relative address byte of this instruction.

See Also: JZ

JZ – Jump if Accumulator Zero
Syntax: JZ *relAddr*

Instructions	OpCode	Bytes	Cycles	Flags
JZ *relAddr*	0x60	2	2	None

JZ branches to the address indicated by *relAddr* if the accumulator contains the value 0. If the value of the accumulator is non-zero, program execution continues with the instruction following the JZ instruction.

Please see "Calculating Relative Addresses" at the beginning of this appendix for information regarding the calculation of the relative address byte of this instruction.

See Also: JNZ

LCALL – Long Call
Syntax: LCALL *address16*

Instructions	OpCode	Bytes	Cycles	Flags
LCALL *address16*	0x12	3	2	None

LCALL calls a subroutine. LCALL increments the program counter by 3 (to point to the instruction following LCALL) and pushes that value onto the stack, low-byte first, high-byte second. The program counter is then set to the 16-bit value *address16*, causing program execution to continue at that address.

See Also: ACALL, RET

<u>LJMP</u> – Long Jump
Syntax: LJMP *address16*

Instructions	OpCode	Bytes	Cycles	Flags
LJMP *address16*	0x02	3	2	None

LJMP jumps unconditionally to the specified *address16*.

See Also: AJMP, SJMP, JMP

<u>MOV</u> – Move Memory into/out of Accumulator
Syntax: MOV *operand1, operand2*

Instructions	OpCode	Bytes	Cycles	Flags
MOV A,#*data8*	0x74	2	1	None
MOV A,@R0	0xE6	1	1	None
MOV A,@R1	0xE7	1	1	None
MOV @R0,A	0xF6	1	1	None
MOV @R1,A	0xF7	1	1	None
MOV A,R0	0xE8	1	1	None
MOV A,R1	0xE9	1	1	None
MOV A,R2	0xEA	1	1	None
MOV A,R3	0xEB	1	1	None
MOV A,R4	0xEC	1	1	None
MOV A,R5	0xED	1	1	None
MOV A,R6	0xEE	1	1	None
MOV A,R7	0xEF	1	1	None
MOV A,*direct*	0xE5	2	1	None
MOV R0,A	0xF8	1	1	None
MOV R1,A	0xF9	1	1	None
MOV R2,A	0xFA	1	1	None
MOV R3,A	0xFB	1	1	None
MOV R4,A	0xFC	1	1	None
MOV R5,A	0xFD	1	1	None
MOV R6,A	0xFE	1	1	None
MOV R7,A	0xFF	1	1	None
MOV *direct*,A	0xF5	2	1	None

MOV copies the value of *operand2* into *operand1*. The value of *operand2* is not affected.

See Also: MOVC, MOVX, XCH, XCHD, PUSH, POP

MOV – Move into/out of Carry Bit
Syntax: MOV *bit1,bit2*

Instructions	OpCode	Bytes	Cycles	Flags
MOV C,*bitAddr*	0xA2	2	1	C
MOV *bitAddr*,C	0x92	2	2	None

MOV copies the value of *bit2* into *bit1*. The value of *bit2* is not affected. Either *bit1* or *bit2* must refer to the Carry bit.

MOV – Move into/out of Internal RAM
Syntax: MOV *operand1,operand2*

Instructions	OpCode	Bytes	Cycles	Flags
MOV @R0,#*data8*	0x76	2	1	None
MOV @R1,#*data8*	0x77	2	1	None
MOV @R0,*direct*	0xA6	2	2	None
MOV @R1,*direct*	0xA7	2	2	None
MOV R0,#*data8*	0x78	2	1	None
MOV R1,#*data8*	0x79	2	1	None
MOV R2,#*data8*	0x7A	2	1	None
MOV R3,#*data8*	0x7B	2	1	None
MOV R4,#*data8*	0x7C	2	1	None
MOV R5,#*data8*	0x7D	2	1	None
MOV R6,#*data8*	0x7E	2	1	None
MOV R7,#*data8*	0x7F	2	1	None
MOV R0,*direct*	0xA8	2	2	None
MOV R1,*direct*	0xA9	2	2	None
MOV R2,*direct*	0xAA	2	2	None
MOV R3,*direct*	0xAB	2	2	None
MOV R4,*direct*	0xAC	2	2	None
MOV R5,*direct*	0xAD	2	2	None
MOV R6,*direct*	0xAE	2	2	None
MOV R7,*direct*	0xAF	2	2	None
MOV *direct*,#*data8*	0x75	3	2	None
MOV *direct*,@R0	0x86	2	2	None
MOV *direct*,@R1	0x87	2	2	None
MOV *direct*,R0	0x88	2	2	None
MOV *direct*,R1	0x89	2	2	None
MOV *direct*,R2	0x8A	2	2	None
MOV *direct*,R3	0x8B	2	2	None
MOV *direct*,R4	0x8C	2	2	None
MOV *direct*,R5	0x8D	2	2	None
MOV *direct*,R6	0x8E	2	2	None
MOV *direct*,R7	0x8F	2	2	None
MOV *direct1,direct2*	0x85	3	2	None

MOV copies the value of *operand2* into *operand1*. The value of *operand2* is not affected.

-☼- **NOTE**: In the case of "MOV *direct1,direct2*", the two auxiliary bytes of the instruction are stored in reverse order. That is, the instruction consisting of the bytes `85h 20h 50h` means "Move the contents of internal RAM location 20h to internal RAM location 50h whereas the opposite would be generally presumed.

See Also: MOVC, MOVX, XCH, XCHD, PUSH, POP

MOV DPTR – Move value into DPTR
Syntax: MOV DPTR,#*data16*

Instructions	OpCode	Bytes	Cycles	Flags
MOV DPTR,#*data16*	0x90	3	2	None

Sets the value of the data pointer (DPTR) to the value *data16*.

See Also: **MOVX, MOVC**

MOVC – Move Code Byte to Accumulator
Syntax: MOVC A,@A+register

Instructions	OpCode	Bytes	Cycles	Flags
MOVC A,@A+DPTR	0x93	1	2	None
MOVC A,@A+PC	0x83	1	1	None

MOVC moves a byte from code memory into the accumulator. The code memory address from which the byte will be moved is calculated by summing the value of the accumulator with either DPTR or the Program Counter (PC). In the case of the Program Counter, PC is first incremented by 1 before being summed with the accumulator.

See Also: MOV, MOVX

MOVX – Move Data to/from External RAM
Syntax: MOVX *operand1,operand2*

Instructions	OpCode	Bytes	Cycles	Flags
MOVX @DPTR,A	0xF0	1	2	None
MOVX @R0,A	0xF2	1	2	None
MOVX @R1,A	0xF3	1	2	None
MOVX A,@DPTR	0xE0	1	2	None
MOVX A,@R0	0xE2	1	2	None
MOVX A,@R1	0xE3	1	2	None

MOVX moves a byte to or from external memory into or from the accumulator.

If *operand1* is @DPTR, the accumulator is moved to the 16-bit External Memory address indicated by DPTR. This instruction uses both P0 (port 0) and P2 (port 2) to output the 16-bit address and data. If *operand2* is DPTR then the byte is moved from external memory into the accumulator.

If *operand1* is @R0 or @R1, the accumulator is moved to the 8-bit external memory address indicated by the specified register. This instruction uses only P0 (port 0) to output the 8-bit address and data. P2 (port 2) is not affected. If *operand2* is @R0 or @R1 then the byte is moved from external memory into the accumulator.

See Also: MOV, MOVC

MUL – Multiply Accumulator by B
Syntax: MUL AB

Instructions	OpCode	Bytes	Cycles	Flags
MUL AB	0xA4	1	4	C, OV

Multiplies the unsigned value in the accumulator by the unsigned value in the "B" register. The least-significant byte of the result is placed in the accumulator and the most-significant-byte is placed in the "B" register.

The **Carry Flag (C)** is always cleared.

The **Overflow Flag (OV)** is set if the result is greater than 255 (if the most-significant byte is not zero), otherwise it is cleared.

See Also: DIV

NOP – No Operation
Syntax: NOP

Instructions	OpCode	Bytes	Cycles	Flags
NOP	0x00	1	1	None

NOP, as it's name suggests, causes no operation to take place for one machine cycle. NOP is generally used only for timing purposes. Absolutely no flags or registers are affected.

ORL – Bitwise OR
Syntax: ORL *operand1,operand2*

Instructions	OpCode	Bytes	Cycles	Flags
ORL *direct*,A	0x42	2	1	None
ORL *direct*,#*data8*	0x43	3	2	None
ORL A,#*data8*	0x44	2	1	None
ORL A,*direct*	0x45	2	1	None

Instructions	OpCode	Bytes	Cycles	Flags
ORL A,@R0	0x46	1	1	None
ORL A,@R1	0x47	1	1	None
ORL A,R0	0x48	1	1	None
ORL A,R1	0x49	1	1	None
ORL A,R2	0x4A	1	1	None
ORL A,R3	0x4B	1	1	None
ORL A,R4	0x4C	1	1	None
ORL A,R5	0x4D	1	1	None
ORL A,R6	0x4E	1	1	None
ORL A,R7	0x4F	1	1	None
ORL C,bitAddr	0x72	2	2	C
ORL C,/bitAddr	0xA0	2	1	C

ORL does a bitwise "OR" operation between *operand1* and *operand2*, leaving the resulting value in *operand1*. The value of operand2 is not affected. A logical "OR" compares the bits of each operand and sets the corresponding bit in the resulting byte if the bit was set in either of the original operands, otherwise the resulting bit is cleared.

See Also: ANL, XRL

POP – Pop Value from Stack
Syntax: POP *register*

Instructions	OpCode	Bytes	Cycles	Flags
POP *direct*	0xD0	2	2	None

POP "pops" the last value placed on the stack into the *direct* address specified. In other words, POP will load *direct* with the value of the internal RAM address pointed to by the current stack pointer. The stack pointer is then decremented by 1.

NOTE #1: The address of direct must be an internal RAM or SFR address. You cannot POP directly into "R" registers, such as R0, R1, etc. To pop a value off the stack into R0, for example, you must pop the value into the accumulator and then move the value of the accumulator into R0.

NOTE #2: When popping a value off the stack into the accumulator, you must code the instruction as **POP ACC**, not **POP A**. The latter is invalid and will result in an error at assemble-time.

See Also: PUSH

PUSH – Push Value onto Stack
Syntax: PUSH *register*

Instructions	OpCode	Bytes	Cycles	Flags
PUSH *direct*	0xC0	2	2	None

PUSH "pushes" the value of the specified *direct* address onto the stack. PUSH first increments the value of the stack pointer by 1, then takes the value stored in *direct* and stores it in internal RAM at the location pointed to by the incremented stack pointer.

💡 **NOTE #1**: The address of direct must be an internal RAM or SFR address. You cannot PUSH directly from "R" registers, such as R0, R1, etc. To push a value onto the stack from R0, for example, you must move R0 into the accumulator, then PUSH the value of the accumulator onto the stack.

💡 **NOTE #2**: When pushing a value from the accumulator onto the stack into the, you must code the instruction as **PUSH ACC**, not **PUSH A**. The latter is invalid and will result in an error at assemble-time.

See Also: POP

RET – Return from Subroutine
Syntax: RET

Instructions	OpCode	Bytes	Cycles	Flags
RET	0x22	1	2	None

RET is used to return from a subroutine previously called by LCALL or ACALL. Program execution continues at the address that is calculated by popping the top-most 2 bytes off the stack. The most-significant-byte is popped off the stack first, followed by the least-significant-byte.

See Also: LCALL, ACALL, RETI

RETI – Return from Interrupt
Syntax: RETI

Instructions	OpCode	Bytes	Cycles	Flags
RETI	0x32	1	2	None

RETI is used to return from an interrupt service routine. RETI first enables interrupts of equal and lower priorities to the interrupt that is terminating. Program execution continues at the address that is calculated by popping the top-most 2 bytes off the stack. The most-significant-byte is popped off the stack first, followed by the least-significant-byte.

See Also: RET

RL – Rotate Accumulator Left
Syntax: RL A

Instructions	OpCode	Bytes	Cycles	Flags
RL A	0x23	1	1	C

Shifts the bits of the accumulator to the left. The left-most bit (bit 7) of the accumulator is loaded into bit 0.

See Also: RLC, RR, RRC

RLC – Rotate Accumulator Left Through Carry
Syntax: RLC A

Instructions	OpCode	Bytes	Cycles	Flags
RLC A	0x33	1	1	C

Shifts the bits of the accumulator to the left. The left-most bit (bit 7) of the accumulator is loaded into the Carry Flag, and the original Carry Flag is loaded into bit 0 of the accumulator.

See Also: RL, RR, RRC

RR – Rotate Accumulator Right
Syntax: RR A

Instructions	OpCode	Bytes	Cycles	Flags
RR A	0x03	1	1	None

Shifts the bits of the accumulator to the right. The right-most bit of the accumulator is loaded into bit 7.

See Also: RL, RLC, RRC

RRC – Rotate Accumulator Right Through Carry
Syntax: RRC A

Instructions	OpCode	Bytes	Cycles	Flags
RRC A	0x13	1	1	C

Shifts the bits of the accumulator to the right. The right-most bit (bit 0) of the accumulator is loaded into the Carry Flag, and the original Carry Flag is loaded into bit 7.

See Also: RL, RLC, RR

SETB – Set Bit
Syntax: SETB *bitAddr*

Instructions	OpCode	Bytes	Cycles	Flags
SETB C	0xD3	1	1	C
SETB *bitAddr*	0xD2	2	1	None

Sets the specified bit.

If the instruction requires the Carry bit to be set, the assembler will automatically use the 0xD3 opcode. If any other bit is set, the assembler will automatically use the 0xD2 opcode.

See Also: CLR

SJMP – Short Jump
Syntax: SJMP *relAddr*

Instructions	OpCode	Bytes	Cycles	Flags
SJMP *relAddr*	0x80	2	2	None

SJMP jumps unconditionally to the address specified *relAddr*. *RelAddr* must be within -128 or +127 bytes of the instruction that follows the SJMP instruction.

Please see "Calculating Relative Addresses" at the beginning of this appendix for information regarding the calculation of the relative address byte of this instruction.

See Also: LJMP, AJMP

SUBB – Subtract from Accumulator with Borrow
Syntax: SUBB A,*operand*

Instructions	OpCode	Bytes	Cycles	Flags
SUBB A,#*data8*	0x94	2	1	C, AC, OV
SUBB A,*direct*	0x95	2	1	C, AC, OV
SUBB A,@R0	0x96	1	1	C, AC, OV
SUBB A,@R1	0x97	1	1	C, AC, OV
SUBB A,R0	0x98	1	1	C, AC, OV
SUBB A,R1	0x99	1	1	C, AC, OV
SUBB A,R2	0x9A	1	1	C, AC, OV
SUBB A,R3	0x9B	1	1	C, AC, OV
SUBB A,R4	0x9C	1	1	C, AC, OV
SUBB A,R5	0x9D	1	1	C, AC, OV
SUBB A,R6	0x9E	1	1	C, AC, OV
SUBB A,R7	0x9F	1	1	C, AC, OV

SUBB subtracts the value of *operand* from the value of the accumulator, leaving the resulting value in the accumulator. The value *operand* is not affected.

The **Carry Bit** (**C**) is set if a borrow was required for bit 7, otherwise it is cleared. In other words, if the unsigned value being subtracted is greater than the accumulator the carry flag is set.

The Auxiliary **Carry** (**AC**) bit is set if a borrow was required for bit 3, otherwise it is cleared. In other words, the bit is set if the low nibble of the value being subtracted was greater than the low nibble of the accumulator.

The **Overflow** (**OV**) bit is set if a borrow was required for bit 6 or for bit 7, but not both. In other words, the subtraction of two signed bytes resulted in a value outside the range of a signed byte (-128 to 127). Otherwise it is cleared.

See Also: ADD, ADDC, DEC

SWAP – Subtract Accumulator Nibbles
Syntax: SWAP A

Instructions	OpCode	Bytes	Cycles	Flags
SWAP A	0xC4	1	1	None

SWAP swaps bits 0-3 of the accumulator with bits 4-7 of the accumulator. This instruction is identical to executing "RR A" or "RL A" four times.

See Also: RL, RLC, RR, RRC

XCH – Exchange Bytes
Syntax: XCH A,*register*

Instructions	OpCode	Bytes	Cycles	Flags
XCH A,@R0	0xC6	1	1	None
XCH A,@R1	0xC7	1	1	None
XCH A,R0	0xC8	1	1	None
XCH A,R1	0xC9	1	1	None
XCH A,R2	0xCA	1	1	None
XCH A,R3	0xCB	1	1	None
XCH A,R4	0xCC	1	1	None
XCH A,R5	0xCD	1	1	None
XCH A,R6	0xCE	1	1	None
XCH A,R7	0xCF	1	1	None
XCH A,*direct*	0xC5	2	1	None

Exchanges the value of the accumulator with the value contained in *register*.

See Also: MOV

<u>**XCHD**</u> – Exchange Digit
Syntax: XCHD A,*register*

Instructions	OpCode	Bytes	Cycles	Flags
XCHD A,@R0	0xD6	1	1	None
XCHD A,@R1	0xD7	1	1	None

Exchanges bits 0-3 of the accumulator with bits 0-3 of the internal RAM address pointed to indirectly by R0 or R1. Bits 4-7 of each register are unaffected.

See Also: DA

<u>**XRL**</u> – Bitwise Exclusive OR
Syntax: XRL operand1,operand2

Instructions	OpCode	Bytes	Cycles	Flags
XRL *direct*,A	0x62	2	1	None
XRL *direct*,#*data8*	0x63	3	2	None
XRL A,#*data8*	0x64	2	1	None
XRL A,*direct*	0x65	2	1	None
XRL A,@R0	0x66	1	1	None
XRL A,@R1	0x67	1	1	None
XRL A,R0	0x68	1	1	None
XRL A,R1	0x69	1	1	None
XRL A,R2	0x6A	1	1	None
XRL A,R3	0x6B	1	1	None
XRL A,R4	0x6C	1	1	None
XRL A,R5	0x6D	1	1	None
XRL A,R6	0x6E	1	1	None
XRL A,R7	0x6F	1	1	None

XRL does a bitwise "EXCLUSIVE OR" operation between *operand1* and *operand2*, leaving the resulting value in *operand1*. The value of *operand2* is not affected. A logical "EXCLUSIVE OR" compares the bits of each operand and sets the corresponding bit in the resulting byte if the bit was set in either (but not both) of the original operands, otherwise the bit is cleared.

See Also: ANL, ORL

<u>**UNDEFINED**</u> – Undefined Instruction
Syntax: ???

Instructions	OpCode	Bytes	Cycles	Flags
???	0xA5	1	1	C

The "undefined" instruction is, as the name suggests, not a documented instruction. The 8052 supports 255 instructions and OpCode 0xA5 is the single OpCode that is not used by any documented function. Since it is not documented nor defined it is not recommended that it be executed.

However, based on research, executing this undefined instruction takes 1 machine cycle and appears to have no effect on the system except that the carry bit always seems to be set.

See Also: NOP

Appendix C: SFR Quick Reference

Name	Addr	Bit 7	Bit 6	Bit 5	Bit 4	Bit 3	Bit 2	Bit 1	Bit 0
B*	F0	*No bit-level significance*							
ACC*	E0	*No bit-level significance*							
PSW*	D0	CY	AC	F0	RS1	RS0	OV	-	P
IP*	B8	-	-	PT2	PS	PT1	PX1	PT0	PX0
P3*	B0	RD	WR	T1	T0	INT1	INT0	TXD	RXD
IE*	A8	EA	-	ET2	ES	ET1	EX1	ET0	EX0
P2*	A0	A15	A14	A13	A12	A11	A10	A9	A8
SBUF	99	*No bit-level significance*							
SCON*	98	SM0	SM1	SM2	REN	TB8	RB8	TI	RI
P1*	90	P1.7	P1.6	P1.5	P1.4	P1.3	P1.2	P1.1	P1.0
TH1	8D	*No bit-level significance*							
TH0	8C	*No bit-level significance*							
TL1	8B	*No bit-level significance*							
TL0	8A	*No bit-level significance*							
TMOD	89	GATE_1	C/T_1	M1_1	M0_1	GATE_0	C/T_0	M1_0	M0_0
TCON*	88	TF1	TR1	TF0	TR0	IE1	IT1	IE0	IT0
PCON	87	SMOD	-	-	-	GF1	GF0	PD	IDL
DPH	83	*No bit-level significance*							
DPL	82	*No bit-level significance*							
SP	81	*No bit-level significance*							
P0*	80	AD7	AD6	AD5	AD4	AD3	AD2	AD1	AD0

(*) = SFR which is bit-addressable. SFRs without asterisk are *not* bit-addressable.

NOTE: The bit names assigned to P0 and P2 refer to their function at the hardware level. P0.0 is named "AD0" because its function is Address/Data Line 0. However, the names assigned to the bits of P0 and P2 will not be recognized as valid by a standard 8052 assembler unless they are specifically defined as a symbol in the program that references those bits.

Appendix D: SFR Detailed Reference (Alphabetical)

ACCUMULATOR (ACC)

SFR Name: ACC
SFR Address: E0h
Default Value: 00h
Bit-Addressable: Yes
Bit-Definitions:

	Bit 7	Bit 6	Bit 5	Bit 4	Bit 3	Bit 2	Bit 1	Bit 0
NAME	ACC.7	ACC.6	ACC.5	ACC.4	ACC.3	ACC.2	ACC.1	ACC.0
BIT ADDR	E0h	E1h	E2h	E3h	E4h	E5h	E6h	E7h

The accumulator is used in a majority of the instructions in the 8052 instruction set. It is often referred to as just "A" but is also referred to as ACC when modifying specific bits and when referencing the accumulator in PUSH and POP instructions.

B REGISTER (B)

SFR Name: B
SFR Address: F0h
Default Value: 00h
Bit-Addressable: Yes
Bit-Definitions:

	Bit 7	Bit 6	Bit 5	Bit 4	Bit 3	Bit 2	Bit 1	Bit 0
NAME	B.7	B.6	B.5	B.4	B.3	B.2	B.1	B.0
BIT ADDR	F7h	F6h	F5h	F4h	F3h	F2h	F1h	F0h

The "B" register is often used as an additional holding register. It is also used in the MUL and DIV instructions.

DATA POINTER HIGH / DATA POINTER LOW (DPH/DPL)

SFR Name: DPH   DPL
SFR Address: 83h   82h
Default Value: 00h   00h
Bit-Addressable: No   No

The data pointer high (83h) and data pointer low (82h) SFRs together hold the 16-bit value of the data pointer. DPTR is used to access external RAM with the MOVX instruction and code memory with the MOVC instruction. When pushing DPTR onto the stack, DPH and DPL must be pushed onto the stack individually.

INTERRUPT ENABLE (IE)

SFR Name: IE
SFR Address: A8h
Reset Value: 00h
Bit-Addressable: Yes
Bit-Definitions:

	Bit 7	Bit 6	Bit 5	Bit 4	Bit 3	Bit 2	Bit 1	Bit 0
NAME	EA	-	ET2	ES	ET1	EX1	ET0	EX0
BIT ADDR	AFh	AEh	ADh	ACh	ABh	AAh	A9h	A8h

EA – Enable/Disable All Interrupts. 1=Interrupts enabled, 0=interrupts disabled.

ET2 – Enable Timer 2 Interrupt. 1=Interrupt enabled, 0=interrupt disabled.

ES – Enable Serial Interrupt. 1=Interrupt enabled, 0=interrupt disabled.

ET1 – Enable Timer 1 Interrupt. 1=Interrupt enabled, 0=interrupt disabled.

EX1 – Enable External 1 Interrupt. 1=Interrupt enabled, 0=interrupt disabled.

ET0 – Enable Timer 0 Interrupt. 1=Interrupt enabled, 0=interrupt disabled.

EX0 – Enable External 0 Interrupt. 1=Interrupt enabled, 0=interrupt disabled.

INTERRUPT PRIORITY (IP)

SFR Name: IP
SFR Address: B8h
Reset Value: 00h
Bit-Addressable: Yes
Bit-Definitions:

	Bit 7	Bit 6	Bit 5	Bit 4	Bit 3	Bit 2	Bit 1	Bit 0
NAME	-	-	PT2	PS	PT1	PX1	PT0	PX0
BIT ADDR	BFh	BEh	BDh	BCh	BBh	BAh	B9h	B8h

PT2 – Priority Timer 2 Interrupt. 1=High priority interrupt, 0=Low priority interrupt

PS – Priority Serial Interrupt. 1=High priority interrupt, 0=Low priority interrupt

PT1 – Priority Timer 1 Interrupt. 1=High priority interrupt, 0=Low priority interrupt

PX1 – Priority External 1 Interrupt. 1=High priority interrupt, 0=Low priority interrupt

PT0 – Priority Timer 0 Interrupt. 1=High priority interrupt, 0=Low priority interrupt

PX0 – Priority External 0 Interrupt. 1=High priority interrupt, 0=Low priority interrupt

POWER CONTROL (PCON)

SFR Name: PCON
SFR Address: 87h
Reset Value: 00h
Bit-Addressable: No
Bit-Definitions:

	Bit 7	Bit 6	Bit 5	Bit 4	Bit 3	Bit 2	Bit 1	Bit 0
NAME	SMOD	-	-	-	GF1	GF0	PD	IDL

SMOD – Serial Baud Rate Double. 1=Double baud rate in serial modes 1, 2, and 3.

GF1 – General Purpose Flag Bit 1. Available for any use.

GF2 – General Purpose Flag Bit 2. Available for any use.

PD – Power Down. 1=Go to power-down/stop mode. See section 22.2.2.

IDL – Idle. 1=Go to idle mode. See section 22.2.1.

PORT 0 (P0)

SFR Name: P0
SFR Address: 80h
Reset Value: FFh
Bit-Addressable: Yes
Bit-Definitions:

	Bit 7	Bit 6	Bit 5	Bit 4	Bit 3	Bit 2	Bit 1	Bit 0
NAME	AD7	AD6	AD5	AD4	AD3	AD2	AD1	AD0
BIT ADDR	87h	86h	85h	84h	83h	82h	81h	80h

NOTE: These bit names indicate that I/O line's function on the P0 bus when used with external memory (code/RAM). A standard 8052 assembler will not recognize these bits by the given names, rather they'll only be recognized as P0.7, P0.6, etc.

NOTE: Port 0 is only available for general input/output if the project does not use external code memory or external RAM. When such external memory is used, port 0 is used automatically by the microcontroller to address the memory and read/write data from/to said memory.

PORT 1 (P1)

SFR Name: P1
SFR Address: 90h
Reset Value: FFh
Bit-Addressable: Yes
Bit-Definitions:

	Bit 7	Bit 6	Bit 5	Bit 4	Bit 3	Bit 2	Bit 1	Bit 0
NAME	P1.7	P1.6	P1.5	P1.4	P1.3	P1.2	T2EX	T2
BIT ADDR	97h	96h	95h	94h	93h	92h	91h	90h

T2EX – Timer 2 Capture/Reload. Optional external capturing or reloading of timer 2.

T2 – Timer 2 External Input. Optionally used to control timer/counter 2 via external source.

PORT 2 (P2)

SFR Name: P2
SFR Address: A0h
Reset Value: FFh
Bit-Addressable: Yes
Bit-Definitions:

	Bit 7	Bit 6	Bit 5	Bit 4	Bit 3	Bit 2	Bit 1	Bit 0
NAME	A15	A14	A13	A12	A11	A10	A9	A8
BIT ADDR	A7h	A6h	A5h	A4h	A3h	A2h	A1h	A0h

NOTE: These bit names indicate each I/O line's function on the P2 bus when used with external memory (code/RAM). A standard 8052 assembler will not recognize these bits by the given names, rather they'll only be recognized as P2.7, P2.6, etc.

NOTE: Port 2 is only available for general input/output if the project does not use external code memory or external RAM. When such external memory is used, port 2 is used automatically by the microcontroller to address the memory and read/write data from/to said memory.

318

PORT 3 (P3)

SFR Name: P3
SFR Address: B0h
Reset Value: FFh
Bit-Addressable: Yes
Bit-Definitions:

	Bit 7	Bit 6	Bit 5	Bit 4	Bit 3	Bit 2	Bit 1	Bit 0
NAME	RD	WR	T1	T0	INT1	INT0	TXD	RXD
BIT ADDR	B7h	B6h	B5h	B4h	B3h	B2h	B1h	B0h

RD – Read Strobe. 0=External memory read strobe.

WR – Write Strobe. 0=External memory write strobe.

T1 – Timer/Counter 1 External Input. Optionally used to control timer/counter 1 via external source.

T0 – Timer/Counter 0 External Input. Optionally used to control timer/counter 0 via external source.

INT1 – External Interrupt 1. Used to trigger external interrupt 1.

INT0 – External Interrupt 0. Used to trigger external interrupt 0.

TXD – Serial Transmit Data. 8052's serial transmit line (from 8052 to external device).

RXD – Serial Transmit Data. 8052's serial receive line (to 8052 from external device).

> **NOTE**: These bit names indicate that I/O line's function on the P3 bus. A standard 8052 assembler will not recognize these bits by the given names, rather they'll only be recognized as P3.7, P3.6, etc.

PROGRAM STATUS WORD (PSW)

SFR Name: PSW
SFR Address: D0h
Reset Value: 00h
Bit-Addressable: Yes
Bit-Definitions:

	Bit 7	Bit 6	Bit 5	Bit 4	Bit 3	Bit 2	Bit 1	Bit 0
NAME	CY	AC	F0	RS1	RS0	OV	-	P
BIT ADDR	D7h	D6h	D5h	D4h	D3h	D2h	D1h	D0h

CY – Carry Flag. Set or cleared by instructions ADD, ADDC, SUBB, MUL, and DIV.

AC – Auxiliary Carry. Set or cleared by instructions ADD, ADDC.

F0 – Flag 0. General flag available to developer for user-defined purposes.

RS1/RS0 – Register Select Bits. These two bits, taken together, select the register bank which will be used when using "R" registers R0 through R7, according to the following table:

RS1	RS0	Register Bank	Register Bank Address
0	0	0	00h - 07h
0	1	1	08h - 0Fh
1	0	2	10h - 17h
1	1	3	18h - 1Fh

OV – Overflow Flag. Set or cleared by instructions ADD, ADDC, SUBB, and DIV.

P – Parity Flag. Set or cleared automatically by core to establish even parity with the accumulator, such that the number of bits set in the accumulator plus the value of the parity bit will always equal an even number.

RELOAD/CAPTURE TIMER 2 (RCAP2H/RCAP2L)

SFR Name: RCAP2H   RCAP2L
SFR Address: CBh   CAh
Default Value: 00h   00h
Bit-Addressable: No

These bytes hold the value the 16-bit value of the timer 2 reload or capture.

SERIAL BUFFER (SBUF)

SFR Name: SBUF
SFR Address: 99h
Default Value: 00h
Bit-Addressable: No

The serial buffer is used to transmit and receive data to and from the serial port. Writing to SBUF will initiate serial transmission. When serial data has been received, the data that was received may be obtained by reading this SFR.

SERIAL CONTROL (SCON)

SFR Name: SCON
SFR Address: 98h
Reset Value: 00h
Bit-Addressable: Yes
Bit-Definitions:

	Bit 7	Bit 6	Bit 5	Bit 4	Bit 3	Bit 2	Bit 1	Bit 0
NAME	SM0	SM1	SM2	REN	TB8	RB8	TI	RI
BIT ADDR	9Fh	9Eh	9Dh	9Ch	9Bh	9Ah	99h	98h

SM0/SM1– Serial Mode. These two bits, taken together, select the serial mode in which the serial port will operate.

SM0	SM1	Serial Mode	Explanation	Baud Rate
0	0	0	8-bit shift register	Oscillator / 12
0	1	1	8-bit UART	Set by timer 1 or 2*
1	0	2	9-bit UART	Oscillator / 32 *
1	1	3	9-bit UART	Set by timer 1 or 2*

(*) NOTE: The baud rate indicated in this table is doubled if PCON.7 (SMOD) is set.

SM2 – Serial Mode 2 (Multiprocessor Communication). When this bit is set, multiprocessor communication is enabled in modes 2 and 3 causing the RI bit to only be set when the 9th bit of a byte received is set. In mode 1, RI will only be set if a valid stop bit is received. SM2 should be cleared in mode 0.

REN – Received Enable. This bit must be set to enable data reception via the serial port. No data will be received by the serial port if this bit is clear.

TB8– Transmit Bit 8. When in modes 2 and 3, this bit will be the 9th bit that is sent when a byte is written to SBUF.

RB8– Receive Bit 8. When in modes 2 and 3, this is the 9th bit that was received. In mode 1, and if SM2 is set, RB8 holds the value of the stop bit that was received. RB8 is not used in mode 0.

TI – Transmit Interrupt. Set by hardware when the byte previously written to SBUF has been completely clocked out the serial port.

RI – Receive Interrupt. Set by hardware when a byte has been received by the serial port and is available to be read in SBUF.

STACK POINTER (SP)

SFR Name: SP
SFR Address: 81h
Default Value: 07h
Bit-Addressable: No

The stack pointer points to the end of the current stack. A value that is popped off the stack will be taken from the internal RAM address pointed to by SP, then SP will be decremented by one. When a value is pushed onto the stack, SP will first be incremented by one and then the value being pushed will be stored at the internal RAM address pointed to by SP.

TIMER 0 (TH0/TL0)

SFR Name: TH0   TL0
SFR Address: 8Ch   8Ah
Default Value: 00h   00h
Bit-Addressable: No

These bytes hold the value of timer 0 when in timer modes 0 and 1. In timer mode 2 (auto-reload), TH0 holds the reload value and TL0 is the 8-bit value of the timer. In timer mode 3 (split timer), TH0 acts as the 8-bit timer 1 and TL0 acts as the 8-bit timer 0.

TIMER 1 (TH1/TL1)

SFR Name: TH1   TL1
SFR Address: 8Dh   8Bh
Default Value: 00h   00h
Bit-Addressable: No

These bytes hold the value of timer 1 when in timer modes 0 and 1. In timer mode 2 (auto-reload), TH1 holds the reload value and TL1 is the 8-bit value of the timer. In timer mode 3 (split timer), TH1 and TL1 act as two independent 8-bit timers.

TIMER 2 (TH2/TL2)

SFR Name: TH2   TL2
SFR Address: CDh   CCh
Default Value: 00h   00h
Bit-Addressable: No

These bytes hold the value the 16-bit value of timer 2.

TIMER CONTROL (TCON)

SFR Name: TCON
SFR Address: 88h
Reset Value: 00h
Bit-Addressable: Yes
Bit-Definitions:

	Bit 7	Bit 6	Bit 5	Bit 4	Bit 3	Bit 2	Bit 1	Bit 0
NAME	TF1	TR1	TF0	TR0	IE1	IT1	IE0	IT0
BIT ADDR	8Fh	8Eh	8Dh	8Ch	8Bh	8Ah	89h	88h

TF1 – Timer 1 Overflow. Set when timer 1 overflows.

TR1 – Timer 1 Run. 1=Timer 1 is running. 0=Timer 1 is stopped.

TF0 – Timer 0 Overflow. Set when timer 0 overflows.

TR0 – Timer 0 Run. 1=Timer 0 is running. 0=Timer 0 is stopped.

IE1 – External 1 Interrupt Flag. 1=An external 1 interrupt has been triggered.

IT1 – External 1 Interrupt Type. 1=Triggered on High to low transition. 0=Triggered on low level.

IE0 – External 0 Interrupt Flag. 1=An external 0 interrupt has been triggered.

IT0 – External 0 Interrupt Type. 1=Triggered on High to low transition. 0=Triggered on low level.

TIMER MODE (TMOD)

SFR Name: TMOD
SFR Address: 89h
Reset Value: 00h
Bit-Addressable: No
Bit-Definitions:

	Bit 7	Bit 6	Bit 5	Bit 4	Bit 3	Bit 2	Bit 1	Bit 0
NAME	GATE1	C/T1	T1M1	T1M0	GATE0	C/T0	T0M1	T0M0

GATE1 – Gate Timer 1. 1=Timer 1 only runs when INT1 (P3.3) is high. 0=Timer 1 runs regardless of state of INT1.

C/T1 – Counter/Timer 1. 1=Timer 1 counts events on T1 (P3.5). 0=Timer 1 incremented every instruction cycle.

T1M1, T1M0 – Timer 1 Mode. Selects timer mode according to the following table.

TxM1	TxM0	Timer Mode	Description
0	0	0	13-bit timer
0	1	1	16-bit timer
1	0	2	8-bit auto-reload
1	1	3	Split timer

GATE0 – Gate Timer 0. 1=Timer 0 only runs when INT0 (P3.2) is high. 0=Timer 0 runs regardless of state of INT0.

C/T0 – Counter/Timer 0. 1=Timer 0 counts events on T0 (P3.4). 0=Timer 0 incremented every instruction cycle.

T0M1, T0M0 – Timer 0 Mode. Selects timer mode according to the table above.

TIMER 2 CONTROL (T2CON)

SFR Name: T2CON
SFR Address: C8h
Reset Value: 00h
Bit-Addressable: Yes
Bit-Definitions:

	Bit 7	Bit 6	Bit 5	Bit 4	Bit 3	Bit 2	Bit 1	Bit 0
NAME	TF2	EXF2	RCLK	TCLK	EXEN2	TR2	C/T2	CP/RL2
BIT ADDR	CFh	CEh	CDh	CCh	CBh	CAh	C9h	C8h

TF1 – Timer 1 Overflow. Set when timer 2 overflows, triggers timer 2 interrupt. Not set if either TCLK or RCLK are set.

EXF2 – Timer 2 External Flag. Set by a reload or capture caused by a 1-0 transition on T2EX (P1.1), but only when EXEN2 is set. Triggers timer 2 interrupt.

RCLK – Receive Clock. 1=Timer 2 used as receive baud rate generator. 0=Timer 1 used as receive baud rate generator.

TCLK – Transmit Clock. 1=Timer 2 used as transmit baud rate generator. 0=Timer 1 used as transmit baud rate generator.

EXEN2 – Timer 2 External Enable. When set, 1-0 transition on T2EX (P1.1) will cause a capture or reload to occur.

324

TR2 – Timer 2 Run. 1=Timer 2 is running. 0=Timer 0 is stopped.

C/T2 – Counter/Timer 2. 1=Timer 2 counts events on T2 (P1.0). 0=Timer 2 incremented every instruction cycle.

CP/RL2 – Capture/Reload Timer 2. 1=Capture will occur on 1-0 transition of T2EX (P1.1) if EXEN2 set. 0=Reload will occur on overflow, and on 1-0 transition of T2EX if EXEN2 set.

Bibliography

Online Resources

Kalinsky, David. "Introduction to Serial Peripheral Interface". <u>Embedded.com.</u> Ed. Jim Turley
      14 Jan. 2005. <http://www.embedded.com/story/OEG20020124S0116>

"Philips Semiconductors I²C-bus Information". <u>Philips Semiconductors</u>.
      18 Jan. 2005. <http://www.semiconductors.philips.com/markets/mms/protocols/i2c>

"Ultra-High-Speed Flash Microcontroller User's Guide". <u>Maxim Integrated Products</u>. 10 Nov. 2004.
      <http://www.maxim-ic.com/products/microcontrollers/pdfs/DS89c420_user_guide.pdf>

"80C51 Family Architecture". <u>Philips Semiconductors</u>. 10 Nov. 2004.
      <http://www.semiconductors.philips.com/acrobat/various/80C51_FAM_ARCH_1.pdf>

"80C51 Family Hardware Description ". <u>Philips Semiconductors</u>. 10 Nov. 2004.
      <http://www.semiconductors.philips.com/acrobat/various/80C51_FAM_HARDWARE_1.pdf>

"80C51 Family Programmer's Guide and Instruction Set". <u>Philips Semiconductors</u>. 10 Nov. 2004.
      <http://www.semiconductors.philips.com/acrobat/various/80C51_FAM_PROG_GUIDE_1.pdf>

"8052.com Online Resource". <u>Vault Information Services</u>. 10 Oct. 2004 <http://www.8052.com>

"LCD Interfacing Reference Page." <u>Myke Predko</u>. 27 Jul. 2005.
      <http://www.myke.com/lcd.htm>

Books

MacKenzie, Scott I. <u>The 8051 Microcontroller</u>. 2nd ed. Columbus: Prentice Hall, 1995.

Index

44780 standard, 245

accumulator, 20, 21, 65, 82, 88, 157, 315
addition,
    double-byte (16-bit), 100
    single-byte (8-bit), 85, 86
Addressing Modes,
    code indirect, 28
    direct, 25
    external direct, 26
    external indirect, 27
    immediate, 25
    indirect, 26
Address Latch Enable (ALE), 116, 129
Assembly Language,
    characters, 72
    equates, 73
    expressions, 71
    number bases, 71
    operator precedence, 71
    Pinnacle 52 IDE, 150
    setting assemble address, 72
    strings, 72
    syntax, 69
Auxiliary carry flag, 19, 86, 87, 320

baud rate generation, 45, 52-55
bits, 11-13, 16
B register, 20, 22, 65, 87, 88, 315
button, 127

Carry flag, 19, 64, 81, 82, 84-90, 100, 101, 103, 108, 238, 269
ChipProg+, 189
comment, 69
conditional branching, 29

Data Pointer (DPTR), 17, 22, 26-28, 65, 94-97, 114, 116, 133, 158, 315
DB (Data Byte) directive, 74
debouncing keys, 236, 238, 240, 241
device programmer, 4, 187, 189
direct calls, 30

directives, 72
direct jumps, 29
division,
    double-byte (16 bit), 107
    single-byte (8-bit), 88
DW (Data Word) directive, 74

EN (LCD signal line), 245, 247
EPROM eraser, 4
EQU directive, 73, 81, 210, 211, 247, 280
External Access (EA), 117, 127, 128, 143
EXTERN directive, 183, 186

greater than comparison, 84

Harvard (memory architecture), 8, 177
hex file, 153, 174, 189, 193, 197, 205-207, 209

In-Circuit Emulation (ICE),
    description, 187
instruction, 69, 70
In-System Programming,
    Atmel AT89S8252, 141
    Dallas DS89C420, 143
    description, 187, 191, 193
    using, 194
    With 8052.com SBC, 192
Integrated Development Environment (IDE), 149
Inter-IC Protocol (I2C),
    coding, 268
    description, 267, 268
    receiving/transmitting data, 270
    start condition, 269
    stop condition, 269
Interrupts,
    common problems, 66
    configuring, 59
    described, 31, 57
    events that can trigger, 58
    exiting, 61
    external interrupts, 62
    Interrupt Enable (EA), 316
    Interrupt Enable (IE), 59, 157

www.ingramcontent.com/pod-product-compliance
Lightning Source LLC
Chambersburg PA
CBHW080913220326
41598CB00034B/5565